Summation Notation 3.2

Means & Medians 3.3

Standard Deviation 4.4

Probability Ch 5 ~70-97

Probability Distribution 104-114

Binomial Distribution 124-130

Poisson Distribution 131-134

Acceptance Sampling (control) 17.5 352-356

Wilcoxon Test Paired Data 18.3 366-368

Rank-Sum Test 18.4 368-370

Normal Distribution 8.2 142-149

Normal Approximation to Bin Dist 8.3 149-152

Chi Square Test 15.1 302-304

Goodness of fit 15.2 304-307

Contingency Tables 15.3 307-313

Testing Means 11.2 198-207

Estimation of Means 10.3 174-178

t-distribution 10.4 178-183

ANOVA 16.1 & 16.2 321-328

Bivariate back

→ idea

14   14.734 15

2.955
estimate ↑ t

calculate range of difference between means using

find C.I using

No diff

-0.167 0   17.~

no difference   since 0 is between sts   you cannot be sure if a change occurred

HENRICK J. MALIK · KENNETH MULLEN
University of Guelph, Ontario, Canada

APPLIED STATISTICS FOR BUSINESS AND ECONOMICS

ADDISON-WESLEY PUBLISHING COMPANY
Reading, Massachusetts
Menlo Park, California · London · Amsterdam · Don Mills, Ontario · Sydney

ISBN 0-201-04410-2
GHIJKLMNO-MA-898765432

To
Rom, Bina, Renee, Becky
Denise, Siobhan, Dierdre

PREFACE

In developing this text, our aim has been to give business students a good basic course in statistics, one that would not be too mathematically oriented. Such an approach meant that the book would have to be founded on the solid principles of modern statistics while omitting many of its more sophisticated arguments. At the same time we wanted to emphasize the application of statistics to a broad range of business problems. These restrictions necessitated the following considerations:

1. That descriptive statistics be included, since business students and businessmen will almost always meet situations where data have to be arranged into frequency tables, histograms, graphs, etc;

2. That probability be covered, since most statements made in statistics are of a probabilistic nature;

3. That there be a large number of interesting problems covering a broad range of topics, so that the student can see the scope of coverage of statistics;

4. That a chapter on the elementary techniques of sampling be included, to emphasize the importance of this often neglected area;

5. That the classical areas of Statistics, such as Estimation and Hypothesis Testing be discussed, while at the same time introducing more modern areas such as Decision-Making, along with the concepts of Payoff, Loss, Expected Monetary Value, etc;

6. That topics of special interest such as time-series analysis, index numbers, and quality control be included.

No previous statistical background is assumed or necessary, but a fair knowledge of high-school mathematics would be an asset. The emphasis is on clarity and simplicity. Thus, new ideas are introduced by posing questions and then answering them in the ensuing discussion. We realize that we have left ourselves open to the charge of being repetitious, but we are willing to accept this responsibility in the belief that some repetition at this stage of statistical development is necessary. There are a large number of worked examples, and many of these examples are carried over from chapter to chapter to illustrate new or different approaches.

The primary goals of acquainting students with the ideas and areas of application of statistics are rounded off by including a large number of exercises at the end of each chapter, in accordance with the authors' belief that no student in statistics can progress far without self-testing through problem solving.

The text contains sufficient material to allow for flexibility in the length of the course. The first four chapters are devoted to descriptive statistics; Chapters 5, 6, 7, and 8 introduce the concepts and applications of probability. Chapter 9 discusses sampling and sampling techniques, while estimation and hypothesis-testing are covered in Chapters 10 and 11. In short, the first part of the book is divided into descriptive statistics (Chapters 1–4) and inferential statistics (Chapters 5–11). The second part of the book is devoted to regression and correlation, time-series analysis, index numbers, the chi-square test, analysis of variance, quality control, nonparametric statistics, and decision-making under uncertainty.

The book can be used for several different kinds of introductory statistics courses, for students in business, public administration, or economics. Depending on the time available, it is possible to design courses covering one or two semesters, or one or two quarters. For example, a one-semester course might consist of descriptive statistics (Chapters 1–4), selected sections on probability, selected sections on correlation and regression, time series, and index numbers. This could be done without heavy involvement in probability and statistical inference. Courses which place more emphasis on probability, or on statistical inference, or on any of the other topics in the latter part of the book, could of course be designed. Use of the book is intended to be extremely flexible.

We wish to thank all those who helped us in the preparation of this textbook. First of all, we express our deep gratitude to original thinkers of the past and the present, and to the scholars who have written textbooks on probability and statistics. We gratefully acknowledge the contributions of Professors George Heitmann of Pennsylvania State University, William Pinney of the University of Texas, Jerome Herniter of Boston University, George M. Brooker of Dean Junior College, George W. Summers and Robert St. Louis of Arizona State University, Robert L. Winkler, Indiana University, Jack B. ReVelle, University of Nebraska, David. D. Krueger, St. Cloud State University, Manfred W. Hopfe, Georgia State University, and Delbert Hastings of the University of Minnesota, who read the manuscript and provided comments, criticisms, and suggestions that have proved helpful in preparation of the final draft. The authors accept sole responsibility for any errors or omissions which may be discovered. We are particularly thankful to Patrick Murray and Roger Trudel for reading the final manuscript, to Mrs. Linda Selby for her careful and accurate typing, to Addison-Wesley Publishing Company for its great help and cooperation, and to our wives and children for their patience and encouragement.

We are indebted to Professor E. S. Pearson and the Biometrika Trustees for permission to use Tables from *Biometrika Tables for Statisticians, Vol. 1.*

H.J.M.

K.M.

Guelph, Ontario
January 1975

CONTENTS

INTRODUCTION 1

1.1. THE ROLE OF STATISTICS IN BUSINESS DECISIONS

One of the cornerstones of modern business theory is a free and adequate flow of information. Decisions are made and actions are taken based on the nature of this information. Thus, a company planning to market a new brand of soap needs to know, among other things, the market potential, the expected sales, and what its competitors are doing. Much of this information is quantitative in nature; its analysis, therefore, must be performed with those tools and techniques specifically developed and designed to handle quantitative information. Further, the information is usually incomplete, in that the full market potential is rarely completely known, the expected sales are calculated based on assumptions only, and one cannot always know what one's competitors are doing. In short, the businessman must make decisions in the face of uncertainty.

Statistics is a discipline which has been developed by business experts, mathematicians, economists, psychologists, and many others (including biologists and zoologists) to extract the relevant facts from a large body of information and to help people make decisions when uncertainty exists concerning the information.

In general, in order to answer a question or problem arising in a business situation, one has to gather information or data. This data must be collected properly, analyzed efficiently, presented in an orderly and coherent fashion, and interpreted correctly. All of these functions fall into the realm of statistics.

Broadly speaking, statistics is the study of a problem from a numerical point of view. This implies that a problem must be capable of being reduced to a measurable scale or to numerical quantities, in order to use the tools and techniques of statistics. Some examples of problems whose solutions or partial solutions depend on statistics or statistical analysis follow.

1

Example 1 (Planning) An oil company is considering expanding its production capacity for regular gasoline; this is a long-term project requiring much planning. In order to decide how much new capacity will be needed in ten years' time, the company must try to predict the sales of regular gasoline in the future. To do this, data is needed on such things as the growth rate in the past, the number of cars on the road in ten years, whether these cars will be bigger or smaller than now, the effect of antipollution legislation, and many others. Since all of these effects can be expressed in numerical terms, the question of how much capacity will be needed in ten years can be partially answered with the aid of statistics.

Example 2 (Market research) The sales of a particular toothpaste are affected by its flavor. The manufacturer wishes to know if sales can be increased by doubling the amount of flavoring that goes into each tube; the company believes that increased production costs will be more than offset by increased sales. To test this idea, a small amount of the more highly flavored toothpaste is produced, and given to a sample of 10,000 families for their reaction. Based on the number of favorable reactions to the new product, the manufacturer will be helped in making his decision to market the new brand of toothpaste. The problem of how to select the 10,000 families who are to get the experimental brand is answered by the statistical technique called sampling.

Example 3 (Industry) Iron used by a steel company can be obtained from two regions (A and B). Of particular importance is the melting point of the iron. In the table below are shown ten determinations of melting points of iron from a smelter in region A and six from a smelter in region B.

Region A	1493,	1519,	1518,	1512,	1512,	1514,	1489,	1508,	1508,	1494
Region B	1509,	1494,	1512,	1483,	1507,	1491				

Statistics provides a means for saying whether the two regions differ in the melting points of their iron, and if so which is higher. It also enables one to calculate how reliable the decision is.

Example 4 (Quality control) Machine parts turned out by an automatic process are known to be one percent defective, on the average. In order to judge each day's run, a quality control expert examines 200 similar parts each day. He rejects the whole day's run if three or more parts in the 200 are found to be defective. Since rejecting the whole run is expensive, one might reasonably ask whether he just happened to have bad luck in getting three or more defective parts. In other words, there are two possible reasons for obtaining such a result:

a) There are indeed too many defective parts in the run, and it should be rejected;

b) Luck was against the quality-control expert and he just happened to choose the wrong parts.

Example 5 (Qualitative measurements) A cigarette company wants to see if small changes in flavor can be detected by the public. In order to test this, a panel of people is chosen (this is sometimes a panel of expert tasters and sometimes a panel chosen from the public) and asked to compare several slightly different flavors. They compare the flavors two at a time and conclude for each pair that either (i) there is no difference between them or (ii) there *is* a difference. Thus, even though flavor cannot be measured on a scale (we say that it is a *qualitative* variable), we will, given a large enough panel, be able to see if a majority of the panel can detect the difference between any two flavors. Determination of the size of the panel needed to detect specific differences can be obtained from the study of statistics.

Example 6 (Management) A company wishes to relocate its offices. It needs to know the prevailing wage patterns in its prospective new location; in particular, what the average wage is for (say) skilled technicians. A sample of skilled technicians in the new location must be chosen and their average wages calculated. The central problem is to ensure that the sample of wage earners is representative of the whole area. If, for instance, the wage earners in the sample are all from the same plant, then the results will not give a general picture of the new location, but only of that particular plant. The problem of how to choose a sample to be representative is one which can be answered by statistical methods.

We could go on giving similar examples but it should be clear by now that statistics can be used to solve or help solve a wide variety of problems. We have stated that statistics deals with a problem from a numerical point of view, that is, by looking at the data pertaining to the problem. There is one binding quality of all of these data, and that is, that true relationships in the data are obscured by variation or variability due to unknown or uncontrollable causes. Thus, in Example 1, the true relationship between regular gasoline sales and other factors such as past growth, number of cars on the road in ten years, and pollution legislation, are unknown. The best that can be done is to estimate the true relationship from the data. In Example 2, the only way to truly find out the popularity of the new toothpaste is to test it on everyone in the country. Since this is impractical, one has to be satisfied with a sample of 10,000 families, and hope that the results of the sample can be generalized to the whole country. That is, an element of uncertainty has been introduced into the problem. In Example 3, we would like to compare the two regions and see if they are different in spite of the variability of results. In truth, to know which region had the highest iron melting point would necessitate measuring all the iron in that region. Clearly that is impossible, and one must be satisfied with the sample results, even though they introduce an element of uncertainty.

We shall see how to deal with and measure this uncertainty in due course, but it will be convenient to divide the study of statistics into two parts, namely (i) descriptive statistics and (ii) inductive statistics.

1.2. DESCRIPTIVE STATISTICS

Descriptive statistics is concerned with the problem of describing a mass of data in a concise, clear, useful, and informative way. This is done by considering such techniques as graphing, tabular presentation, and calculation of averages. In the past, elementary statistics courses dealt principally with descriptive statistics; and indeed the layman's idea of what statistics is is still largely confined to these concepts.

We shall eventually discuss what to do when presented with a large number of data, including methods for organizing them, for presenting them, and finally for describing them in a simple and informative way. Chapters 2, 3, and 4 deal with descriptive statistics.

1.3. INDUCTIVE STATISTICS

Once the data have been collected and adequately sorted and described, the statistician is only beginning his task. Generally, he wishes to draw conclusions from the data, and try to generalize these conclusions. For instance, in Example 2, he wishes to draw conclusions as to the attitude of all toothpaste users to his new product, based on the results of his sample of 10,000 families. In Example 4, conclusions about the whole day's run of machine parts are based on a sample of 200 pieces.

The reasoning used to generalize the results of a sample is called inductive reasoning, and inductive statistics is that part of the study of statistics which allows us to reason inductively. Inductive statistics is a much larger area of study than descriptive statistics, and hence will occupy much of our effort. Basic to the development of inductive statistics is the study of probability.This is because when we make statements about certain populations on the basis of the samples, we will not be sure if our statements are completely correct. We would, of course, like to make correct statements and decisions all the time, but sometimes the nature of the variation will obscure the results, so that errors in judgment will result. We want to keep the chance or probability of these errors small, but we will not be able to eradicate them completely. Thus although we cannot make correct statements with certainty, we can make statements which have a high probability of being correct. From this it follows that we must start the study of inductive statistics with preliminary ideas of probability.

1.4. SOME COMMON TERMS USED IN STATISTICS

All areas of work and study have their own special words, special meanings, and special vocabulary. This is simply because there are things and ideas that are peculiar to that area of work or study. Statistics is no exception, and in order that the reader can fully comprehend the ideas being discussed, it is necessary for us to say what we mean when we use particular words. Many of these words will become clear as the reader progresses, but the following words will be necessary at the beginning.

Population (or **Universe**)

The total group under discussion or the group to which the results will be generalized is called the population.

Characteristic

A variable which can be measured for each member of the population is called a characteristic of the population.

Sample

A sample consists of a group of objects chosen from the population because measurements on the entire population cannot or will not be made. In rare instances, the sample will consist of the entire population.

Example 7 In Example 2, the population might be all toothpaste users in the country (or in a particular geographic area), while the sample is the group of 10,000 people on which the data were obtained. The characteristic measured is simply their opinion of the new toothpaste.

Parameter

A constant which describes the *population* (e.g., its average, dispersion, maximum, etc.) is called a parameter.

Statistic

A statistic is a number which describes the *sample* (e.g., its average, center, dispersion, etc.).

Data

By data, we mean recorded observations made on the sample.

From data we can compute a statistic (or sometimes more than one) that can be used to reach conclusions about the parameter of the population from which it came.

Example 8 In Example 2 (discussed in Example 7), the data are the number of people in favor and the number not in favor of the new brand of toothpaste. The parameter might be the number of people in the population who would use the toothpaste, while the statistic is the number of people in the sample who would use the toothpaste.

Example 9 In Example 4, the population consists of all machine parts made on a particular day; the sample consists of 200 chosen for inspection. The characteristic might be the width of the parts; and the parameter might be the maximum width of parts in the population, while the data are the 200 widths from the sample, from which we would obtain the maximum width (the statistic).

Note, in Example 9, that the parameter will never be known unless the whole population is measured. Nevertheless, based on the 200 observations, we would like to draw some conclusions about the population; that is, we would like to infer from the sample to the population. There is always a chance (hopefully small) that the conclusion or inference will be wrong; and the chance of being wrong is measured in terms of probability.

SUMMARY

This chapter starts with the premise that many business problems may be expressed in quantitative terms, which in turn must be analyzed with the techniques and ideas of statistics. There are two main branches of statistics: (i) descriptive statistics, whereby data are collected, analyzed, and presented in a clear, concise, and useful form; and (ii) inductive statistics, whereby conclusions are drawn from collected data by means of analysis and interpretation. The conclusions are then generalized from the sample to the population.

Various important terms and words are introduced and a number of examples are given, to demonstrate how problems can be at least partially solved by statistical ideas.

Words to Remember

Variability	Population	Statistic
Descriptive statistics	Characteristic	Data
Inductive statistics	Sample	
Probability	Parameter	

EXERCISES

1. In Example 3, discuss what constitutes the population, the sample, the parameter, the statistic, and the characteristic.

. In Example 5, discuss how it would be possible to draw inferences from the panel results to the population of all smokers. Would the inferences be different if the panel were expert tasters or simply a sample from the general public? What might be the purpose of a panel of experts?

. Take a problem in your particular area of interest which can be expressed quantitatively. Write a short note defining the population, sample, parameter, statistic, and characteristic. How might the sample be chosen?

ORGANIZATION
OF DATA 2

.1. INTRODUCTION

Generally the collection of information to solve problems in business and industry leads to a large mass of data, and in recent years business decisions have come to depend on such masses of data more and more. In their original form, these data usually appear meaningless; they need to be condensed or rearranged into a more easily understandable form before they can be of much use.

In this chapter we discuss how to organize, summarize and analyze a large number of business data in such a way that their important characteristics can be seen at a glance.

Example 1 (Marketing) Suppose that a sample of 200 filling stations is selected in a study to determine the optimum conditions under which filling stations can be operated, and each is judged as large (L) or small (S), based on weekly sales of regular gasoline. Instead of presenting the results as a sequence of 200 letters, L, S, L, S, etc., one may simply say how many values of L and S there are. Suppose that there are 50 L's and 150 S's; then the results may be summarized by saying that the *proportion* of large stations is 0.25 (or 25%), and that the *ratio* of small to large stations is three to one.

In this example, there are only two possible values for each filling station, L or S. In situations where the number of possible values for each response is large —for example, if the actual weekly gas sales were recorded, they could be anywhere from zero gallons per week upwards—then the data can be most easily summarized if they are grouped into a relatively small number of *classes* or intervals. We may then present them pictorially or graphically. Note that this summarization of the data is not mandatory for subsequent discussion and analysis, but can be done at the discretion of the individual, in order to provide clarity and simplicity.

Before proceeding it is advisable to distinguish between various types of data, since often the choice of a statistical procedure depends on the nature of the data. The four main types of data that one will observe in business are:

a)   Continuous data or measurements, such as weight, length, pressure, temperature, rates, returns, profits, etc.

Continuous Data

If a variable can take on any value within certain limits, it is called continuous, and we speak of the data as *continuous data*.

Example 2 The amount of gasoline sold per day in gallons by a particular filling station is a continuous variable, since it can be any value between zero and some reasonable upper value.

Example 3 The projected return on an investment is a continuous variable since it can take on any value within a certain or particular interval; that is, it is not limited to integer values, for instance.

b)   Data which portray the number of automobile accidents in a given time period, the number of employees absent each week, the number of executives owning pleasure boats, etc., are called *discrete data*. In general we shall call data discrete if it is *counted* rather than measured; and in this book we will consider discrete data to be able to take on *only integer values*.

Discrete Data

If a variable can take on only integer values, then it is called a discrete variable and we speak of the data as discrete.

Example 4 In Example 1, each station was classified as large or small, so that there were only two values for the variable "size of station." Size of station was therefore a discrete variable.

Example 5 The number of workers reporting sick in any particular week is a discrete variable with possible values of $0, 1, 2, \ldots$

c)   Ranked or subjective data, such as data which can be classified into ordered categories; for example, very good, good, fair, poor, very poor.

Ranked Data

If the possible values of a variable can be arranged into a set of ordered categories or classes, then the variable will be called ranked, and we speak of the data as *ranked data*.

Example 6 Suppose that the factory workers in a particular plant have their work judged as (a) good, (b) fair, or (c) poor, during a quality-control study; then the variable "quality of work" is a ranked variable.

Nominal Data

Discrete data on a variable which cannot be ordered is called nominal data.

Example 7 Suppose that, in a particular city, there are five large retail outlets, A, B, C, D, and E; and that we wish to compare the number of employees in each outlet. Our data would consist of the number of observations in each of the five groups. We could not think of the outlets as being in any "correct" order; that is, when we present the data, we may arrange the outlets in any order we like, and probably alphabetical order is as easy as any other.

In general there are no problems associated with grouping discrete, ranked, or nominal data, since the data are already arranged into categories. Thus our main concern is with continuous data.

2.2. FREQUENCY DISTRIBUTIONS

After data have been collected, they need to be organized in such a way that they can be used efficiently for purposes of description or analysis. When dealing with large masses of data, as many business people must do at present, one can gain much information from the data simply by organizing them into what is called a *frequency distribution*. This is a system of grouping the data into number of classes or intervals. To illustrate the need for and use of frequency tables, consider the following example concerning the amount of time that 136 television viewers spent in front of their sets.

Example 8 An advertising agency obtained the following 136 observations in a study of the number of hours per day (between 5 P.M. and 11 P.M., weekdays) that a group of television viewers spent in front of their sets.

2.34	3.32	3.28	3.27	3.46	3.55	3.14	2.32
3.14	3.48	2.61	3.62	3.68	3.29	2.53	2.76
3.51	2.43	3.79	3.09	1.96	1.57	4.23	3.40
2.51	3.31	2.53	2.66	3.56	3.57	2.82	2.26
3.12	2.27	3.53	3.49	1.83	2.97	4.05	3.26
2.84	1.52	2.14	3.71	2.94	5.17	2.49	5.85
2.65	3.35	3.25	4.14	1.22	2.07	3.07	3.46
3.82	1.61	2.74	3.25	1.18	4.94	4.06	3.96
2.78	4.28	1.74	4.37	3.87	1.63	2.71	2.52
4.75	3.42	3.49	3.17	2.08	3.18	3.26	4.76
3.17	1.23	4.58	3.88	3.34	2.28	2.91	2.01
4.42	3.53	3.91	2.31	2.65	3.26	3.89	1.38
2.13	1.85	4.32	3.30	3.26	2.15	3.79	2.93
4.51	3.62	3.48	2.77	3.75	3.83	1.47	1.36
3.90	3.35	2.86	3.53	2.37	3.72	2.88	4.59
4.93	5.00	3.46	3.04	3.33	4.46	3.42	2.62
3.64	4.13	5.13	4.25	3.17	0.72	4.07	1.76

From the data, it would be very difficult to draw intelligent conclusion about the duration of time spent by the average viewer in front of his set. Th difficulty would be magnified for larger sets of data.

Let us now put the data into the form of a *frequency distribution* and see i we can obtain a better overall view of the main characteristics. A frequenc distribution is a way of arranging a mass of data into a number of classes so tha each datum falls into exactly one of the classes. The resulting distribution show the number of observations in each class.

Frequency Distribution

A frequency distribution is a system for classifying data, usually in the form of a table, where the observations in a sample of size N are grouped into classes or intervals, so that the frequency of observations in each class can be ascertained.

The construction of a frequency distribution requires three major steps:

1. Choosing the classes into which the data are to be grouped,
2. Actually putting the data into these classes (called tallying); and
3. Counting the number in each class.

The first step, that of choosing the classes, is done as follows:

1. Find the *range* of the data (that is, the value of the largest observation *minu* the value of the smallest observation). There will be automatically *no value* *outside* of these limits.

2. The number of classes, although arbitrary, should rarely be less than six or greater than twenty. A rough rule of thumb is to let the number of classes be approximately equal to the *square root* of the number of observations.

3. Make sure that the observations belong to *only one class*.

4. Try to make classes of *equal width*, so that each will cover the same range of values.

Example 9 Continuing with Example 8, let us form a *frequency distribution*.

1. The range of the data is $5.85 - 0.72 = 5.13$.

2. Since there are 136 observations, the number of classes is approximately $\sqrt{136}$ or 12.

3. If each class is of equal width, then that width should be approximately the *range* divided by the *number* of classes, or $5.13/12 = 0.42$. A more convenient width would be 0.40, but this would call for the use of 13 classes. Starting with 0.70 and a class width of 0.40, the first class will be between 0.70 and 1.10; the second class will be from 1.10 to 1.50, etc.

4. Note that some ambiguity will arise above. Into which class do we put the observation 1.10? This difficulty is easily overcome by measuring the class limits one-half unit beyond the accuracy of the observations. Thus the successive class limits would be 0.705, 1.105, 1.505, etc.

5. The frequency distribution is formed by listing the class limits down one side of a page, and marking a slash (/) on any line for which an observation is assigned. It saves time if every *fifth* entry in a class is denoted by a diagonal line *across* the preceding four ($\cancel{||||}$). The center of a class is usually called the *class mark*. Any observation placed in a class loses its individuality, so to speak, and it is assumed that all values in the same class take the value of the class mark.

For instance, take the first observation, 2.34, which lies between 2.305 and 2.705 (the fifth class). It is assumed to be represented by the class mark of that class, namely 2.505. Other observations are treated similarly. Classes, class marks, tallies, and class frequencies are shown in Table 2.1.

In summary then, if a number of observations are to be arranged in a frequency distribution, one first selects the *number* of classes. The smallest and largest values of each class are called the *class limits*. The *center* of the class is called the *class mark*, and the number of observations in a class is called the *class frequency*. The difference between the largest and smallest values of a class is called the *class width*. Class widths should, if possible, be equal for all classes. In our subsequent discussions we shall assume that class widths are *always* equal.

Sometimes it is preferable to present the data in what is called a *cumulative frequency distribution*, which shows how many observations are *less than* various

TABLE 2.1

Class number	Class limits	Class mark (Y)	Tally marks	Class frequency (f)
1	0.705–1.105	0.905	/	1
2	1.105–1.505	1.305	ⅼⅼⅼⅼ /	6
3	1.505–1.905	1.705	ⅼⅼⅼⅼ ///	8
4	1.905–2.305	2.105	ⅼⅼⅼⅼ ⅼⅼⅼⅼ	10
5	2.305–2.705	2.505	ⅼⅼⅼⅼ ⅼⅼⅼⅼ ⅼⅼⅼⅼ	15
6	2.705–3.105	2.905	ⅼⅼⅼⅼ ⅼⅼⅼⅼ ⅼⅼⅼⅼ /	16
7	3.105–3.505	3.305	ⅼⅼⅼⅼ ⅼⅼⅼⅼ ⅼⅼⅼⅼ ⅼⅼⅼⅼ ⅼⅼⅼⅼ ⅼⅼⅼⅼ ///	33
8	3.505–3.905	3.705	ⅼⅼⅼⅼ ⅼⅼⅼⅼ ⅼⅼⅼⅼ ⅼⅼⅼⅼ //	22
9	3.905–4.305	4.105	ⅼⅼⅼⅼ ⅼⅼⅼⅼ	10
10	4.305–4.705	4.505	ⅼⅼⅼⅼ //	7
11	4.705–5.105	4.905	ⅼⅼⅼⅼ	5
12	5.105–5.505	5.305	//	2
13	5.505–5.905	5.705	/	1
Total				136

values. By successively adding the frequencies in Table 2.1, we obtain the cumulative frequency distribution for viewing hours of our 136 TV watchers; this will be further discussed in Example 10.

TABLE 2.2

Class number	Class limits	Class mark (Y)	Class frequency (f)	Cumulative frequency up to upper class limit
1	0.705–1.105	0.905	1	1
2	1.105–1.505	1.305	6	7
3	1.505–1.905	1.705	8	15
4	1.905–2.305	2.105	10	25
5	2.305–2.705	2.505	15	40
6	2.705–3.105	2.905	16	56
7	3.105–3.505	3.305	33	89
8	3.505–3.905	3.705	22	111
9	3.905–4.305	4.105	10	121
10	4.305–4.705	4.505	7	128
11	4.705–5.105	4.905	5	133
12	5.105–5.505	5.305	2	135
13	5.505–5.905	5.705	1	136

Example 10 We may construct the cumulative frequency distribution for Examples 8 and 9 by considering the frequency distribution of Example 9. The results are shown in Table 2.2. The cumulative frequency for class 5, for instance, is $1 + 6 + 8 + 10 + 15$, or 40. Other cumulative frequencies are similarly constructed.

This table is interpreted as follows: one person watches television for less than 1.105 hours, seven for less than 1.505 hours, etc.

Sometimes one prefers to show the *percentage* of observations in each class, instead of the actual numbers in that class. To convert a frequency (or cumulative frequency) into a percentage (and consequently, to form a *percentage distribution*), divide the class frequency by the total number of observations, and then multiply by 100. For example, by referring to Example 9, Table 2.1, we note that the fifth class contains 15 observations, which, as a percentage, is $(15/136) \cdot 100 = 11\%$.

So far we have been considering the construction of frequency distributions only for *continuous* (or *measurement*) *data*. For discrete data, ranked data, and nominal data, the construction of a frequency distribution is very similar, except that we do not usually have to worry about the class limits, since they are often self-evident. In some cases there are difficulties in defining a class exactly, that is, in saying exactly *what* that class *contains*. For this reason, when in doubt, one should use the class definitions given by government agencies (such as the Census Bureau). A further reference on this subject is *Government Statistics for Business Use*, by P. M. Hauser and W. Leonard (John Wiley and Sons, New York).

2.3. SOME GENERAL COMMENTS ON GRAPHICAL PRESENTATIONS

Once the observations have been grouped, the next step is usually to present the results *pictorially*, so that their salient features may be made clear at a glance. Such pictorial presentation should not be thought of as a substitute for statistical treatment, but rather as an added tool for bringing clarity to the data. Indeed, it might be added that graphical presentation alone can often be very misleading, especially if the person presenting it has a partisan point of view. The statistician has a responsibility to be objective. Some simple guidelines for eliminating subjectivity from graphical presentation are given below.

a) Try not to eliminate the *zero point* from the vertical axis; that is, don't present just the top parts of a graph, for instance. To do so creates an impression of greater variability than is actually present.

b) Try to make the height of the *maximum* point (maximum frequency) on the vertical axis about *three-quarters* the length of the horizontal axis. This rule, while somewhat arbitrary, prevents subjectivity (in the choice of axes) from creeping in.

c) *Label* graphs clearly, so that they may explain themselves as far as possible. One should include information on what the measurements are, what material or subjects were used, and what restrictions apply. The horizontal axis should be clearly labelled, showing what is measured and in what units.

2.4. GRAPHICAL PRESENTATION OF A FREQUENCY DISTRIBUTION

The two most common and convenient ways to present a frequency distribution graphically are by means of (a) a *histogram* and (b) a *frequency polygon*. For both, one employs an appropriate piece of graph paper and writes, on a *horizontal scale*, the class limits and corresponding class marks. The *class frequencies* are plotted on the *vertical* scale. One then plots points defined by the *intersection* of the class mark and its corresponding frequency. The *frequency polygon* is constructed by joining each adjacent pair of points by a straight line. It is usual to "complete" the polygon by drawing lines from its right- and left-hand ends to points at the center of the base of the next class interval. The *histogram* is constructed by placing vertical bars with width equal to the class width (interval) and centered at the class mark, and with height equal to the frequency.

Frequency Polygon

A frequency polygon is a graphical representation of a frequency distribution in which class frequencies are plotted against class marks. These are then joined by straight line segments.

Histogram

A histogram is a graphical representation of a frequency distribution in which class frequencies are plotted against class marks. The class frequencies are then represented by vertical bars, centered at the class marks, whose area is *proportional to the frequency*.

In both the polygon and the histogram, the total areas under the graphs are proportional to the total number of observations, assuming that we have equal class widths. In our discussion we have tacitly assumed that all class intervals have the same width. While this is not absolutely necessary, the interpretation of graphs and later calculations are complicated if the class widths vary. For this reason we shall deal only with *equal class widths*.

Example 11 The observations of Example 8, having been put in a frequency table in Example 9, can now be represented either by a histogram as in Fig. 2.1, r by a frequency polygon as in Fig. 2.2. In both cases, the horizontal and ertical axes are plotted in the same way. From Table 2.2 the highest class ·equency is seen to be 33. Thus the vertical axis need go only from 0 to 33. On ıe horizontal axis, the class marks are plotted. For the first class interval, the lass mark is 0.905, with frequency 1, and a dot is recorded on the graph ¡presenting that point. Other points are similarly recorded. The frequency olygon is then completed by joining, with a straightedge, each adjacent pair of oints. The polygon is completed by joining the endpoints to the base at the ınter of the next class interval. The histogram is completed by drawing a ɔrizontal line, through each dot, the length of the class interval. Vertical lines ɾe then dropped to the base line.

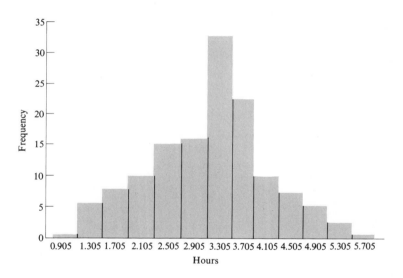

1 A histogram showing the number of hours spent watching television (between 5 and P.M.).

5. GRAPHICAL PRESENTATION OF DISCRETE, RANKED, AND NOMINAL ATA

ɔr discrete, ranked, and nominal data, the class intervals are already suggested ⸗ the nature of the variable being measured. The data are most simply ːesented by means of a bar graph, where the height of each bar is proportional the frequency of that class interval. Each bar is usually of the same width.

cample 12 (Discrete data) In a large plant the number of workers reporting :k on each working day (Monday through Friday) are 24, 8, 6, 6, 18. The bar aph showing the data is given in Fig. 2.3.

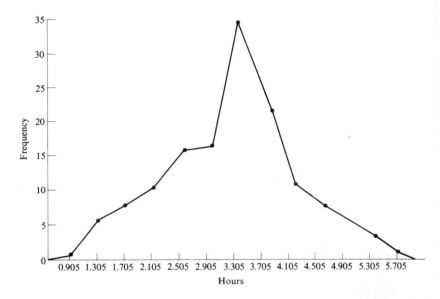

2.2 A frequency polygon showing the number of hours spent watching television (between 5 and 11 P.M.).

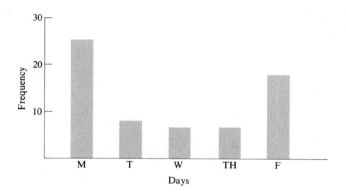

2.3 Number of workers reporting sick each working day (Monday through Friday).

Example 13 (Ranked data) Fifty factory workers, making similar machine parts had their work judged as (a) very good, (b) good, (c) fair, (d) poor, (e) very poor during a quality-control study. The number in each category were respectively 8 20, 14, 6, 2. They are shown in the bar graph of Fig. 2.4.

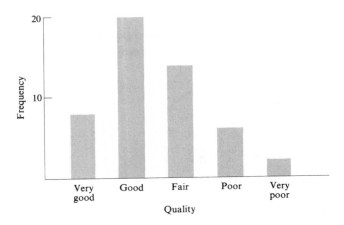

2.4 Ranked quality-control data, as judged by the number of workers whose work falls into various classes.

Example 14 (Nominal data) Consider that five retail outlets A, B, C, D, and E, have, respectively, 350, 420, 270, 140, and 380 employees. The bar graph showing this data is given in Fig. 2.5.

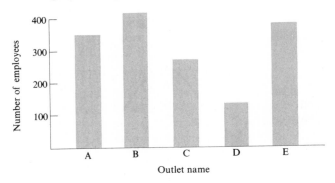

2.5 The number of employees in each of five retail outlets.

There is no particular reason why all of the bar graphs are drawn with vertical bars; they could as well have been horizontal. Consider the data of Example 14 presented in a horizontal bar graph (Fig. 2.6).

2.6. OTHER METHODS OF GRAPHICAL PRESENTATION

We present below a number of examples of different pictorial devices for presenting data graphically. It should be emphasized that most of the devices present nothing more, numerically, than has already been discussed; they are, in effect, eye-catchers.

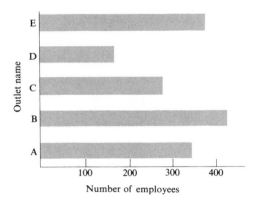

2.6   The number of employees in each of five retail outlets.

Example 15 The numbers of new houses built in states A, B, C, D, E, and F during 1970 were as follows:

State	A	B	C	D	E	F
New houses	70,409	15,310	40,257	27,108	19,217	35,615

We may present the data pictorially as in Fig. 2.7. Here each house (⌂) represents 10,000 houses. This method is imprecise, but gives a useful overall impression to the viewer who is not interested in details.

Example 16 The data of Example 15 could be presented in a pie chart (see Fig. 2.8, where different frequencies are represented by the areas of the segments). Pie charts have the advantage of being visually attractive and precise, because the *actual numbers* of houses are printed on the chart.

Example 17 The data of Example 12 may be presented in a *component bar graph*, as shown in Fig. 2.9. This graph indicates the total number of sick reports for the week, but tends to obscure day-to-day comparison.

Example 18 To show the relationship between two variables we may employ the line graph. To illustrate this graph, consider the table below, which compares the number of years that a (semiskilled) worker has been with a company, and his hourly earnings. These data are plotted in Fig. 2.10.

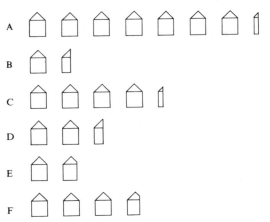

2.7 The number of new houses built in various states during 1970.

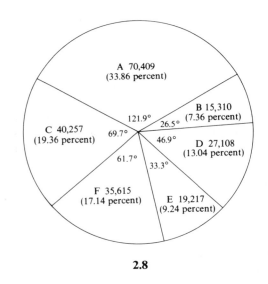

2.8

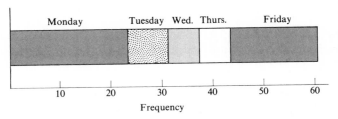

2.9 The number of workers reporting sick (per day).

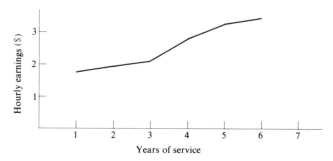

2.10   The relationship between years of service and earnings.

Number of years with the company	Hourly earnings
1	$1.75
2	1.87
3	2.04
4	2.78
5	3.21
6	3.35

The use of computers eliminates much of the tedious work formerly associated with the procedure of drawing histograms and graphs.

SUMMARY

This chapter has been concerned with organizing, summarizing, and presenting graphically large masses of business data. Of the four types of data discussed, discrete, ranked, and nominal data present no special problems. Continuous data are adequately handled by means of a frequency distribution, the construction of which is outlined in Section 2.2. The methods of graphical presentation are then discussed, with a few general comments on unbiased representation of data. Finally, several examples of miscellaneous devices for pictorially presenting different kinds of data are given.

Words to Remember

Grouped data   Class mark
Ungrouped data   Class frequency
Continuous data   Class width
Discrete data   Tally mark

Ranked data	Cumulative frequency
Nominal data	Histogram
Frequency distribution	Frequency polygon
Range	Bar graph
Class interval	Pie chart
Class limit	Line graph

Symbols to Remember

N	Number of observations
X_i	Value of an observation
Range	Largest observation minus smallest observation
Class width	Range/number of classes
f_i	Class frequency
Y_i	Class mark
Σf_i	Total number of frequencies in a distribution

EXERCISES

1. The personnel department of Company X wants to know something about the age distribution of its 72 employees. The data are as follows:

42.7	47.1	43.9	39.0	42.4	39.7
24.1	34.4	50.1	33.0	53.7	56.7
35.0	46.7	32.3	57.1	27.2	57.3
52.9	54.7	29.2	59.2	33.3	55.0
49.2	50.5	61.1	48.0	40.0	30.5
42.7	61.0	44.2	31.4	54.2	37.2
47.1	32.8	30.0	51.3	37.8	65.7
31.3	46.3	19.3	56.7	45.3	40.8
47.7	27.8	44.8	32.7	34.9	31.6
44.1	46.1	40.3	55.5	37.8	52.8
51.3	36.3	47.3	46.0	45.1	41.9
40.2	37.3	37.8	45.4	26.4	42.2

Make a frequency table of the data, and plot the histogram.

2. The diameter of machine parts turned out by a manufacturing process varies slightly with time. The quality-control engineer, in order to study this variation, takes a sample of forty-four parts. The data are given below (in inches):

0.512	0.502	0.507	0.504	0.513	0.511
0.505	0.513	0.508	0.509	0.502	0.507
0.517	0.516	0.521	0.516	0.504	0.512
0.512	0.510	0.511	0.512	0.512	0.503
0.518	0.510	0.504	0.511	0.511	0.508
0.514	0.514	0.513	0.515	0.513	0.515
0.505	0.502	0.511	0.512	0.505	0.506
0.520	0.501				

a) Make a frequency table of this data.

b) Superimpose, on one graph, the histogram and the frequency polygon.

3. Compare the areas of the histogram and the corresponding frequency polygon in Exercise 1.

4. Construct a cumulative-frequency distribution for Exercise 1.

5. Construct the cumulative-frequency distribution for Exercise 2.

6. The weekly receipts for the transit company of city A are:

Week	Receipts (thousands of dollars)
1	217
2	342
3	147
4	284
5	420
6	290
7	195
8	217
9	233
10	271

Draw a bar chart comparing receipts for the ten weeks.

7. The monthly sales for five food stores are:

Store	Sales (thousands of dollars)
A	121
B	432
C	94
D	252
E	77

Construct (a) a bar graph, and (b) a pie chart, to compare the sales of the five stores.

8. The following is a list of the salaries (in dollars) of members of a professional football team.

23,500	36,000	37,000	38,000	38,000
23,000	36,000	30,000	37,000	12,000
23,000	30,000	51,000	22,500	13,500
22,500	34,500	24,000	24,700	14,200
20,500	130,000	25,000	23,300	12,500

Construct a frequency distribution, and plot the histogram.

9. Of 1060 people who bought new cars in 1971, the distribution of the lengths of time they had owned their previous cars is given below (in years):

Length of time	Number
Less than 1 year	180
1 year and under 2	151
2 years and under 3	243
3 years and under 4	167
4 years and under 5	117
5 years and under 6	67
6 years and under 7	62
7 years and under 8	73

Plot a bar graph comparing the figures.

10. The monthly rent of 30 houses (in dollars) is given below:

150	130	100	140	170	120
170	160	130	200	150	160
180	170	170	210	180	130
190	200	190	130	250	270
120	210	300	160	200	150

Construct a frequency distribution, and plot the histogram.

11. The cost of construction of a house in a certain city in the U.S.A. is given below:

Land	$5,000
Labor	$10,000
Materials	$15,000

Draw a suitable diagram to compare the three costs.

12. The production of a manufacturing plant for the last six months is given below:

Month	Production (units)
January	12,000
February	11,000
March	10,000
April	13,000
May	14,000
June	12,000

Draw a bar graph to represent the above data.

13. The monthly expenditures, profits, and the number of units produced by three factories is given in the table at the top of page 26.

	Factory A	Factory B	Factory C
Wages	$3,000	$3,500	$4,000
Materials	4,000	4,200	4,500
Miscellaneous	500	400	600
Profits	1,500	1,200	2,000
Units produced	1,000	1,200	1,500

Draw a suitable diagram to compare the three factories. Compare also the cost and profit per unit.

14. The monthly amounts of take-home pay of two families, family A and family B, are $500 and $700, respectively. The monthly expenditures of the two families are given below:

Item	Family A	Family B
Groceries	$120	$150
Mortgage	200	175
Car	100	75
Heat and light	30	50
Miscellaneous	50	100
Savings	0	150
Total	500	700

a) Compare the expenditures of the two families by a suitable diagram.

b) Find the percentage spent on each item, relative to their take-home pay.

MEASURES
OF LOCATION 3

3.1. INTRODUCTION

In the previous chapter we learned how to organize a mass of business data graphically for inspection; in particular, we discussed the formation of a frequency distribution and the resulting histogram or polygon. We would now like to go further than this, and try to describe the data in a quantitative way. We shall confine our attention to continuous data, both grouped and ungrouped, but the methods are readily applicable to discrete data.

The student may ask why we need to calculate anything at all. Why not just present the histogram (or polygon) and let the viewer interpret the results? The answer to this lies in our purpose in collecting data. Often we want to compare two or more sets of data, or compare an observed set of data with some preconceived theoretical values. This is difficult to do unless we can describe the data simply and numerically.

If we look at the histogram of Example 11 of Chapter 2, we see that the data cluster around some central value, and that, as we move away from the central value in either direction, the frequency of observations decreases to zero. In many situations, it seems reasonable to use the *central value* as the typical value, or the value which summarizes the data. This value is usually in the vicinity of the bulk of the data. Where the center actually lies is a bit difficult to define (as will be seen below) and our first interest will be in discussing this central value.

Measure of Location

By measure of location, we mean a number which describes, in some manner, the "center" of a set of observations.

We put the word *center* in quotation marks to show that there are various ways of defining it.

Example 1 The center of a billiard cue may be (a) the point halfway between the ends or (b) the point at which the cue balances. These are usually not the same.

The calculations for measures of location are slightly different for grouped and ungrouped data; hence they will be discussed separately. Since we are concerned with organizing a large mass of data and since the calculations for measures of location and other statistics (to be discussed later) frequently involve adding large numbers of terms, a compact or simplified notation is needed to indicate *summation*.

3.2. THE SUMMATION NOTATION

In many business problems, there are various types of decisions based on the sum of quantities. For example, total annual income and expenditures of a firm, total sales revenue, total inventory on hand, total supply of money, total number of apartment units in a particular city, etc. In order to differentiate between different quantities, items, individuals, or observations, we let X_1 (read "X sub one") represent the first quantity, individual item, or measurement, X_2 (read "X sub two") represent the second, and so on, X_N (read "X sub N") representing the Nth quantity, item, or individual. To find the sum of a large number of items $X_1, X_2, \ldots, X_N$, we need a shortcut notation to indicate summation.

Summation Notation

Let the symbol X_i (read "X sub i") denote the N values $X_1, X_2, \ldots, X_N$. Then the symbol

$$\sum_{i=1}^{N} X_i$$

(read "summation of X sub i" or "sigma X sub i") is used to denote the sum of all the X_i's from $i = 1$ to $i = N$.

The Greek capital letter Σ denotes the sum, and the subscript i assumes values $1, 2, \ldots, N$. Thus,

$$\sum_{i=1}^{N} X_i = X_1 + X_2 + \cdots + X_N.$$

Example 2 Let $X_1 = \$8$, $X_2 = \$6$, $X_3 = \$9$, $X_4 = \$10$ be the daily wage of four workers. Then $\sum_{i=1}^{4} X_i$ means the sum of the four wages, that is,

$$\sum_{i=1}^{4} X_i = X_1 + X_2 + X_3 + X_4$$

$$= \$8 + \$6 + \$9 + \$10 = \$33.$$

Example 3 Express each of the following summations in expanded form.

a) $\displaystyle\sum_{i=1}^{5} X_i$ b) $\displaystyle\sum_{i=1}^{4} X_i^2$ c) $\displaystyle\sum_{i=1}^{6} f_i X_i$

d) $\displaystyle\sum_{i=1}^{3} (X_i - 3)^2$ e) $\displaystyle\sum_{i=1}^{4} f_i(X_i - 3)^2$ f) $\left(\displaystyle\sum_{i=2}^{4} X_i\right)^2$

Solution

a) $\displaystyle\sum_{i=1}^{5} X_i = X_1 + X_2 + X_3 + X_4 + X_5$

b) $\displaystyle\sum_{i=1}^{4} X_i^2 = X_1^2 + X_2^2 + X_3^2 + X_4^2$

c) $\displaystyle\sum_{i=1}^{6} f_i X_i = f_1 X_1 + f_2 X_2 + f_3 X_3 + f_4 X_4 + f_5 X_5 + f_6 X_6$

d) $\displaystyle\sum_{i=1}^{3} (X_i - 3)^2 = (X_1 - 3)^2 + (X_2 - 3)^2 + (X_3 - 3)^2$

e) $\displaystyle\sum_{i=1}^{4} f_i(X_i - 3)^2 = f_1(X_1 - 3)^2 + f_2(X_2 - 3)^2 + f_3(X_3 - 3)^2 + f_4(X_4 - 3)^2$

f) $\left(\displaystyle\sum_{i=2}^{4} X_i\right)^2 = (X_2 + X_3 + X_4)^2.$

Example 4 If $X_1 = 2$, $X_2 = 4$, $X_3 = 6$, $X_4 = 8$, find:

a) $\displaystyle\sum_{i=1}^{4} X_i$

b) $\left(\displaystyle\sum_{i=1}^{4} X_i\right)^2$

c) $\displaystyle\sum_{i=1}^{4} X_i^2$

d) $\displaystyle\sum_{i=1}^{4} (X_i - 5)^2$

Solution

a) $\sum_{i=1}^{4} X_i = X_1 + X_2 + X_3 + X_4$

   $= 2 + 4 + 6 + 8$

   $= 20$

b) $\left(\sum_{i=1}^{4} X_i \right)^2 = (X_1 + X_2 + X_3 + X_4)^2$

   $= (2 + 4 + 6 + 8)^2$

   $= 20^2 = 400$

c) $\sum_{i=1}^{4} X_i^2 = X_1^2 + X_2^2 + X_3^2 + X_4^2$

   $= 2^2 + 4^2 + 6^2 + 8^2$

   $= 4 + 16 + 36 + 64$

   $= 120$

d) $\sum_{i=1}^{4} (X_i - 5)^2 = (X_1 - 5)^2 + (X_2 - 5)^2 + (X_3 - 5)^2 + (X_4 - 5)^2$

   $= (2 - 5)^2 + (4 - 5)^2 + (6 - 5)^2 + (8 - 5)^2$

   $= 9 + 1 + 1 + 9$

   $= 20.$

Three properties which could be used to simplify operations using summation notation are given below:

Property 1

The sum of a constant multiplied by a variable is equal to the constant multiplied by the sum of the values of the variable. That is, if c is a constant,

$$\sum_{i=1}^{N} cX_i = c \sum_{i=1}^{N} X_i.$$

Example 5 If each worker in Example 2 works 5 days in a week, find the total weekly wage of the four workers.

Solution

Worker	Daily wage X_i
1	$40
2	30
3	45
4	50
	$\sum\limits_{i=1}^{4} X_i = 165$

Thus, by Property 1, the total weekly wage of four workers is

$$\sum_{i=1}^{4} 5X_i = 5 \sum_{i=1}^{4} X_i = 5(\$33) = \$165.$$

Property 2

The summation of a constant is equal to the number of terms in the summation multiplied by the constant. Thus, if c is a constant,

$$\sum_{i=1}^{N} c = Nc.$$

Example 6 If each worker in Example 2 works 5 days in a week, their total number of working days is

Worker	Number of days c
1	5
2	5
3	5
4	5
Total	20

Therefore, by Property 2, the total number of working days is

$$\sum_{i=1}^{4} c = \sum_{i=1}^{4} 5 = 4(5) = 20.$$

Property 3

The summation of the sum (or difference) of two variables is equal to the sum (or difference) of the summations of the individual variables. That is,

$$\sum_{i=1}^{N} (X_i + Y_i) = \sum_{i=1}^{N} X_i + \sum_{i=1}^{N} Y_i$$

and

$$\sum_{i=1}^{N} (X_i - Y_i) = \sum_{i=1}^{N} X_i - \sum_{i=1}^{N} Y_i.$$

Example 7 The daily wages of four workers, called X_i, and the overtime wages, called Y_i, are as listed below:

Worker	Wages, X_i	Overtime wages, Y_i	Total, $X_i + Y_i$
1	$40	$15	$55
2	30	5	35
3	45	10	55
4	50	10	60
	$165	$40	$205

Thus, using Property 3, the total wage of the four workers is:

$$\sum_{i=1}^{4} (X_i + Y_i) = \sum_{i=1}^{4} X_i + \sum_{i=1}^{4} Y_i$$

$$= \$165 + \$40$$

$$= \$205.$$

3.3. THE ARITHMETIC MEAN

The measure of location with which the student is probably most familiar and which is most widely used, is the arithmetic mean or, simply, the *mean*. In common usage, people often speak of the "average" when the *mean* is understood; but the word "average" has other meanings (as in an *average person*, an *average day*, etc.), so we shall usually use the word "mean." In general, N ungrouped observations will be represented by $X_1, X_2, \ldots, X_N$.

Mean

The mean of a set of observations is the sum of their values divided by the number of observations.

For ungrouped data, the mean, represented by $\overline{X}$ (read "X-bar"), is written as:

$$\overline{X} = \frac{X_1 + X_2 + \cdots + X_N}{N} = \frac{\sum_{i=1}^{N} X_i}{N}.$$

The use of the letter X is completely arbitrary; we could just as well have used the letters W, Z, or any other letter we desired. For *grouped data*, the class marks for h classes will be represented by $Y_1, Y_2, \ldots, Y_h$, and the class frequencies by $f_1, f_2, \ldots f_h$.

For grouped data, the mean represented by $\overline{Y}$ is computed as follows:

$$\overline{Y} = \frac{f_1 Y_1 + f_2 Y_2 + \cdots + f_h Y_h}{f_1 + f_2 + \cdots + f_h} = \frac{\sum_{i=1}^{h} f_i Y_i}{\sum_{i=1}^{h} f_i}.$$

Note that $\sum f_i$ is the sum of the observations in each class; thus it is the *total* number of observations, and hence may be replaced by N. Thus

$$\overline{Y} = \frac{\sum_{i=1}^{h} f_i Y_i}{N}.$$

Note also that the formula for grouped data reduces to that for ungrouped data when $h = N$, $f_i = 1$, and $Y_i = X_i$.

Example 8 (Ungrouped data) The wages of five workers (in dollars per hour) are 2.73, 4.02, 2.50, 2.15, and 1.40. Their arithmetic mean is:

$$\bar{X} = \frac{2.73 + 4.02 + 2.50 + 2.15 + 1.40}{5} = \frac{12.80}{5} = 2.56.$$

Example 9 (Ungrouped data) Reconsider Example 8 of Chapter 2. There are 136 observations of TV viewing time. The arithmetic mean of these observations is

$$\bar{X} = \frac{2.34 + 3.14 + \cdots + 1.76}{136} = \frac{430.35}{136} = 3.16.$$

Example 10 (Grouped data) When those same data have been grouped into a frequency table, their arithmetic mean is:

$$\bar{Y} = \frac{(1)(0.905) + (6)(1.305) + \cdots + (1)(5.705)}{136}$$

$$= \frac{431.480}{136} = 3.17.$$

The seeming discrepancy between the two values (3.16 and 3.17) is due to the *approximation* introduced by grouping the data. That is, instead of the observations in a particular class having their individual values, they are *represented* by the class mark.

The use of the mean to describe the *center* of a set of data is widespread because:

1. It can be calculated for all and any sets of data;
2. For any set of the data there is only one mean (we can say that it is *unique*);
3. It can be manipulated simply (for instance, the mean of two sets of data can be easily found from the two individual means);
4. It can be shown to be more reliable than other measures of location (this will be discussed below); and
5. The mean takes into account all of the observations.

Two useful properties of the mean are as follows:

$$\text{Property } 1(a): \quad \Sigma(X_i - \bar{X}) = 0$$
$$\text{Property } 1(b): \quad \Sigma f_i(Y_i - \bar{Y}) = 0.$$

Property 1(a) simply says that the *sum of the differences* between each value and the mean is *zero* for ungrouped data. Property 1(b) says the same thing for *grouped* data. In later chapters, where we may have to calculate $X_i - \overline{X}$ for further work, we can check our calculations by seeing whether the sum of the differences is zero.

Example 11 In Example 8, the mean of the five workers' hourly wages (2.73, 4.02, 2.50, 2.15, 1.40) is 2.56.

Worker	Wage (X_i)	$X_i - \overline{X}$
1	2.73	0.17
2	4.02	1.46
3	2.50	−0.06
4	2.15	−0.41
5	1.40	−1.16
		$\Sigma(X_i - \overline{X}) = 0.00$

A second property (mentioned above) is that if we have two sets of data, the first having N_1 observations with mean $\overline{X}_1$ and the second having N_2 observations with mean $\overline{X}_2$, then the *combined* mean of the two sets of data (often called the *weighted mean*), denoted by $\overline{X}$, is

$$\overline{X} = \frac{N_1\overline{X}_1 + N_2\overline{X}_2}{N_1 + N_2}.$$

Example 12 Suppose that a secretary types ten short reports and two long ones. The mean of the time to type a short report is 30 minutes, while that for a long report is 60 minutes. The (combined) mean of the time to type a report (large or small) is

$$\overline{X} = \frac{(10)(30) + (2)(60)}{12} = \frac{420}{12} = 35 \text{ minutes.}$$

The formula is easily verified by reasoning that the total time taken to do all ten short reports is (10)(30) or 300 minutes, while the total time taken to do the two long reports is (2)(60) or 120 minutes. The total time taken to do all twelve reports is 300 + 120, or 420 minutes, giving an arithmetic mean of 420/12 = 35 minutes.

3.4. THE MEDIAN

In some situations, the mean is a poor figure with which to characterize the center of a set of data. For example, suppose that we have the values 1, 2, 3, 4, and 100. The mean is 110/5 or 22. Note that there are four values less than the mean and only *one* greater than it. In this case, it appears that the mean is *not a very good way* of describing the center of the observation, since four values lie below it and only one above it.

A better way to describe such data is to calculate the middle observation (that is the observation which has half the number of observations less than it and the other half greater than it). Such a value is called the *median* of a set of data. If there is an *odd number* of observations, the median is always the observation whose value is *in the middle*, when all observations have been arranged in order of magnitude. In general, for N observations (where N is odd) the median is the $((N+1)/2)$th largest observation. (For $N = 35$, the median is the 18th largest observation. For $N = 999$, the median is the 500th largest observation.) For an *even number* of observations, since there is not one but two "middle" observations, the median is defined as the *mean of the middle two values*.

Median

That value in a set of observations which, after ordering according to magnitude, has an equal number of observations above and below it, is called the *median*.

Example 13 Consider the five observations of wages of Example 8; they are 2.73, 4.02, 2.50, 2.15, and 1.40. In increasing order of magnitude, they are 1.40, 2.15, 2.50, 2.73, 4.02. Clearly the middle value is 2.50. We say that the median wage is $2.50.

Example 14 If there had been six observations (for instance, 1.40, 2.15, 2.50, 2.70, 2.73, 4.02), then the median would be $(2.50+2.70)/2$, or 2.60.

For data which have been grouped, the median is defined as that value which divides the area of their histogram in half.

To find the median of a set of data arranged in a frequency distribution, we find the class interval that contains the median and then linearly interpolate within that interval. The following example will illustrate the procedure.

Example 15 Suppose that we wish to find the median for the data of Example 9 of Chapter 2. There are 136 observations, so the histogram is divided in half by the value which has $136/2 = 68$ observations above it and 68 below it. Now from Example 10 of Chapter 2 on the cumulative-frequency distribution, we see that there are 56 observations through Class 6 and 89 observations through Class 7. Thus the median lies in Class 7; that is, it lies *between* 3.105 and 3.505. The class width is 0.40 and the class frequency is 33. If we divide the class interval into 33 tiny intervals each of length 0.40/33, then we can imagine that for every unit increase in cumulative frequency from 56 to 89 we move a distance of 0.40/33 hours upward from 3.105. We then ask how far we must move to correspond to the frequency 68. There are twelve units from 56 to 68 (68 − 56 = 12), each corresponding to a distance 0.40/33. Thus the total distance is

$$(12)(0.40)/33 = 0.145.$$

Add this to the upper limit of the previous class interval, namely, 3.105, and we have the answer:

$$3.105 + 0.145 = 3.25.$$

More briefly,

$$\text{Median} = 3.105 + \frac{0.40}{33}(68 - 56) = 3.25.$$

For the general case, if

a) there are N observations,

b) a and b are the lower and upper class limits of the class containing the median,

c) the class width is Δ,

d) N_1 and N_2 are the class frequencies up to a and b, respectively, then:

$$\text{Median} = a + \frac{(N/2 - N_1)(\Delta)}{N_2 - N_1}.$$

Note that $N_2 - N_1$ is the class frequency of the class containing the median.

The properties of the median are:

1. It can be calculated for any and all sets of data;

2. It is unique for any set of data;

3. It can be used to define the center of a set of ranked observations (suppose that observations cannot be measured but can be ranked as best, second best, etc., down to the worst; then the median observation is the one in the middle of this ranking);

4. It is less sensitive to extreme values, that is, to values very different from the majority of the observations.

With regard to point (4), one may say that it is affected more by the *number* of observations than by the *size* of their values. For instance, the median of the numbers 1, 2, 3, 4, and 5 is 3, while the median of the numbers 1, 2, 3, 4, 5, and 1000 is 3.5. On the negative side, the median can be shown to be less reliable than the mean (that is, estimates of the median will tend to vary more than estimates of the mean); for large sets of data, the ordering of the values needed in order to locate the median can be tedious; and, finally, the median is not so easy to manipulate as the mean (for instance, the median of the combination of two sets of data, each with its own median, is not easy to define). If there are N observations $X_1, X_2, \ldots, X_N$, the median is usually denoted by $\tilde{X}$.

3.5. THE MODE

Another quantity which is sometimes used to measure the center of a set of data is the mode, which is simply that value which *occurs most often* or has the highest frequency.

Mode

The most frequently occurring value in a set of data is called the *mode*.

Thus if more people at a football game are 22 than any other age, then 22 is the *modal age*. For ungrouped data, the mode is easy to determine, and can be found almost by inspection. However, for grouped data, although the *class* containing the mode is easy to find, it is difficult to locate the mode within that class. Its main disadvantages are (1) sometimes the mode does not exist and (2) it is not always unique.

Example 16(a) The following figures are the times taken, in minutes, by eleven different workers to perform the same job: 1, 1, 2, 2, 2, 3, 3, 5, 7, 8, 8. The mode of the data is 2, since 2 occurs most often.

Example 16(b) There is no mode to the following six observations: 5, 3, 1, 8, 12, 4, since each value occurs just once.

Example 16(c) The following set of data has two modes (and is called *bimodal*): 1, 1, 2, 2, 2, 3, 3, 5, 5, 5, 8.

3.6. COMPARISON OF THE MEAN, MEDIAN AND MODE

The properties of the mean, median and mode have already been discussed in detail. It should be apparent from this discussion that the *mean* is the most useful measure of the center of a set of data. It always exists, is unique, reliable, takes into account all of the observations, and is easy to manipulate algebraically. For these reasons, the mean is the most common and widely used measure of the center of a set of data.

The median comes second in its usefulness. It always exists, is unique, is not oversensitive to extreme values, and can be used with ranked data. However, it is less reliable than the mean, can be tedious to calculate, and is not easy to manipulate algebraically.

The mode is the least useful of the three measures. It is handy when a quick or rough estimate of the middle of a set of data is required, but in general it does not always exist, is not always unique, cannot be found for grouped data, and is not easy to manipulate. It has little value in modern business statistics.

Finally note that, for a symmetric frequency distribution, the mean, median, and mode coincide, whereas for a skewed frequency distribution, the mean, median, and mode will usually be arranged as shown in Fig. 3.1.

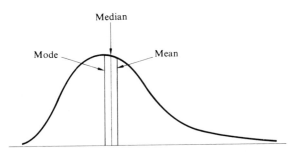

3.1   A typical representation of the mean, median, and mode of a frequency distribution.

3.7. THE GEOMETRIC MEAN

Although the mean, median, and mode are the three most important and widely used measures of the center of a set of data, under certain circumstances other more complex measures are required. We mention just one of them here, the geometric mean, denoted by G, and defined as the Nth root of the product of N positive numbers.

Geometric Mean

The geometric mean, denoted by G, is the Nth root of the product of N positive observations $X_1, X_2, \ldots, X_N$; that is,

$$G = \sqrt[N]{X_1 \cdot X_2 \cdot X_3 \cdots X_N} = (X_1 \cdot X_2 \cdot X_3 \cdots X_N)^{1/N}.$$

Example 17 Consider the costs of four items:

Item	Cost ($)
1	2
2	4
3	4
4	8

The geometric mean of the costs is

$$G = (2 \cdot 4 \cdot 4 \cdot 8)^{1/4} = (256)^{1/4} = 4.$$

A more convenient way of calculating the geometric mean is to use logarithms. Using the fact that

$$\log(X_1 X_2 \cdots X_N) = \log X_1 + \log X_2 + \cdots + \log X_N,$$

we can write:

$$\log G = \log\left\{(X_1 X_2 \cdots X_N)^{1/N}\right\}$$

$$= \frac{1}{N}\left\{\log X_1 + \log X_2 + \cdots + \log X_N\right\}$$

$$= \frac{1}{N}\sum \log X_i.$$

G is then found by using antilogarithms.

Example 18 Using the hourly wages of the five workers of Example 8, namely 2.73, 4.02, 2.50, 2.15, and 1.40, we find the geometric mean as follows:

Worker	Wage (X_i)	$\log(X_i)$
1	2.73	0.4362
2	4.02	0.6042
3	2.50	0.3979
4	2.15	0.3324
5	1.40	0.1461
		$\sum \log X_i = 1.9168$

$$\log G = \tfrac{1}{5}(1.9168) = 0.3834,$$

and

$$G = 2.42 \text{ (from antilogs).}$$

The geometric mean is sometimes useful for averaging ratios, percentages and index numbers and for computing average rates of change. It can also be used with data whose dispersion appears to increase proportionally to its mean. It has certain disadvantages which have limited its use. These are (1) it is not defined for observation values which are negative or zero, and (2) it is difficult to interpret. Thus, for instance, profit and loss data could not be averaged using the geometric mean. The geometric mean is the correct measure to use if we require (1) the mean of ratios and (2) an average percentage rate of change.

Example 19 Consider the following population figures for a particular town.

Year	Population	% of previous period	122.7% × population
1940	5000		
1950	6000	120	6135
1960	7800	130	7528
1970	9204	118	9237
	Mean	368/3 = 122.7	

The average rate of change for the years 1950, 1960, and 1970 is 122.7%, but if each year's figure is multiplied by 122.7, the 1970 figure of 9237 does not agree with the actual figure of 9204, indicating that the average rate of change 122.7%) does not reflect the growth situation accurately. The *geometric mean* is

$$G = (120 \times 130 \times 118)^{1/3} = 122.56.$$

When this is multiplied by each previous year's figure, it gives a 1970 population which agrees with the actual result, indicating that the overall rate of change is best expressed in terms of the *geometric mean*.

SUMMARY

After data have been organized and graphically presented, we wish to describe them in terms of a few quantitative variables. In this chapter we look at three ways of measuring the "center" of the data. These three measures of location are (a) the arithmetic mean, (b) the median, and (c) the mode. Each measure is defined both for grouped and ungrouped data, and a number of examples are given.

The arithmetic mean is the sum of the observations divided by the number of observations, and the sum of deviations of observations from their mean is zero. The median is the point which divides the data in two, while the mode is the value which occurs most frequently.

A comparison of the three measures is made. The mode is almost never used; the median is useful under certain circumstances, but the arithmetic mean is the most useful and most used measure of central location.

A passing reference to the geometric mean is made.

Words to Remember

Measure of location

Arithmetic mean

Weighted mean

Median

Mode

Geometric mean

Symbols to Remember

Y_i	Class mark
$\overline{X}$	Arithmetic mean (ungrouped data)
$\overline{Y}$	Arithmetic mean (grouped data)
Median	Middle observation
Mode	Most frequent observation
G	Geometric mean

EXERCISES

1. Express each of the following summations in expanded form:

   a) $\sum_{i=1}^{10} f_i X_i^2$ b) $\sum_{i=6}^{9} X_i^2$ c) $\left[\sum_{i=1}^{4} X_i(X_i - 2) \right]^2$

2. Write each of the following expressions, using a summation sign:

   a) $X_1 + X_2 + X_3 + X_4 + X_5$

   b) $f_1 X_1 + f_2 X_2 + \cdots + f_N X_N$

   c) $X_2^2 + X_3^2 + X_4^2 + X_5^2$

   d) $(X_1 - 2)^2 + (X_2 - 2)^2 + (X_3 - 2)^2 + (X_4 - 2)^2$

   e) $f_1(X_1 - 4)^2 + f_2(X_2 - 4)^2 + \cdots + f_{10}(X_{10} - 4)^2$.

3. If $X_1 = 5$, $X_2 = 2$, $X_3 = -2$, $X_4 = 3$, find

   a) $\sum_{i=1}^{4} X_i$ b) $\sum_{i=1}^{4} (X_i - 2)$ c) $\sum_{i=1}^{4} (X_i - 2)^2$.

4. If $X_1 = 2$, $X_2 = 7$, $X_3 = 6$, $X_4 = 5$, show that $(\sum_{i=1}^{4} X_i)^2 \neq \sum_{i=1}^{4} X_i^2$.

5. The amounts of money spent on purchases by ten drugstore customers during a particular one-hour period were (in dollars) 1.73, 2.57, 10.05, 0.93, 4.17, 3.30, 4.85, 0.47, 2.75, 4.00. Calculate (a) the arithmetic mean, and (b) the median.

6. In Exercise 5, verify that $\sum (X_i - \overline{X}) = 0$.

7. The times taken by a secretary to type eight letters are (in minutes) 3.4, 1.2, 4.7, 4.0, 2.7, 3.1, 3.8, 4.9. Calculate the arithmetic mean of these times.

8. A sample survey of the yearly incomes of families in a particular neighborhood yielded the following results: $12,700,   8,200,   15,500,   97,300,   9,200,   10,000, 7,400. Calculate (a) the arithmetic mean, and (b) the median. Discuss the relative merits of each as a measure of location of this neighborhood's yearly income.

9. In Exercise 8, suppose that one observation was added later. It is $13,300. Recalculate the median and the arithmetic mean. Comment on the results.

10. For the data of the age distribution of 72 employees of Company X, given in Exercise 1 of Chapter 2, calculate the arithmetic mean when (a) the data are ungrouped, (b) when the data are grouped into a frequency table. Explain any discrepancies between the two results.

11. For the data on age distribution of 72 employees of Company X, given in Exercise 1 of Chapter 2, calculate the median when (a) the data are ungrouped, and (b) when the data are grouped into a frequency table. Explain any discrepancies between the two results.

12. For the data on machine parts turned out by a manufacturing process, given in Exercise 2 of Chapter 2, calculate the arithmetic mean and the median for the ungrouped data.

13. For the data on machine parts turned out by a manufacturing process, given in Exercise 2 of Chapter 2, calculate the arithmetic mean and the median for the grouped data.

14. In Exercise 6 of Chapter 2, calculate the arithmetic mean and the median of weekly receipts for the transit company of City A.

15. For Exercise 7 of Chapter 2, calculate the arithmetic mean and the median of weekly sales for five food stores.

16. What is the arithmetic mean of the salaries of the members of the pro football team of Exercise 8 of Chapter 2?

17. Treating the data of Exercise 9 of Chapter 2 as grouped data, calculate the arithmetic mean and the median length of time of ownership of the previous car.

18. What is the arithmetic mean and the median rent for the data on thirty houses, given in Exercise 10 of Chapter 2?

19. What is the arithmetic mean and median production level (in units) for the data of Exercise 12 of Chapter 2?

20. For the data on age distribution of 72 employees of Company X, given in Exercise 1 of Chapter 2, which class interval contains the mode?

21. In which class interval does the mode lie for the data on machine parts, given in Exercise 2 of Chapter 2?

22. Find the class interval which contains the modal rent value for the data on thirty houses, given in Exercise 10 of Chapter 2.

23. Calculate the geometric mean for the data on weekly sales for five food stores, given in Exercise 7 of Chapter 2.

24. What is the geometric mean of the number of units of production for Exercise 12 of Chapter 2?

25. The mean weekly wage of a group of 100 workers is found to be $120.00. In another group of 200 workers the mean weekly wage is found to be $125.00. What is the mean weekly wage of the combined group of 300 workers?

26. Given the following frequency distribution of the wages of 500 employees in a certain factory:

Daily wage (in dollars)	Number of employees
16–20	2
21–25	7
26–30	101
31–35	180
36–40	120
41–45	50
46–50	25
51–55	10
56–60	5
Total	500

Find (a) the mean wage, and (b) the median wage.

27. The following table gives the life length of 200 light bulbs of type A made by a certain manufacturer.

Life (in hours)	Number of light bulbs
0–500	5
500–1000	13
1000–1500	57
1500–2000	85
2000–2500	26
2500–3000	8
3000–3500	6
Total	200

Calculate (a) the mean life, and (b) the median life.

28. The following table gives the sizes of 300 pairs of shoes sold during the week at a certain shoe store.

Size of shoes	Number of pairs
6.0	2
6.5	3
7.0	68
7.5	72
8.0	85
8.5	40
9.0	16
9.5	11
10.0	3
Total	300

Find (a) the mean size, and (b) the median size.

MEASURES OF VARIABILITY AND DISPERSION 4

4.1. INTRODUCTION

In almost any imaginable situation, the set of data generated by that situation or used to describe it will invariably differ from each other. In short, they will not all have the same value. We may say that they *vary among themselves*, and it is this same variation which can be measured and used to describe the data. In the two preceding chapters, we were concerned with the problem of describing a set of data. First, a frequency distribution was established, where the observations were sorted into a number of classes. Then the mean was calculated and designated as a typical value. These two devices are useful and important but they do not describe all the characteristics of the data. Another tool is needed to show how the data vary about their mean, and this variation is as important a consideration as the mean itself.

In Fig. 4.1, we see two histograms representing the wages of the workers in industries A and B. Although both industries have the same mean, the data for industry B are more spread out or dispersed than those for industry A, indicating that industry B contains more higher-paid and lower-paid workers than does industry A.

As another example, consider the situation where one must choose between two brands of light bulbs on the basis of their life length. If both brands have equal mean life but the first is less variable than the second (in that it has fewer very long- or very short-burning bulbs), then it is more reliable; i.e., intuitively, we would have greater faith in that brand.

It is clear, then, that in order to adequately describe data we need, in addition to a measure of the *center* of the data, like the mean, a measure of the *variation* of the data about this center. The concept of variation is basic to the study of statistical analysis of data, since all statistical methods (including estimation, testing hypotheses, and making forecasts, all to be discussed later are techniques of studying variation, in particular, *chance* variation. As an example, consider that a penny is thrown 100 times and we obtain 20 tails and 80 heads. Is there any evidence to suppose that the penny is not a fair coin? Even if it were fair, we would be unlikely to get *exactly* 50 heads and 50 tails

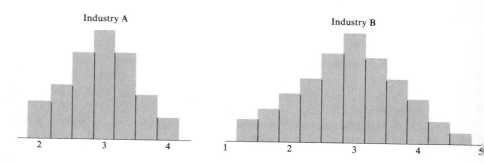

4.1 Wages (in dollars) per hour, for workers in two industries.

In fact we will show later that the number of tails could quite easily vary between 40 and 60, but that values outside of that range are rare. Hence, 20 tails represents too large a variation from that which we might expect by chance, and we would conclude that the penny was unfair. This heuristic discussion will be looked at more critically in later chapters. Hence, the main objective of this chapter is to study different measures of variation (or *dispersion*) that measure the degree of variability in the data. The following measures of variability are discussed:

1. The range
2. The mean deviation
3. The variance
4. The standard deviation
5. The quartile deviation
6. The coefficient of variation

4.2. THE RANGE

We have already introduced one measure of variability of a set of data in the previous chapter, when we were dealing with constructing a frequency distribution. This was the *range*, which is the difference between the values of the largest and the smallest observation. Thus, in Example 8 of Chapter 2, the range of the time spent by viewers in front of their television sets is $5.85 - 0.72 = 5.13$. Clearly the range is a very simple measure of variability of data, and it is easy for the layman to interpret. Despite this, the range has limited value because, being based on the two extreme (largest or smallest) values, it tells us nothing about the variability of the data lying *between* those two values. Consider the following three sets of data, each with the same *range* of $11 - 1 = 10$.

Set 1: 1, 2, 3, 4, 5, 6, 7, 8, 9, 10, 11

Set 2: 1, 6, 6, 6, 6, 6, 6, 6, 6, 6, 11

Set 3: 1, 1, 1, 1, 1, 1, 1, 1, 1, 1, 11

Intuitively we would feel that, even though the range is the same in each case, the three sets of data are not equally dispersed. For this reason the range is used as a measure of dispersion only when (a) quick results are needed, and (b) those who wish to know about the dispersion do not have the training to calculate more sophisticated measures of dispersion. One of the main uses for the range has been in the area of statistical quality control, where the sample sizes are small (in which case the range is a quite adequate measure of dispersion) and quick results are needed to check the variability of manufactured products.

4.3. THE MEAN DEVIATION

Another measure of dispersion or variability that has a certain intuitive appeal is the sum of differences of observations from their arithmetic mean, that is, for the observations $X_1, X_2, \ldots, X_N$, we would calculate

$$\sum_{i=1}^{N} (X_i - \overline{X}).$$

But it can be shown that *this is always zero* no matter how close or disperse the observations are. Now the reason that the sum of differences from the mean is zero is that some differences $X_i - \overline{X}$ are positive and some are negative. When added together the negatives and positives balance each other, making the sum zero. Since, in order to describe the dispersion of data, we don't care whether the difference $X_i - \overline{X}$ is positive or negative, we might add the *absolute values* of the differences; that is, we could ignore the plus or minus signs and simply take *all* differences as positive.

It is usual to denote the absolute value of a number Y by $|Y|$. Thus $|+2| = 2$, $|-3| = 3$, etc. The sum of absolute values of differences from the arithmetic mean is denoted by:

$$\sum_{i=1}^{N} |X_i - \overline{X}| = |X_1 - \overline{X}| + |X_2 - \overline{X}| + \cdots + |X_N - \overline{X}|.$$

When this sum is divided by the number of observations, the result is called the *mean deviation*.

Mean Deviation (ungrouped data)

The mean deviation of a set of data, denoted by M.D., is a measure of dispersion of the data and is the *sum of absolute values* of the differences of observations from their arithmetic mean, divided by the *number* of observations. For N observations, this is

$$\text{M.D.} = \frac{1}{N} \sum_{i=1}^{N} |X_i - \overline{X}|.$$

Example 1 (Refer to Example 8, Chapter 3.) The wages of the five workers are (in dollars) 2.73, 4.02, 2.50, 2.15, and 1.40. The arithmetic mean of these amounts is \$2.56, and the mean deviation is calculated from Table 4.1:

TABLE 4.1.

Wage (X_i)	$X_i - \overline{X}$	$|X_i - \overline{X}|$
2.73	0.17	0.17
4.02	1.46	1.46
2.50	−0.06	0.06
2.15	−0.41	0.41
1.40	−1.16	1.16

$$\sum |X_i - \overline{X}| = 3.26$$

$$\text{M.D.} = \frac{\$3.26}{5} = \$0.65.$$

In the first column of Table 4.1 are the wages or X-values. The value of $\overline{X} = 2.56$ is subtracted from each and recorded in the second column. The third column shows the absolute values of the differences, which are summed at the foot of the column. For grouped data arranged in a frequency table we have the following.

Mean Deviation (grouped data)

In a frequency table with class marks $Y_1, Y_2, \ldots, Y_h$ and class frequencies $f_1, f_2, \ldots, f_h$, the mean deviation is

$$\text{M.D.} = \frac{1}{N} \sum_{i=1}^{h} f_i |Y_i - \overline{Y}|,$$

where $\overline{Y}$ is the arithmetic mean and $N = \sum_{i=1}^{h} hf_i$.

Example 2 (Refer to Chapter 2, Example 8.) The data referring to the number of hours spent watching television are arranged into a frequency distribution in Table 2.1 of Chapter 2. The arithmetic mean $\overline{Y}$ is found (in Example 10 of Chapter 3) to be 3.17. The calculation of the mean deviation is shown in Table 4.2.

TABLE 4.2

Class mark (Y)	$Y_i - \overline{Y}$	$|Y_i - \overline{Y}|$	Class frequency (f)	$f_i |Y_i - \overline{Y}|$
0.905	−2.265	2.265	1	2.265
1.305	−1.865	1.865	6	11.190
1.705	−1.465	1.465	8	11.720
2.105	−1.065	1.065	10	10.650
2.505	−0.665	0.665	15	9.975
2.905	−0.265	0.265	16	4.240
3.305	0.135	0.135	33	4.455
3.705	0.535	0.535	22	11.770
4.105	0.935	0.935	10	9.350
4.505	1.335	1.335	7	9.345
4.905	1.735	1.735	5	8.675
5.305	2.135	2.135	2	4.270
5.705	2.535	2.535	1	2.535

$$\sum f_i |Y_i - \overline{Y}| = 100.440$$

$$\text{M.D.} = \frac{100.440}{136} = 0.7385.$$

The mean deviation has some value as a measure of dispersion since the greater the dispersion of a set of data, the greater is the mean deviation. It is easy to calculate and easy to define and, unlike the range, uses all of the observations. Its main disadvantage is that it is not suited for algebraic manipulations. For instance, if the mean deviations of two sets of data are known, there is no way of combining them without going back to the original sets of data. The main purpose of discussing the mean deviation at all is not for its practical uses, but in order to explain intuitively the development of the concept of a "good" measure of dispersion.

4.4. THE VARIANCE AND THE STANDARD DEVIATION

In the previous section we considered the differences $X_i - \overline{X}$. We had to take their *absolute* values; otherwise their sum would have been zero. Another way of converting the differences $X_i - \overline{X}$ into positive quantities is to square them and add the squared differences. This sum, when divided by N, the number of observations, is called the *variance* of the observations.

For the purposes of estimation (which we'll discuss in Chapter 10), one can show that a better and commonly used way of calculating the variance is to divide by $(N-1)$ rather than by N.

Variance (ungrouped data)

The variance of a set of observations $X_1, X_2, \ldots, X_N$, denoted by s^2, is the sum of the squares of deviations of observations from their arithmetic mean, divided by $(N-1)$, or

$$s^2 = \frac{1}{N-1} \sum_{i=1}^{N} \left(X_i - \overline{X} \right)^2.$$

Example 3 (Refer to Example 8, Chapter 3.) The wages of the five workers (in dollars) are $2.73, 4.02, 2.50, 2.15$, and 1.40. Their arithmetic mean is \$2.56. The variance is calculated with the aid of Table 4.3:

TABLE 4.3

Wages (X_i)	$X_i - \overline{X}$	$(X_i - \overline{X})^2$
2.73	0.17	0.0289
4.02	1.46	2.1316
2.50	−0.06	0.0036
2.15	−0.41	0.1681
1.40	−1.16	1.3456

$$\sum (X_i - \overline{X})^2 = 3.6778$$

$$s^2 = \frac{3.6778}{4} = 0.9195.$$

The first column shows the five wages; the differences are shown in the second column, while the squares of these differences occur in the third column. These are summed at the foot of the column.

Variance (grouped data)

In a frequency table with class marks $Y_1, Y_2, \ldots, Y_h$ and class frequencies $f_1, f_2, \ldots, f_h$, the variance is

$$s^2 = \frac{1}{N-1} \sum_{i=1}^{h} f_i (Y_i - \overline{Y})^2.$$

Example 4 (Refer to Chapter 2, Example 8.) The data refer to the number of hours spent watching television and are arranged into a frequency distribution in Table 2.2 of Chapter 2. The arithmetic mean is found (in Example 10 of Chapter 3) to be 3.17. The calculation of the variance uses Table 4.4.

TABLE 4.4

Class mark (Y)	$Y_i - \overline{Y}$	$(Y_i - \overline{Y})^2$	Class frequency (f)	$f_i(Y_i - \overline{Y})^2$
0.905	−2.265	5.130225	1	5.130225
1.305	−1.865	3.478225	6	20.869350
1.705	−1.465	2.146225	8	17.169800
2.105	−1.065	1.134225	10	11.342250
2.505	−0.665	0.442225	15	6.633375
2.905	−0.265	0.070225	16	1.123600
3.305	0.135	0.018225	33	0.601425
3.705	0.535	0.286225	22	6.296950
4.105	0.935	0.874225	10	8.742250
4.505	1.335	1.782225	7	12.475575
4.905	1.735	3.010225	5	15.051125
5.305	2.135	4.558225	2	9.116450
5.705	2.535	6.426225	1	6.426225

$$\sum f_i(Y_i - \overline{Y})^2 = 120.978600$$

$$s^2 = \frac{120.978600}{135} = 0.8961.$$

The variance is a commonly used measure of dispersion in statistics. It is simple to calculate and simple to define and it uses all of the data. Sets of observations which are more dispersed than others will have a correspondingly larger variance. Since it is a sum of squares, the units in which the variance is expressed are the *squares* of the original units. For instance, if the original units are in inches, the variance will be expressed in square inches. For many purposes we would like the measure of dispersion to be in the *same units* as the original. Such a measure of dispersion is obtained from the variance, by taking the square root of the variance. The resulting quantity is called the *standard deviation*, and denoted by s.

Standard Deviation

The positive square root of the variance is called the standard deviation; that is,

$$s = \sqrt{\frac{1}{N-1} \sum_{i=1}^{N} (X_i - \bar{X})^2} \qquad \text{(Ungrouped data)};$$

$$s = \sqrt{\frac{1}{N-1} \sum_{i=1}^{h} f_i (Y_i - \bar{Y})^2} \qquad \text{(Grouped data)}.$$

Example 5 (Refer to Example 3.) The standard deviation of the five workers' wages is

$$s = \sqrt{0.9195} = 0.96.$$

Example 6 (Refer to Chapter 2, Example 8.) The data refer to the number of hours spent watching television. The variance is found in Example 4 to be $s^2 = 0.8961$. The standard deviation is then

$$s = \sqrt{\text{Variance}} = \sqrt{0.8961} = 0.947.$$

For some purposes it is preferable to express the variance in alternative forms. For ungrouped data, two such forms are

$$s^2 = \frac{1}{N-1} \left\{ \sum_{i=1}^{N} X_i^2 - N\bar{X}^2 \right\}$$

and

$$s^2 = \frac{1}{N-1} \left\{ \sum_{i=1}^{N} X_i^2 - \frac{\left(\sum_{i=1}^{N} X_i \right)^2}{N} \right\}.$$

For grouped data, the two corresponding forms are

$$s^2 = \frac{1}{N-1} \left\{ \sum_{i=1}^{h} f_i Y_i^2 - N\overline{Y}^2 \right\}$$

and

$$s^2 = \frac{1}{N-1} \left[\sum_{i=1}^{h} f_i Y_i^2 - \frac{\left(\sum_{i=1}^{h} f_i Y_i \right)^2}{N} \right],$$

where

$$N = \sum_{i=1}^{h} f_i.$$

We indicate below how the first of these alternative formulas is found. By the basic definition of s^2 for ungrouped data,

$$s^2 = \frac{1}{N-1} \sum_{i=1}^{N} \left(X_i - \overline{X} \right)^2$$

$$= \frac{1}{N-1} \sum_{i=1}^{N} \left(X_i^2 - 2X_i\overline{X} + \overline{X}^2 \right)$$

$$= \frac{1}{N-1} \left\{ \sum_{i=1}^{N} X_i^2 - 2\overline{X} \sum_{i=1}^{N} X_i + N\overline{X}^2 \right\}.$$

But from the definition of $\overline{X}$ we have that

$$\frac{\sum_{i=1}^{N} X_i}{N} = \overline{X}$$

or

$$\sum_{i=1}^{N} X_i = N\overline{X}.$$

Replacing $\sum_{i=1}^{N} X_i$ above by $N\overline{X}$, we have

$$s^2 = \frac{1}{N-1} \left\{ \sum_{i=1}^{N} X_i^2 - 2\overline{X} \cdot N\overline{X} + N\overline{X}^2 \right\} = \frac{1}{N-1} \left\{ \sum_{i=1}^{N} X_i^2 - N\overline{X}^2 \right\}.$$

The standard deviation is found by taking the square root of the variance.

Example 7 (Refer to Example 8, Chapter 3.) The wages of the five workers are 2.73, 4.02, 2.50, 2.15, and 1.40. The variance is calculated in Table 4.5, using the formula:

$$s^2 = \frac{1}{N-1}\left\{ \sum_{i=1}^{N} X_i^2 - N\bar{X}^2 \right\}.$$

TABLE 4.5

Worker	Wage (X_i)	X_i^2
1	$2.73	7.4529
2	4.02	16.1604
3	2.50	6.2500
4	2.15	4.6225
5	1.40	1.9600
Total	12.80	36.4458

$$\bar{X} = \frac{12.80}{5} = 2.56$$

$$\bar{X}^2 = (2.56)^2 = 6.5536$$

$$N\bar{X}^2 = (5)(6.5536) = 32.7680$$

$$s^2 = \frac{1}{4}\{36.4458 - 32.7680\} = 0.9195.$$

It may appear at first glance that the alternative way of calculating s^2 is more difficult than the direct way; but in actual practice, with realistically large and complex sets of data (as opposed to the small textbook example above), the alternative forms are usually more efficient. This is because, with the use of a good desk calculator, we can accumulate $\sum X_i$ and $\sum X_i^2$ fairly simply and quickly.

An important property of the variance is that the variances of two independent sets of data may be combined to get a joint or pooled variance. More specifically, if one group of M observations has a variance of s_1^2 and a second group of N observations has a variance of s_2^2, then the combined (or *weighted*) variance, denoted by s^2, is

$$s^2 = \frac{(M-1)s_1^2 + (N-1)s_2^2}{M+N-2}.$$

Example 8 Suppose that a secretary types ten short reports and two long ones; the variance of the time taken to type the short reports is 8 minutes, while that for the long reports is 2 minutes. The variance of the time taken to type all the reports is

$$s^2 = \frac{(9)(8)+(1)(2)}{10} = \frac{74}{10} = 7.4 \text{ minutes.}$$

The variance and the standard deviation are by far the most useful and widely used measures of dispersion. Their advantages are that they use all of the data; they vary with the amount of dispersion; and, finally, they can be manipulated algebraically. In particular, the variance of two sets of data whose individual variances are known can be easily found.

Before proceding with the next section, we mention briefly a measure of variation related to the standard deviation called the *coefficient of variation*. The coefficient of variation (C.V.) is the standard deviation expressed as a percentage of the mean, or

$$\text{C.V.} = (s/\overline{X}) \cdot 100.$$

In Example 5, the standard deviation of the five workers' wages is $s = 0.96$; their mean is 2.56. Thus, the coefficient of variation is

$$\text{C.V.} = \frac{0.96}{2.56} \times 100 = 37.5\%.$$

One use of the coefficient of variation is in comparing the *relative variability* of two distributions which are not expressed in the same units. For example, is the distribution of wages more or less variable than the daily gallonage of gasoline? One could not compare the standard deviations, since one is in dollars and one in gallons. The coefficients of variation *are* comparable since they do not have any units. Note, however, that the coefficients of variation can also be used to compare the relative variability of distributions expressed in the *same* units; for instance, the variability of morning and evening sales at several large department stores.

4.5. QUARTILES AND QUARTILE DEVIATIONS

To end the discussion on measures of variability, we mention a class of such measures that is often of interest to those dealing with business data. This class of measures could be called *positional measures* of a distribution. They are closely related to the median mentioned in the previous chapter. In the discussion which follows, we shall deal only with grouped data, that is, with data which has been formed into a frequency distribution.

Whereas the median of a set of data was that point which divided the data in half (we could thus say that 50% of the data was less than or equal to the median), we define the three quartiles Q_1, Q_2, and Q_3 of a set of data as follows: 25% of the data are less than or equal to the first quartile Q_1; 50% are less than or equal to the second quartile Q_2; and 75% are less than or equal to the third quartile Q_3. (Note that Q_2 corresponds to the median.)

The quantity $(Q_3 - Q_1)/2$ is called the quartile deviation (or sometimes the interquartile range) and is sometimes used as a measure of variability. The quantity $(Q_3 - Q_1)/(Q_3 + Q_1)$, called the *coefficient of quartile variation*, is unitless.

We could similarly define nine deciles $D_1, D_2, \ldots, D_9$, so that 10% of the data were less than or equal to D_1; 20% less than or equal to D_2, etc. Finally, we could extend the discussion to ninety-nine percentiles $P_1, P_2, \ldots, P_{99}$, where 1% of the data are less than or equal to P_1, 2% less than or equal to $P_2, \ldots$

For the data of Chapter 2, Example 8, we have already shown that the median is 3.25. The reader will be asked in Exercise 25 to show that the 3rd decile is 2.725 and that the 8th decile is 3.865.

SUMMARY

In this chapter we develop the concept of how to describe the dispersion of a set of data by showing that the measure of the center (usually the mean) is not sufficient by itself to adequately represent the data. Four measures of dispersion are mentioned. The range is dismissed from general use because of its limited information about the variation of the data. The mean deviation, while intuitively appealing, is not algebraically tractable. The variance and the standard deviation are intuitively appealing, simple to calculate, and algebraically tractable; for these reasons, they are the generally used measures of dispersion in modern business statistics. Also mentioned are the quartiles, deciles, percentiles, quartile deviation, coefficient of variation, and coefficient of quartile variation.

Words to Remember

Measure of dispersion	Variance
Range	Standard deviation
Mean deviation	Pooled variance
Quartile	Decile
Percentile	Quartile deviation
Coefficient of variation	

Symbols to Remember

M.D.	Mean deviation
s^2	Variance of the observations
s	Standard deviation of the observations

Q_1 First quartile
Q_2 Second quartile (or median)
Q_3 Third quartile

EXERCISES

1. For the data of the age distribution of 72 employees of Company X, given in Exercise 1 of Chapter 2, calculate the mean deviation for (a) the ungrouped data, (b) the grouped data.

2. For the data of the age distribution of 72 employees of Company X, given in Exercise 1 of Chapter 2, calculate the variance of (a) the grouped data, (b) the ungrouped data.

3. For the data on machine parts turned out by a manufacturing process, given in Excercise 2 of Chapter 2, calculate the mean deviation for (a) the ungrouped data, (b) the grouped data.

4. For the data on machine parts given in Exercise 2 of Chapter 2, calculate the variance and standard deviation of (a) the ungrouped data, (b) the grouped data.

5. In Exercises 1, 2, 3, and 4, explain any discrepancies between the results for the grouped and the ungrouped data.

6. Calculate, for the data of Exercise 6 of Chapter 2, (a) the mean deviation, (b) the variance, and (c) the standard deviation, for weekly receipts.

7. Calculate the coefficient of variation for the data of Exercise 1, Chapter 2.

8. The table below gives the number of earners found per family in a study that was carried out in a large city.

Number of earners	Number of families
0	93
1	1327
2	471
3	232
4	68
5	24
6	3
7	2
Total	2220

Calculate the mean deviation and the standard deviation of the number of earners per family.

9. Calculate the variance of weekly sales for the five stores in Exercise 7 of Chapter 2.

10. What is the mean deviation, the variance, and the standard deviation of the earnings of members of the pro football team in Exercise 8 of Chapter 2?

11. Show that $\sum_{i=1}^{N}(X_i - \bar{X})^2 = \sum_{i=1}^{N} X_i^2 - \left(\left(\sum_{i=1}^{N} X_i \right)^2 / N \right)$.

12. Consider a set of data $X_1, X_2, \ldots, X_N$ with variance s^2 and standard deviation s. Show for a constant c $(c \neq 0)$ that $cX_1, cX_2, \ldots, cX_N$ has variance $c^2 s^2$ and standard deviation cs.

13. Find the variance and standard deviation for the length of time that owners kept their previous cars, in Exercise 9 of Chapter 2.

14. What is the variance of the amount of rent paid in Exercise 10 of Chapter 2?

15. The ages of the ten top executives of the PQR Company are: 53, 49, 64, 45, 51, 57, 50, 56, 61, 58. Calculate the variance and standard deviation of their ages.

16. What is the standard deviation of monthly production for the data in Exercise 12 of Chapter 2?

17. What is the variance of ten identical observations?

18. If we get a variance of zero, what does that indicate about the observations?

19. Can the variance ever be negative? Give reasons for your answer.

20. Can the mean deviation ever be negative? Give reasons for your answer.

21. The standard deviation of wages of a group of 100 workers is $2.50. In another group of 200 workers, the standard deviation of wages is found to be $3.00. What is the standard deviation of the wages of the combined group of 300 workers?

22. Calculate the variance and standard deviation of the wages of 500 employees given in Exercise 26 of Chapter 3.

23. Calculate the variance and standard deviation of the life length of 200 light bulbs, given in Exercise 27 of Chapter 3.

24. Calculate the variance and standard deviation of the sizes of 300 pairs of shoes given in Exercise 28 of Chapter 3.

25. Using the data of Example 8, Chapter 2, show that the quartiles and deciles are as follows:

Quartiles: 2.545, 3.250, 3.741

Deciles: 1.835, 2.364, 2.725, 3.065, 3.250, 3.415, 3.618, 3.865, 4.385.

ELEMENTARY
PROBABILITY 5

5.1. INTRODUCTION

In Chapter 1, we pointed out that in order to study observed business data it is necessary to collect them, and to collect them in such a way that valid inferences can be made from them. In Chapters 2 to 4, we discussed descriptive statistics: measures concerned with the problem of describing a mass of data in a concise, clear, useful, and informative way. In Chapter 1, we pointed out, in several examples, how a sales manager is often confronted with the need to make decisions in the face of chance or uncertainty. The theory of probability is very helpful in making economic analysis and business decisions in the face of uncertainty. For example, a sales manager may want to know the probability of doubling the sales of a particular product; management may want to know the probability of the success of a new product; a quality-control engineer may want to know the probability of a machine's needing resetting; a marketing manager may want to know the probability of success of a plant at a new location, etc. In order to make correct statements and decisions in the face of uncertainty, we must learn the fundamentals of probability.

A probability is a number between 0 and 1, inclusive, and can be calculated or estimated in three different ways. The first approach is classical and can be easily understood when applied to gambling games. In such situations, the probability of an event is simply the number of favorable outcomes of an event divided by the number of possible outcomes, where each outcome is equally likely to occur. The second approach is based on relative frequency, and the probability in this case is interpreted as the relative frequency of the event in a large number of repeated trials. The third approach is called *personal probability* or *subjective probability*. In many business problems, it may not be possible to calculate probability by using the classical approach or the relative frequency approach, since the events may not be repeatable. Personal probability or subjective probability is purely a subjective number and can be determined by experienced businessmen based on their degree of belief and experience. In this chapter, we shall discuss the first two approaches and basic probability laws, and give some applications to business problems. In Chapter 19, however, we shall discuss the use of subjective probability in making business decisions.

5.2. THE NOTION OF A SET

The term *set* is regarded as one of the basic ideas of the mathematics necessary for the study of probability and statistics. The idea of a set is common in business. We speak of a set of customers of a product, a set of sales, a set of firm's inventories, a set of personnel in an organization, etc. A set is merely an *aggregate* or *collection* of objects of any sort. Anything that is a member of a set is called an *element* of the set. If a set has a finite number of elements, then it is called a *finite set*; otherwise, it is called an infinite set.

Set

A set is a well-defined collection of objects from a specified universe or population.

By well-defined we mean that for each particular object we must be able to decide whether it does or does not belong to the set.

Example 1 The following are some examples of sets.

1. The set of directors on the board of ABC, Inc.
2. The set of stocks listed on the New York Stock Exchange.
3. The set of employees of a firm.
4. The set of transactions of a firm at the end of a first year in business.

It is customary to denote whole *sets* by capital letters such as A, B, C, and their *elements* by small letters such as a, b, c.

In common practice, there are two ways to describe a set. First, if the set has a finite number of elements, we may *list its members*, enclosing them in braces. Second, a set may be described by enclosing in braces a *defining property* that any object must meet in order to be a member of the set.

Example 2 Let Scott, Wilson, Phillips, and Miller be the directors on the board of ABC Company. Then the set D consisting of the directors on the board is finite, with four elements, and may be written

$$D = \{\text{Scott, Wilson, Phillips, Miller}\}.$$

Example 3 Consider the previous set D of Scott, Wilson, Phillips, and Miller, the directors of ABC Company. The set D may also be described as follows:

$$D = \{X \,|\, X \text{ is a director of ABC Company}\}.$$

Empty Set

A set with no elements is called an empty (or *null*) set. We denote this set by the symbol $\emptyset$.

The empty set plays a role in set theory similar to that of zero in the number system.

Equality of Two Sets

Two sets A and B are said to be equal (or *identical*) if and only if they have *exactly the same elements*.

If two sets A and B are equal, then every element that belongs to A also belongs to B, and every element that belongs to B also belongs to A. Note that the *order* in which we list the elements of a set is immaterial, and that the sets

$$D = \{\text{Scott, Wilson, Phillips, Miller}\}$$

and

$$E = \{\text{Wilson, Phillips, Scott, Miller}\}$$

are equal.

In any particular discussion of sets, it is important to have the *universe* (or the *universal set*) clearly determined. We shall denote the universal set by U; we might then select special subsets from it.

Subset

A set A is a subset of a set B if every element of a set A is also an element of a set B. This is written as $A \subseteq B$ (A is contained in or is equal to B).

Example 4 Let the set

$$D = \{\text{Scott, Wilson, Phillips, Miller}\}$$

and the set

$$E = \{\text{Scott, Wilson, Phillips}\}$$

Then $D \neq E$, since Miller is a member of set D and not a member of set E, but $E \subseteq D$.

5.3. OPERATIONS ON SETS

A useful device, when considering sets and operations on sets of a given universal set, is a Venn diagram. The rectangle U in Fig. 5.1 represents the universal set U, and the elements of U are represented by the points inside and on the rectangle. A subset A of U is represented by the region within and on a circle drawn inside the rectangle.

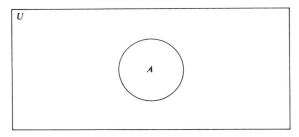

5.1   Venn diagram.

Intersection of Two Sets

The intersection of two sets A and B is the set of all elements of U that are members of both A and B. We denote the intersection of A and B by $A \cap B$ (read "A intersection B").

The intersection of the two sets A and B is represented by the shaded region in Fig. 5.2.

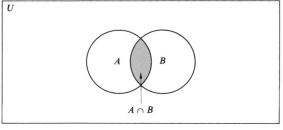

5.2 A intersection B.

Example 5 Let

$$P = \{\text{Scott, Wilson, Lang, Ryan}\}$$

represent the board of directors of PQR company and

$$D = \{\text{Scott, Wilson, Phillips, Miller}\}$$

represent the board of directors of ABC Company. Then their intersection, $P \cap D$, represents the persons who are members of the board of directors of both ABC Company and PQR Company. Thus,

$$P \cap D = \{\text{Scott, Wilson}\}.$$

Mutually Exclusive Sets

Two sets A and B are disjoint, or mutually exclusive, if they have no elements in common. Symbolically, $A \cap B = \emptyset$.

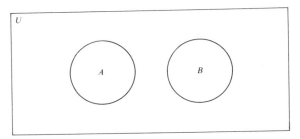

5.3   Mutually exclusive sets.

Example 6 Let $P = \{\text{Scott, Wilson, Lang, Ryan}\}$ represent the partners of ABC Company and $E = \{\text{George, Smith, Johnston}\}$ represent the executives of ABC Company. Then their intersection, $P \cap E$, contains the individuals who are *both* partners and executives of ABC Company. Thus, $P \cap E = \emptyset$.

Union of Two Sets

The union of two sets A and B is the set of all elements of U that belong either to A or to B or to both. We denote the union of A and B by $A \cup B$ (read "A union B").

The union of two sets A and B is represented by the hatched region in Fig. 5.4.

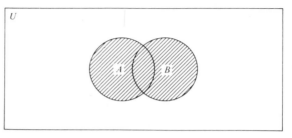

5.4 A union B.

Example 7 Let U be the set of all smokers in a particular district. Let M be the set of male smokers who prefer Brand A and let F be the set of female smokers who prefer Brand A; then the union $M \cup F$ is the set of all smokers in a particular district who prefer Brand A.

Example 8 Let $P = \{$Scott, Wilson, Lang, Ryan$\}$ represent the partners of ABC Company and let $D = \{$Scott, Wilson, Phillips, Miller$\}$ represent the directors of ABC Company. Then the union, $P \cup D$, represents the individuals who are directors *or* partners, *or both*, of ABC Company. Thus,

$$P \cup D = \{ \text{Scott, Wilson, Lang, Ryan, Phillips, Miller} \}.$$

Complement of a Set

The complement of a set A is the set of all elements in U that are not contained in A. We denote the complement of A by A'.

The complement of A is represented by the hatched region in Fig. 5.5.

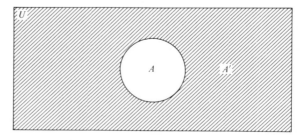

5.5   Complement of A.

Example 9 Let the universal set U be the set of families who purchase coffee. Let A be the set of families who purchase Brand X of coffee. Then the *complement* of the set A is the set of families who do *not* purchase Brand X of coffee.

5.4 EXPERIMENT, SAMPLE SPACE, AND EVENT

The word "experiment" is used in probability and statistics in a much broader sense than in every-day conversation. For example, to observe demand for a particular type of soft drink in a supermarket during a given week is an experiment; to count the number of defective items in a production process involving one or more machines is another experiment. The experiments may not yield the same results when repeated under the same set of conditions. That is, their outcomes are determined by chance. In most business problems, uncertainty exists and the decisions are made in the face of uncertainty. Some examples of such experiments are given below:

1. An advertising campaign for a new product is launched and the number of sales is observed.

2. A poll is taken across the nation, asking whether a certain television program was seen on a particular evening.

3. Items are manufactured on a machine, and the number of defective items during a specified period is counted.

In all of the above examples, experiments can generate a set of possible outcomes. The set of all possible outcomes of an experiment is called the sample space of an experiment.

Sample Space

A sample space S of an experiment is a set of all possible outcomes. An element in a sample space is called a *sample point*.

The idea of a sample space is fundamental in the study of probability and statistics. We shall explain this concept with the help of the following examples.

Example 10 Consider the experiment of picking two items, one at a time, at random, from a box containing defective and nondefective items. Let D represent the defective item and N represent the nondefective item. Then the sample space for this experiment is

$$S = \{ NN, ND, DN, DD \},$$

where NN means both items are nondefective. A *tree diagram* showing all possible outcomes of the above experiment is shown in Fig. 5.6.

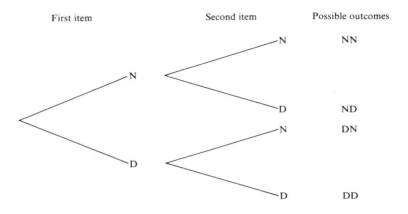

First item   Second item   Possible outcomes

5.6 Tree diagram for selection of two items.

Example 11 Four directors of a company are considered as candidates for a two-member delegation to represent the company at an international meeting. Let A, B, C, and D, denote the directors of a company. Since the order of selection does not count, the number of sample points in this sample space is simply the number of combinations of four different objects, taken two at a time. The six sample points may be listed as

$$S = \{AB, AC, AD, BC, BD, CD\}.$$

Example 12 Suppose an item is manufactured on any one of the three machines, M_1, M_2, M_3. Let D represent a defective item and N represent a nondefective item. Suppose we are interested to know whether the item is defective or nondefective, and the type of machine on which it is manufactured. The sample space for this experiment is

	N	D
M_1	$M_1 N$	$M_1 D$
M_2	$M_2 N$	$M_2 D$
M_3	$M_3 N$	$M_3 D$

A tree diagram showing six possible outcomes of the above experiment is shown in Fig. 5.7.

A sample space S serves as the universal set for all questions related to an experiment. An event E with respect to a particular sample space S is simply a set of possible outcomes favorable to E. For example, consider the experiment of Example 10, picking two items, one at a time, at random, from a box containing defective and nondefective items; the set

$$S = \{NN, ND, DN, DD\}$$

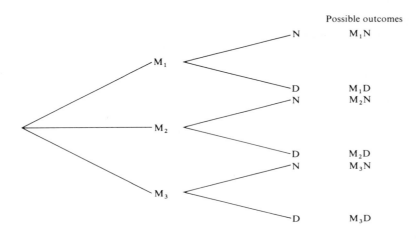

5.7 Tree diagram.

is the sample space for this experiment. For each outcome of the experiment we can determine whether a given event does or does not occur. We may be interested in the event "only one item is defective," we find that this event occurs if the experiment results in an outcome corresponding to an element of the set

$$E = \{ ND, DN \}.$$

We note that E is a *subset* of the *sample space S*.

Event

An "event" is a subset of the sample space of an experiment.

Since an event is a subset of the sample space S, this means that S itself and the empty set Ø are events. The event S is called the certain event, which always occurs, while Ø is the impossible event, which can never occur.

Example 13 The following are some examples of events.

1. In the experiment of picking two items, one at a time, at random, from a box containing defective and nondefective items, "both items are defective" is an event, "both items are nondefective" is another event.

2. In Example 12, "an item is manufactured on a particular machine" is an event; "an item is defective" is another event.

In many business problems we are interested not only in single events but also in *combinations* of events. For example, we might be interested in the event "the first item is nondefective and the second item is nondefective," or we might be interested in the event "an item is manufactured on machine M_1 and is defective." We define below three basic combinations of two events.

Combination of Events

1. Event "A or B". If A and B are events, "A or B" is the event which occurs if A, *or* B, *or both* occur.
2. Event "A and B". If A and B are events, "A and B" is the event which occurs if A *and B both* occur.
3. Event "Not A". If A is an event, "Not A" is the event which occurs if A does not occur.

5.5 RULES OF COUNTING

There are many factors affecting business decisions that are subject to counting or measurement (for example, level of inventory at the end of a given month, number of cars sold during a given week, number of production runs on a given machine in a 24-hour period, etc.). Business decisions depend to a considerable extent on numerous statistics. An efficient way of counting is necessary to handle large masses of statistical data and for an understanding of probability and statistics. In this section we shall develop a few counting techniques. Such techniques will enable us to count the number of ways, the number of samples, or the number of outcomes, without having to *list* all of them.

Addition Rule for Counting

If two operations are *mutually exclusive*, and if the first operation can be performed in N_1 ways and the second operation can be performed in N_2 ways, then one *or* the other can be performed in $(N_1 + N_2)$ ways.

Example 14 The board of directors of XYZ Company has five men and three women as members. A delegate is to be chosen to represent the company at an international meeting. In how many ways can the company choose a delegate?

Solution: There are five possible ways in which a man can be chosen, and three ways in which a woman can be chosen. Therefore, $5 + 3 = 8$ is the number of ways in which the delegate can be chosen.

The addition rule may be extended to any finite number of operations.

Addition Rule for K Operations

If there are K operations, and if the first operation can be performed in N_1 ways, and the second can be performed in N_2 ways, etc., then one of the K operations can be performed in $N_1 + N_2 + \ldots + N_K$ ways, assuming that *no two operations* are performed together.

Example 15 A customer entering a men's clothing store finds that the dealer has 3 white, 4 red, and 6 blue shirts. In how many ways can he buy a white *or* a red *or* a blue shirt?

Solution: Since a white shirt can be chosen in 3 ways, a red shirt in 4 ways, and a blue shirt in 6 ways, the total number of ways in which a white *or* a red *or* a blue shirt can be chosen is

$$3 + 4 + 6 = 13.$$

Multiplication Rule for Counting (Two Operations)

If there are N_1 ways of performing the first operation and, after it is performed in any one of these ways, a second can be performed in N_2 ways, then the two operations can be performed together in $(N_1 \times N_2)$ ways.

Example 16 Two prizes are to be awarded by the ABC Company in a contest of sales of its new product. The first prize is a TV set, and the second prize is a stereo set. In how many ways can the two prizes be distributed among five salesmen if no one can win more than one prize?

Solution: A first prize can be awarded in $N_1 = 5$ ways. After the first contestant has been chosen, a second prize can be awarded in $N_2 = 4$ ways. By the multiplication rule, the total number of ways is $N_1 \times N_2 = (5)(4) = 20$.

The multiplication rule may be extended to any finite number of operations.

Multiplication Rule for Counting (K Operations)

If there are K operations, and if the first operation can be performed in N_1 ways, and if, no matter how the first operation is performed, a second operation can be performed in N_2 ways, and if, no matter how operations one and two are performed, a third operation can be performed in N_3 ways, and so forth for K operations, then the K operations can be performed together in

$$N_1 \times N_2 \times \cdots \times N_K \text{ ways.}$$

Example 17 The life-insurance policies of an insurance company are classified by age of the insured: under 25 years, between 25 and 50 years, and over 50 years old; by sex; and by marital status, single or married. What is the total number of classifications?

Solution: An insurance policy can be classified by age in $N_1 = 3$ ways, by sex in $N_2 = 2$ ways, and by marital status, single or married, in $N_3 = 2$ ways. By the multiplication rule, the total number of classifications is:

$$N_1 \times N_2 \times N_3 = 3 \times 2 \times 2 = 12.$$

Another important counting problem deals with the different arrangements (or *orderings*) of a collection of objects. For example, we might want to know in how many different ways four vice-presidents of a large corporation can be considered for the offices of president and executive vice-president; or we might ask in how many ways five cereal boxes can be arranged on a shelf in a supermarket. The multiplication rule provides a general method for finding the number of arrangements of sets of such objects.

Permutation

An arrangement, or ordering, of all or part of a set is called a permutation.

Example 18 In how many ways can three different machines be arranged on an assembly line?

Solution: The first position can be filled in any one of three ways, the second in any one of two ways, and the last in exactly one way. Hence, by the multiplication rule, the number of possible ways in which three machines can be arranged on an assembly line is:

$$3 \times 2 \times 1 = 6.$$

The six possible permutations or arrangements of three machines M_1, M_2, M_3 in the line may be demonstrated by the tree diagram in Fig. 5.8.

In general, suppose we have n different objects, and we wish to arrange these objects in a line. The first place can be filled in n different ways. The second place must then be filled; this can be done in $(n-1)$ ways. Then there are $(n-2)$ choices for the third place, and so on. By the multiplication rule, we find that there are

$$n(n-1)(n-2)\cdots(2)(1)$$

	First position	Second position	Third position	Possible permutations

M_2 — M_3 — $M_1M_2M_3$

M_1 — M_3 — M_2 — $M_1M_3M_2$

M_1 — M_3 — $M_2M_1M_3$

M_2 — M_3 — M_1 — $M_2M_3M_1$

M_1 — M_2 — $M_3M_1M_2$

M_3 — M_2 — M_1 — $M_3M_2M_1$

5.8 Tree diagram showing the six permutations of the three machines, M_1, M_2, M_3.

different ways. This product is usually denoted by the symbol $n!$ and called "n factorial". In general, we define

$$n! = n(n-1)(n-2)\cdots(2)(1)$$

whenever n is a positive integer. We also define $1! = 1$ and $0! = 1$. Thus the number of permutations of n different objects, taken altogether, is $n!$

Example 19 In how many ways can the supermarket manager display five brands of cereal on a shelf?

Solution: The number of possible displays is:

$$5! = 5 \times 4 \times 3 \times 2 \times 1 = 120.$$

Now consider the permutation of n different objects in which only *some* of the objects are used.

Example 20 In how many ways can the supermarket manager display five brands of cereal in *three spaces* on a shelf?

Solution: The first space can be filled in five ways. After this has been done, the second space can be filled in four ways. Similarly, the third space can be filled in three ways. By the multiplication rule, the three spaces can be filled in

$$5 \times 4 \times 3 = 60 \text{ ways.}$$

In terms of factorial symbols, we have:

$$5 \times 4 \times 3 = \frac{5 \times 4 \times 3 \times 2 \times 1}{2 \times 1} = \frac{5!}{2!} \, .$$

Let the symbol $^5P_3$ denote the number of permutations of five different objects, taken three at a time. Then,

$$^5P_3 = \frac{5!}{2!} = \frac{5!}{(5-3)!} \, .$$

In general, suppose we have n different objects and r spaces to be filled. By the multiplication rule, the r spaces can be filled in

$$n(n-1)(n-2) \cdots (n-r+1) \text{ ways.}$$

Let ^nP_r denote the number of permutations of n different objects, taken r at a time. Then,

$$^nP_r = n(n-1)(n-2) \cdots (n-r+1).$$

In terms of factorial symbols,

$$^nP_r = \frac{n(n-1)(n-2) \cdots (n-r+1)(n-r)!}{(n-r)!}$$

$$= \frac{n!}{(n-r)!} \, .$$

Number of Permutations

The number of permutations of n different objects taken r at a time, denoted by ^nP_r, is

$$^nP_r = \frac{n!}{(n-r)!} \, .$$

Example 21   In how many ways can the manager of a department store arrange a window display with two different colors, choosing from a set of *three* different colors: green, yellow, and red?

Solution: The number of displays or combinations of two colors from a set of three different colors $\{G, Y, R\}$ is

$$GY, \qquad GR, \qquad YR.$$

Now consider the example in which the manager is interested in permutations of three different colors, taken two at a time. The number of permutations of colors is

$$^3P_2 = \frac{3!}{(3-2)!} = \frac{3!}{1!}$$

$$= 3 \times 2 \times 1 = 6.$$

The six possible permutations are listed in the following array:

$$GY \quad GR \quad YR$$
$$YG \quad RG \quad RY$$

The reader may already have observed that the two permutations listed in the first column represent the *same set of colors* $\{GY\}$. Similarly, colors given in the second column represent the same set $\{GR\}$; and the permutations given in the third column represent the same set $\{YR\}$. Thus, the number of *combinations* of three different colors, taken two at a time, is 3, whereas the number of *permutations* of three different colors, taken two at a time, is 6. In other words, the number of *sets or combinations* of three different colors taken two at a time is equal to the number of *permutations* of three different colors taken two at a time divided by 2 (the number of permutations of colors in the selection).

This example illustrates the difference between a permutation and a combination. That is, in a permutation, *order counts*; in a combination, *order does not count*.

In general, the number of combinations of n objects taken r at a time, denoted by $\binom{n}{r}$, is equal to the number of permutations of the r objects divided by $r!$. That is,

$$\binom{n}{r} = {}^nP_r \div r!$$

$$= \frac{n!}{r!(n-r)!}.$$

Number of Combinations

The number of combinations of n different objects, taken r at a time, denoted by $\binom{n}{r}$, is

$$\binom{n}{r} = \frac{n!}{r!(n-r)!}.$$

Example 22 An auto firm decides to advertise its latest models of cars in the following media: two radio stations, one television network, five magazines, and four newspapers.

a) In how many ways can four advertisers be selected from the 12 considered?

b) In how many ways can four advertisers be selected if *no two advertisers* are selected from the *same medium?*

Solution: a) The number of combinations of 12 advertisers taken four at a time is

$$\binom{12}{4} = \frac{12!}{4!\,8!} = \frac{12 \cdot 11 \cdot 10 \cdot 9 \cdot 8!}{4 \cdot 3 \cdot 2 \cdot 1 \cdot 8!}$$

$$= 495.$$

b) If no two advertisers are to be selected from the same medium, this means each medium *must be represented* in the advertising campaign. A radio station can be selected in $\binom{2}{1}$ ways, a television network can be selected in $\binom{1}{1}$ way, a magazine can be selected in $\binom{5}{1}$ ways, and a newspaper can be selected in $\binom{4}{1}$ ways. Hence, by the *multiplication rule*, the number of combinations of advertisers representing all four media is:

$$\binom{2}{1}\binom{1}{1}\binom{5}{1}\binom{4}{1} = (2)(1)(5)(4) = 40.$$

5.6. PROBABILITY OF AN EVENT

In this section we shall develop the concept of probability with equally likely outcomes. The theory of probability has always been associated with games of chance such as dice, cards, etc. For example, let's assume we have a die with six sides, numbered 1 to 6, and that we want to know the chance of rolling a 3. Since there are six possible outcomes we regard each outcome as *equally likely to occur*, and we say that the outcome 3 has a chance of $\frac{1}{6}$ of occurring. Suppose we draw a card from a well-shuffled deck of 52 cards. Since each card has the same chance of being drawn as every other card, the chance of drawing a particular card is $1/52$.

When an experiment is performed, we set up a sample space S of all possible outcomes. In a sample space of N equally likely outcomes, we assign a chance or weight of $1/N$ to each sample point. We define probability for such a sample space as follows:

Probability of an Event

If an experiment can produce N different equally likely outcomes, and if exactly M of these outcomes are favorable to the event A, then the probability of event A, denoted by $P(A)$, is:

$$P(A) = \frac{\text{Number of favorable outcomes}}{\text{Number of possible outcomes}} = \frac{M}{N}.$$

Example 23 Consider the experiment of picking one item at random, from a box containing an *equal number* of defective and nondefective items. The sample space for this experiment is

$$S = \{D, N\}.$$

Since there are two possible outcomes and only one of these is favorable to the event "an item is defective," the probability of the event D, is

$$P(D) = \frac{1}{2}.$$

Similarly, the probability of the event N that "an item is nondefective," is:

$$P(N) = \frac{1}{2}.$$

Since there are only two possibilities, we note that:

$$P(D) + P(N) = 1.$$

If there are no outcomes favorable to an event, then the probability of that event is equal to zero. If all possible outcomes are favorable to an event, then the probability of that event is equal to 1. We list below three basic properties of the probability of an event.

Properties of Probability

The probability of an event satisfies the following three properties.

1. $0 \leqslant P(\text{Event}) \leqslant 1$;

2. $P(\text{Impossible event}) = P(\emptyset) = 0$;

3. $P(\text{Certain event}) = P(S) = 1$.

Example 24 From a box consisting of the same number of defective and nondefective items, two items are drawn, one at a time, at random, with replacement. What is the probability that:

a)   Both items are defective?

b)   Both items are nondefective?

c)   One item is defective and one item is nondefective?

Solution: The sample space S, the set of all possible outcomes, is:

$$S = \{NN, ND, DN, DD\}.$$

a)   There is only one sample point favorable to the event "both items are defective." Therefore,

$$P(DD) = \frac{1}{4}.$$

b)   There is only one sample point favorable to the event "both items are nondefective." Therefore,

$$P(NN) = \frac{1}{4}.$$

c)   There are *two* sample points favorable to the event "one item is defective and one item is nondefective." Therefore,

$$P(\text{One is defective and one is nondefective}) = \frac{2}{4} = \frac{1}{2}.$$

Example 25   (Refer to Example 11.) Four directors of a company are considered as candidates for a two-member delegation to represent the company at an international meeting.

a)   What is the probability that director A is selected?

b)   What is the probability that directors A and D are selected?

c)   What is the probability that director A *or* director D is selected?

d)   What is the probability that director A is *not* selected?

Solution: The sample space S, a set of all possible outcomes, is:

$$S = \{AB, AC, AD, BC, BD, CD\}.$$

a)   There are three sample points favorable to the event "director A is selected." Therefore,

$$P(A) = \frac{3}{6} = \frac{1}{2}.$$

b)   There is only *one* sample point favorable to the event "directors A and D are selected." Therefore,

$$P(\text{A and D}) = \frac{1}{6}.$$

c)   There are *five* sample points favorable to the event "director A *or* D is selected." Therefore,

$$P(\text{A or D}) = \frac{5}{6}.$$

d)   There are *three* sample points favorable to the event "director A is *not* selected." Therefore

$$P(\textit{Not } \text{A}) = \frac{3}{6} = \frac{1}{2}.$$

5.7 RELATIVE FREQUENCY APPROACH TO PROBABILITY

The concept of probability discussed in Section 5.6. is sufficient to define probability in games of chance or in experiments with equally likely outcomes. But the assumption of equally likely outcomes may not be reasonable for many real-life experiments. In actual situations, the outcomes may not be equally likely. For example, what is the probability that an electron tube is defective, or what is the probability of the sale of a particular item, or what is the probability that the president of a company would live to age 70? To calculate probabilities in situations of the above type, we assign *weights* to the events after observing the outcome of an experiment. That is, proportions or relative frequencies may be considered as *weights* or probabilities of events. This method of *proportion*, the number of favorable outcomes divided by the total number of trials, is called the *relative-frequency approach* (or the *empirical approach*) to probability.

Probability as Relative Frequency

If an experiment is performed a sufficient number of times, then in the long run, the relative frequency of an event is called the probability of the occurrence of that event.

Example 26 Suppose a supermarket manager is interested in studying the sales of a particular item. The record of the number of units sold of a particular item in the last 100 sales days is as shown in Table 5.1.

TABLE 5.1

Number of units sold (Events)	Days (f)	Relative frequency
1	4	0.04
2	6	0.06
3	25	0.25
4	35	0.35
5	19	0.19
6	11	0.11
Total	100	1.00

The last column of Table 5.1. gives the relative frequencies of the number of units sold; and consequently, the supermarket manager may use relative frequencies as *weights* for the six mutually exclusive events. From Table 5.1. we find:

$$P(1)=0.04, \quad P(2)=0.06 \quad P(3)=0.25,$$
$$P(4)=0.35, \quad P(5)=0.19, \quad P(6)=0.11.$$

Furthermore, the sum of the weights of the six mutually exclusive events, as shown in the last column of Table 5.1, is 1. The supermarket manager may use these weights to find the probability of making one or more sales. For example, the probability of making a single sale is 0.04; the probability of making either 3 or 4 sales is $0.25+0.35=0.60$; since the two events are mutually exclusive, we *add* the probabilities.

Example 27 The director of personnel of ABC Company is interested in knowing the number of skilled personnel needed by the company for the next year. Table 5.2 gives the probabilities of the number of additional skilled personnel required.

TABLE 5.2

Number of skilled personnel	Probability
Under 30	0.20
30–49	0.30
50–79	0.25
80–99	0.10
100 or over	0.15
Total	1.00

What is the probability that the company will need for the next year:

a) 50 or more additional lled personnel?

b) At least 30 but not more than 79 additional skilled personnel?

Solution: Let us define the following events:

$P(E_1)$ = probability that the additional skilled personnel needed is *under* 30.

$P(E_2)$ = probability that the additional skilled personnel needed is from 30 through 49.

$P(E_3)$ = probability that the additional skilled personnel needed is from 50 through 79.

$P(E_4)$ = probability that the additional skilled personnel needed is from 80 through 99.

$P(E_5)$ = probability that the additional skilled personnel needed is 100 or over.

a) The probability of 50 or more additional skilled personnel is:

$$P(50 \text{ or more}) = P(50\text{–}79) + P(80\text{–}99) + P(100 \text{ or over})$$

$$= P(E_3) + P(E_4) + P(E_5)$$

$$= 0.25 + 0.10 + 0.15$$

$$= 0.50.$$

b) The probability of at least 30 but *not more than* 79 is

$$P(\text{At least 30 but not more than 79}) = P(30\text{–}49) + P(50\text{–}79)$$

$$= P(E_2) + P(E_3)$$

$$= 0.30 + 0.25 = 0.55.$$

5.8 RULES OF PROBABILITY

You may recall that, in Section 5.4, we introduced the concept of "event" and various *combinations* of events. In Sections 5.6 and 5.7, we have learned how to calculate the probability of single events. We now turn to the problem of how to calculate probabilities of various *combinations* of events, such as the event "*A* or *B*", the event "*A* and *B*", and the event "Not *A*". In this section we shall discuss three basic rules for calculating probabilities of related or combined events.

Consider the Venn diagram in Fig. 5.9.

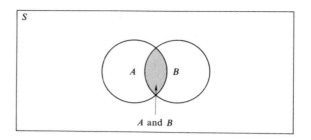

A and B

5.9 Events not mutually exclusive.

The probability of the event "A or B" is the sum of the probabilities of the outcomes favorable to the event A, or B, or both. Now $P(A) + P(B)$ is the *sum* of the probabilities of outcomes favorable to the event A plus the *sum* of the probabilities of outcomes favorable to the event B. It is to be noted that the probability $P(A)$ and the probability $P(B)$ both include the probability of the event "A and B", the *intersection*. That is, by adding $P(A)$ and $P(B)$, we have added $P(A$ and $B)$ twice. If we subtract $P(A$ and $B)$ once, we get the probability of the event "A or B". That is,

$$P(A \text{ or } B) = P(A) + P(B) - P(A \text{ and } B).$$

Rule of Addition: Events not Mutually Exclusive

If A and B are any two events, then:

$$P(A \text{ or } B) = P(A) + P(B) - P(A \text{ and } B).$$

Example 28 (Refer to Example 25.) Four directors of a company are being considered as candidates for a two-member delegation to represent the company at an international meeting. What is the probability that director A or director D is selected?

Solution: The sample space S, a set of all possible outcomes, is :

$$S = \{AB, AC, AD, BC, BD, CD\}.$$

To determine the probability that director A or director D will be selected, let A be the event that director A is selected, and D be the event that director D is selected. Then:

$$A = \{AB, AC, AD\}; \qquad D = \{AD, BD, CD\}; \qquad A \text{ and } D = \{AD\}.$$

Thus, by the rule of addition, the probability of the event "A or D" (that director A or director D is selected) is:

$$P(A \text{ or } D) = P(A) + P(D) - P(A \text{ and } D)$$

$$= \frac{3}{6} + \frac{3}{6} - \frac{1}{6}$$

$$= \frac{5}{6}.$$

In situations where the two events are *mutually exclusive*, that is, when the two events cannot occur at the same time, the probability $P(A \text{ and } B)$ is zero. Thus, the probability of the event "A or B" is simply equal to the sum of the probabilities of the event A and the event B. That is,

$$P(A \text{ or } B) = P(A) + P(B).$$

The Venn diagram for two mutually exclusive events might look as in Fig. 5.10.

Rule of Addition: Mutually Exclusive Events

If A and B are two mutually exclusive events, then:

$$P(A \text{ or } B) = P(A) + P(B).$$

Example 29 (Refer to Example 26.) Let us assume that the probabilities of the number of units sold are as follows:

TABLE 5.3

Number of units sold (Events)	Probability
1	0.04
2	0.06
3	0.25
4	0.35
5	0.19
6	0.11
Total	1.00

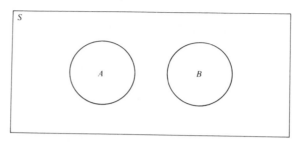

5.10 Mutually exclusive events.

What is the probability that the sales will be three units or more?

Solution: Let us identify the events and their probabilities that will meet the three-unit criterion;

$P(E)$ = probability that the sales will be three units or more.
$P(E_3)$ = probability that the sales will be three units.
$P(E_4)$ = probability that the sales will be four units.
$P(E_5)$ = probability that the sales will be five units.
$P(E_6)$ = probability that the sales will be six units.

From Table 5.3, we find that the events E_3, E_4, E_5, and E_6 are *mutually exclusive*. Thus, by the rule of *addition* for mutually exclusive events,

$$P(E) = P(E_3 \text{ or } E_4 \text{ or } E_5 \text{ or } E_6)$$

$$= P(E_3) + P(E_4) + P(E_5) + P(E_6)$$

$$= 0.25 + 0.35 + 0.19 + 0.11$$

$$= 0.90.$$

Two events are said to be *complementary* if the occurrence of one event precludes the occurrence of the other event. For instance, the events *success* and *failure* are complementary events; the events *defective* and *nondefective* are complementary events. By the rule of addition for mutually exclusive events, the probability of the event A plus the probability of the event "Not A" is equal to one. That is,

$$P(A) + P(\text{Not } A) = 1$$

or

$$P(\text{Not } A) = 1 - P(A).$$

Rule of Complementation

If A and "Not A" are two complementary events, then:

$$P(\text{Not } A) = 1 - P(A).$$

Example 30 (Refer to Example 25.) Four directors of a company are considered as candidates for a two-member delegation to represent the company at an international meeting. What is the probability that director A is *not* selected?

Solution: The sample space S, a set of *all possible outcomes*, is:

$$S = \{AB, AC, AD, BC, BD, CD\}.$$

The probability that director A is not selected is

$$P(\text{Not } A) = 1 - P(A)$$

$$= 1 - \frac{3}{6}$$

$$= \frac{1}{2}.$$

5.9. CONDITIONAL PROBABILITY

In the present section we shall study *conditional probability*. In many business decisions, the manager is interested in knowing the probability that an event A will take place when it is known that another event, B, has occurred. For example, if B is the event "first item is defective," and if A is the event "second item is defective," given the event "first item is defective"; or, if B is the event "a person watches an advertisement of a certain product on television," and if A is the event "he buys the product," then we might ask what the probability is of the event "he buys the product," given the event "a person watches an advertisement of a certain product on television." Thus, the probability of an event A, given the additional information that another event B has occurred, is called the *conditional probability* of A given B. In terms of symbols, this may be written as $P(A|B)$; the vertical line means "given that B has occurred." We shall illustrate the concept of conditional probability with the help of the following example.

Example 31. Suppose two items are drawn at random, one at a time, without replacement, from a box containing three defective and seven nondefective items. Let D denote the event "item is defective," and N denote the event "item is nondefective." Then the two joint events may be read as follows:

NN: Both items are nondefective.

ND: A first item is nondefective and a second item is defective.

DN: A first item is defective and a second item is nondefective.

DD: *Both* items are defective.

The probabilities of the outcomes of the first item and the second item respectively, are shown in the tree diagram Fig. 5.11.

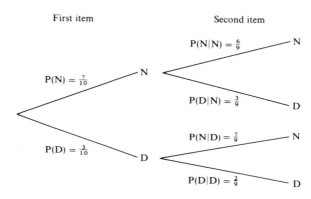

First item · Second item

$P(N) = \frac{7}{10}$ · N · $P(N|N) = \frac{6}{9}$ · N · $P(D|N) = \frac{3}{9}$ · D

$P(D) = \frac{3}{10}$ · D · $P(N|D) = \frac{7}{9}$ · N · $P(D|D) = \frac{2}{9}$ · D

5.11   Tree diagram.

The joint probabilities may now be calculated as follows:

$P(N$ and $N) = P($First item is nondefective$)P($Second item is nondefective$|$ First item is nondefective$)$

$$= P(N)P(N|N)$$

$$= \left(\frac{7}{10}\right)\left(\frac{6}{9}\right) = \frac{42}{90}.$$

Similarly,

$$P(N \text{ and } D) = P(N)P(D|N) = \left(\frac{7}{10}\right)\left(\frac{3}{9}\right) = \frac{21}{90};$$

$$P(D \text{ and } N) = P(D)P(N|D) = \left(\frac{3}{10}\right)\left(\frac{7}{9}\right) = \frac{21}{90};$$

$$P(D \text{ and } D) = P(D)P(D|D) = \left(\frac{3}{10}\right)\left(\frac{2}{9}\right) = \frac{6}{90}.$$

The above information concerning the joint probabilities of two events may be summarized as in Table 5.4; this is known as a *joint probability table*.

TABLE 5.4   Joint Probability Table

		Second item		Probability of outcomes on first item
		N	D	
First item	N	$\dfrac{42}{90}$	$\dfrac{21}{90}$	$\dfrac{63}{90}$
	D	$\dfrac{21}{90}$	$\dfrac{6}{90}$	$\dfrac{27}{90}$
Probability of outcomes on second items		$\dfrac{63}{90}$	$\dfrac{27}{90}$	1.00

From the table we note that:

$$\frac{P(N \text{ and } D)}{P(D)} = \frac{21/90}{27/90} = \frac{21}{27},$$

which is the same as $P(N|D)$. Similarly,

$$\frac{P(N \text{ and } N)}{P(N)} = \frac{42/90}{63/90} = \frac{42}{63},$$

which is the same as $P(N|N)$.

We now generalize this example and introduce a definition of conditional probability.

Conditional Probability

If A and B are two events, and if $P(B)$ is not equal to zero, then the conditional probability of the event A, given the event B, is given by the formula:

$$P(A|B) = \frac{P(A \text{ and } B)}{P(B)}.$$

That is, the conditional probability of the event A, given the event B, is equal to the *joint probability* of A and B, divided by the probability of the event B.

The conditional probability of the event A, given the event B, is:

$$P(A|B) = \frac{P(A \text{ and } B)}{P(B)}.$$

If we multiply both sides by $P(B)$, we get

$$P(A \text{ and } B) = P(A|B)P(B).$$

This equation is called the *rule of multiplication* for dependent events.

Rule of Multiplication

If A and B are two events, then:

$$P(A \text{ and } B) = P(A|B)P(B).$$

That is, the joint probability of A and B is equal to the conditional probability of A given B, times the probability of B.

The above rule of multiplication may also be written as

$$P(A \text{ and } B) = P(B|A)P(A).$$

Example 32 Of all smokers in a particular district, 40 percent prefer Brand A. Of all smokers who prefer Brand A, 30 percent are females. What is the probability that a randomly selected person is a female and prefers Brand A?

Solution: Let

$P(F)$ = probability of a female smoker

$P(F|A)$ = probability of a female smoker, given the smoker who prefers brand A.

Then, by the rule of multiplication,

$$P(\text{Female and Brand A}) = P(\text{Brand A})P(\text{Female}|\text{Brand A})$$

$$= P(A)P(F|A)$$

$$= (0.4)(0.3)$$

$$= 0.12.$$

Example 33 Three items are drawn at random, one at a time, without replacement, from a box containing three defective and seven nondefective items. What is the probability that *all three items* are nondefective?

Solution: Let the possible outcomes be:

N_1: The first item is nondefective.

N_2: The second item is nondefective.

N_3: The third item is nondefective.

Then, by the rule of multiplication,

$$P(N_1 \text{ and } N_2 \text{ and } N_3) = P(N_1)P(N_2|N_1)P(N_3|N_1 \text{ and } N_2)$$

$$= \left(\frac{7}{10}\right)\left(\frac{6}{9}\right)\left(\frac{5}{8}\right)$$

$$= \frac{210}{720}.$$

5.10. STATISTICAL INDEPENDENCE

Two events are said to be *independent* if the occurrence of one event does not influence the occurrence of the second event. For example, if the stock-market prices of one stock do not affect the stock-market prices of another stock (that is, if the selection of one stock gives us *no information* about the second stock), then the two stock prices are said to be *statistically independent*; or, if a product is manufactured on two separate machines and stored in separate places, and if we pick up one item from a lot manufactured on each of these machines, then the event "item manufactured on first machine" is *independent* of the event "item manufactured on the second machine." In other words, the conditional probability, $P(A|B)$, of the event A given the event B, is equal to the unconditional probability of A. That is,

$$P(A|B) = P(A).$$

We know from previous sections that the rule of multiplication for dependent events is

$$P(A \text{ and } B) = P(A|B)P(B).$$

Substituting $P(A|B) = P(A)$ in this equation, the rule of multiplication for independent events becomes

$$P(A \text{ and } B) = P(A)P(B).$$

Independent Events

Two events A and B are said to be independent if and only if $P(A \text{ and } B) = P(A)P(B)$.

We must note that the independence of two events does not mean that the two events are mutually exclusive.

Example 34 (Refer to Example 13). Two items are drawn at random, one at a time, with replacement, from a box consisting of 3 defective and 7 nondefective items. Let D denote the event "item is defective," and N denote the event "item is nondefective."

The probabilities of the outcomes of the first event and the second event respectively, are shown in the tree diagram Fig. 5.12.

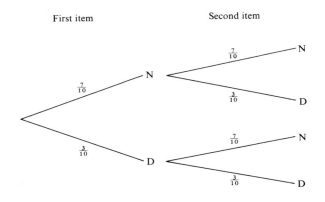

5.12 Tree diagram.

Since we put the first item back in the box, after noting whether it is defective or nondefective, before we draw the second item, the event "first item is defective," and the event "second item is defective" are independent. The joint probabilities may now be calculated as follows:

$$P(N \text{ and } N) = P(N)P(N) = \left(\frac{7}{10}\right)\left(\frac{7}{10}\right) = \frac{49}{100};$$

$$P(N \text{ and } D) = P(N)P(D) = \left(\frac{7}{10}\right)\left(\frac{3}{10}\right) = \frac{21}{100};$$

$$P(D \text{ and } N) = P(D)P(N) = \left(\frac{3}{10}\right)\left(\frac{7}{10}\right) = \frac{21}{100};$$

$$P(D \text{ and } D) = P(D)P(D) = \left(\frac{3}{10}\right)\left(\frac{3}{10}\right) = \frac{9}{100}.$$

Example 35 In a bolt factory, three machines, M_1, M_2, and M_3, manufacture equal numbers of bolts. Of their output, 1, 3, and 2 percent, respectively, are defective bolts. Three bolts are drawn at random, one from each machine's output. What is the probability that *all three* bolts are defective?

Solution: Let the probabilities of the possible events be:

$P(M_1)$ = probability that the bolt manufactured by machine M_1 is defective;

$P(M_2)$ = probability that the bolt manufactured by machine M_2 is defective;

$P(M_3)$ = probability that the bolt manufactured by machine M_3 is defective.

Thus, $P(M_1) = 0.01$, $P(M_2) = 0.03$, $P(M_3) = 0.02$.

Since the three events, "bolt manufactured by machine M_1 is defective," "bolt manufactured by machine M_2 is defective," and "bolt manufactured by machine M_3 is defective," are independent, the probability that *all three* are defective bolts may be calculated as follows:

$$P(M_1 \text{ and } M_2 \text{ and } M_3) = P(M_1)P(M_2)P(M_3)$$

$$= (0.01)(0.03)(0.02)$$

$$= 0.000006.$$

5.11. BAYES' RULE

In Section 5.9, we discussed the application of joint and conditional probabilities to business problems. In this section we shall study how joint and conditional probabilities can be used to *revise* the probability of a particular event in the light of *additional information*. For example, suppose we have two boxes containing defective and nondefective items. One item is picked at random from either one of the boxes and is found defective; and now we might like to know the probability that it came from Box 1. To answer questions of this sort, we use Bayes' Rule, which may be considered an application of conditional probability.

Quite often the businessman has the extra information in a particular event or proposition, either through a personal belief or from the past history of the event. Probabilities assigned on the basis of personal experience, before observing the outcomes of the experiment, are called *prior probabilities*. For example, probabilities assigned to past sales records, to past number of defectives produced by a machine, are examples of *prior probabilities*. When the probabilities are revised with the use of Bayes' rule, they are called *posterior probabilities*. Bayes' rule is very useful in solving practical business problems in the light of additional information.

Although Bayes' rule may be applied to more than two mutually exclusive and exhaustive events, we shall, for the sake of simplicity, state Bayes' rule below only for *two* mutually exclusive and exhaustive events.

Bayes' Rule

If A and B are two mutually exclusive and exhaustive events of the sample space S and if D is any event of S such that $P(D)$ is not equal to zero, then

$$P(A|D) = \frac{P(A \text{ and } D)}{P(A \text{ and } D) + P(B \text{ and } D)}$$

and

$$P(B|D) = \frac{P(B \text{ and } D)}{P(A \text{ and } D) + P(B \text{ and } D)}.$$

Example 36 Box 1 contains three defective and seven nondefective items, and Box 2 contains one defective and nine nondefective items. We select a box at random and then draw one item at random from the box.

a)   What is the probability of drawing a defective item?
b)   What is the probability that Box 1 was chosen, given that a defective item is drawn?

Solution: Let the probabilities of the possible events be:

$P(B_1)=$ probability that Box 1 is chosen;
$P(B_2)=$ probability that Box 2 is chosen;
$P(D)=$ probability that a defective item is drawn;
$P(N)=$ probability that a nondefective item is drawn.

Then the rule of addition gives

$$P(\text{Defective}) = P(\text{Box 1 and defective}) + P(\text{Box 2 and defective}).$$

In terms of symbols

$$P(D) = P(B_1 \text{ and } D) + P(B_2 \text{ and } D)$$

$$= P(B_1)P(D|B_1) + P(B_2)P(D|B_2).$$

The probabilities of the outcomes are shown in the tree diagram of Fig. 5.13.

a)   Using the probabilities given in the tree diagram,

$$P(D) = \left(\frac{1}{2}\right)\left(\frac{3}{10}\right) + \left(\frac{1}{2}\right)\left(\frac{1}{10}\right) = \frac{4}{20}.$$

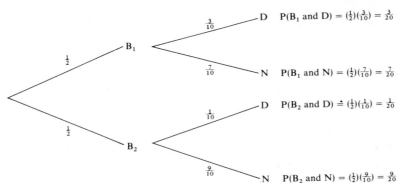

Joint probabilities

5.13 Tree diagram.

b) By Bayes' rule,

$$P(B_1|D) = \frac{P(B_1 \text{ and } D)}{P(D)}$$

$$= \frac{3/20}{4/20} = \frac{3}{4}.$$

$P(B_1)$ and $P(B_2)$ are called *prior probabilities* and $P(B_1|D)$ and $P(B_2|D)$ are called *posterior probabilities*. The above information concerning various probabilities may be summarized as in Table 5.5.

TABLE 5.5.

Event	Prior probability	Conditional probability	Joint probability	Posterior probability
B_1	$\frac{1}{2}$	$\frac{3}{10}$	$\frac{3}{20}$	$\frac{3}{4}$
B_2	$\frac{1}{2}$	$\frac{1}{10}$	$\frac{1}{20}$	$\frac{1}{4}$
Totals	1.0		$\frac{4}{20}$	1.0

Example 37 Of all smokers in a particular district, 40 percent prefer Brand A and 60 percent prefer Brand B. Of those smokers who prefer Brand A, 30

percent are females, and of those who prefer Brand B, 40 percent are females. What is the probability that a randomly selected smoker prefers Brand A, given that the person selected is a female?

Solution: Let the probabilities be:

$P(A)$=probability that a smoker prefers Brand A.

$P(B)$=probability that a smoker prefers Brand B.

$P(M|A)$=probability that a smoker is a male, given brand A.

$P(M|B)$=probability that a smoker is a male, given brand B.

$P(F|A)$=probability that a smoker is a female, given brand A.

$P(F|B)$=probability that a smoker is a female, given brand B.

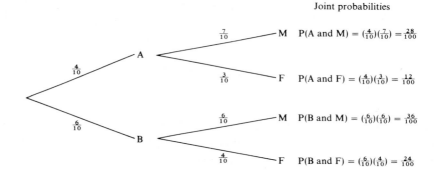

5.14   Tree diagram.

The probabilities of possible outcomes are shown in the tree diagram of Fig. 5.14. By Bayes' rule,

$$P(A|F)=\frac{P(A \text{ and } F)}{P(A \text{ and } F)+P(B \text{ and } F)}.$$

Now, using the probabilities given in the tree diagram, we have:

$$P(A|F)=\frac{12/100}{12/100+24/100}$$

$$=\frac{12/100}{36/100}=12/36.$$

The above information concerning various probabilities may be summarized as in Table 5.6.

TABLE 5.6

Event	Prior probability	Conditional probability	Joint probability	Posterior probability
A	$\dfrac{4}{10}$	$\dfrac{3}{10}$	$\dfrac{12}{100}$	$\dfrac{12}{36}$
B	$\dfrac{6}{10}$	$\dfrac{4}{10}$	$\dfrac{24}{100}$	$\dfrac{24}{36}$
Totals	1.0		$\dfrac{36}{100}$	1.0

Example 38 In a bolt factory, machines M_1, M_2, and M_3 manufacture, respectively, 30, 30, and 40 percent of the total output. Of their output, 1, 3, and 2 percent are defective bolts. A bolt is drawn from a day's output and is found defective. What is the probability that it was manufactured by M_1? by M_2? by M_3?

Solution: Let us identify the probabilities:

$P(D)$ = probability that a defective bolt is drawn.
$P(M_1)$ = probability that a bolt is manufactured by machine M_1.
$P(M_2)$ = probability that a bolt is manufactured by machine M_2.
$P(M_3)$ = probability that a bolt is manufactured by machine M_3.

The data of the problem give the following probabilities:

$$P(M_1) = 0.30, \quad P(D|M_1) = 0.01;$$

$$P(M_2) = 0.30, \quad P(D|M_2) = 0.03;$$

$$P(M_3) = 0.40, \quad P(D|M_3) = 0.02.$$

Now let us calculate $P(M_1 \text{ and } D)$, $P(M_2 \text{ and } D)$, and $P(M_3 \text{ and } D)$.

$$P(M_1 \text{ and } D) = P(M_1)P(D|M_1) = (0.30)(0.01) = 0.003,$$

$$P(M_2 \text{ and } D) = P(M_2)P(D|M_2) = (0.30)(0.03) = 0.009,$$

$$P(M_3 \text{ and } D) = P(M_3)P(D|M_3) = (0.40)(0.02) = 0.008.$$

Using Bayes' rule, we have

$$P(M_1|D) = \frac{P(M_1 \text{ and } D)}{P(M_1 \text{ and } D) + P(M_2 \text{ and } D) + P(M_3 \text{ and } D)}$$

$$= \frac{0.003}{0.003 + 0.009 + 0.008} = \frac{3}{20}.$$

Similarly,

$$P(M_2|D) = \frac{0.009}{0.020} = \frac{9}{20},$$

and

$$P(M_3|D) = \frac{0.008}{0.020} = \frac{8}{20}.$$

The above information concerning various probabilities may be summarized as in Table 5.7.

TABLE 5.7

Event	Prior probability	Conditional probability	Joint probability	Posterior probability
M_1	0.30	0.01	0.003	$\frac{3}{20}$
M_2	0.30	0.03	0.009	$\frac{9}{20}$
M_3	0.40	0.02	0.008	$\frac{8}{20}$
Totals	1.00		0.020	1.00

SUMMARY

The concepts of *experiment, sample space,* and *event* are fundamental in the study of probability and statistics. Several examples of experiments from business and industry are given in Section 5.4. In Section 5.6, the concept of *probability with equally likely outcomes* is discussed. The assumption of equally likely outcomes may not be reasonable for many real-life experiments; in actual situations, the outcomes may not be equally likely. To calculate probabilities in such situations, weights are assigned to the events after observing the outcomes of an experiment. That is, proportions or relative frequencies may be considered as weights or probabilities of events. This approach to probability is called the *relative-frequency approach* (or the *empirical approach*) to probability.

Some rules of probability about events or combinations of events are given in Section 5.8. Several examples showing the application of conditional probability to business problems are presented in Section 5.9. Statistical independence of two or more events is explained in Section 5.10. In Section 5.11 we use the *joint* and *conditional* probabilities to *revise* the probability of a particular event in the light of new information. Probabilities are revised with the help of Bayes' rule, which is very useful in solving practical business problems. More applications of conditional probability and Bayes' rule in making business decisions will be given in a later chapter.

Words to Remember:

Experiment	Probability as relative frequency
Sample space	Conditional probability
Event	Rule of multiplication
Combination of events	Independent events
Probability of an event	Bayes' rule
Mutually exclusive events	Prior probability
Rule of addition	Joint probability
Rule of complementation	Posterior probability

Symbols to Remember

U	Universal set
$\emptyset$	Null set
$\cup$	Union of two sets
$\cap$	Intersection of two sets
A'	Complement of a set A
S	Sample space
$^{n}P_{r}$	Number of permutations of n different objects, taken r at a time
$\binom{n}{r}$	Number of combinations of n different objects, taken r at a time
$P(A)$	Probability of event A
$P(A|B)$	Conditional probability of event A, given that event B has occurred.

EXERCISES

1. Set up the sample space for the experiment of tossing two coins: a nickel and a dime.
2. Set up the sample space for the experiment of tossing three coins: a nickel, a dime, and a quarter.
3. An experiment consists of throwing two dice: a red die and a black die. Let r be the outcome for the red die, and b be the outcome for the black die. Set up the sample space for this experiment.

4. A coin is tossed twice. What is the probability:

   a) of getting two heads?

   b) of getting two tails?

   c) of getting a tail and a head?

5. A die is rolled. What is the probability that an *even number* occurs?

6. A card is drawn from an ordinary deck of 52 cards. What is the probability of getting:

   a) a king?

   b) a queen?

   c) a king or a queen?

   d) a face card?

7. If the letters of the word BUSINESS are arranged at random, what is the probability that:

   a) S and E are together?

   b) The "word" will start with B?

   c) The word will end with S?

   d) The word will start with B and end with S?

8. On the basis of life-expectancy tables, the probability that the president of a company will be alive 25 years hence is $\frac{1}{8}$, and the probability that the vice-president of the same company will be alive 25 years hence is $\frac{3}{5}$. Find the probability that, 25 years hence:

   a) Both will be alive;

   b) At least one of them will be alive;

   c) Only the president will be alive.

9. The office of a company is located in a building which has two elevators; one is waiting at floor 1 twenty-five percent of the time, and the other is waiting there thirty-five percent of the time. Assuming that both elevators operate independently, what is the probability that

   a) Both are waiting at floor 1?

   b) *Neither* is waiting at floor 1?

10. Three copies of a notice are to be sent to three vice-presidents of a company, and three copies of another notice to the directors of the same company. The six copies of the two notices are placed at random in the six envelopes addressed to those six persons. What is the probability that the notices have been correctly dispatched, given that the same person is not both a director and a vice-president?

11. The probability that Harry will be late for work is $\frac{2}{3}$ if he walks, $\frac{1}{4}$ if he takes the bus, and $\frac{1}{6}$ if he drives the car. Assuming that one morning his choice is entirely at random, what is the probability of his being late for work?

12. Let A and B be the events in a sample space S, such that $P(A)=0.6$, $P(B)=0.5$, and $P(A \text{ and } B)=0.2$. Find the probability of the following events:

    a) A or B;   b) Not A.

13. Three salesmen of a company, Smith, Brown, and Jones, make house-to-house sales calls. The probabilities of their making a sale are $\frac{5}{8}$ for Smith, $\frac{3}{8}$ for Brown, and $\frac{1}{3}$ for Jones. If all of them go out for a sales call, calculate the probability that:

    a) A sale will be made;

    b) Only one of them will make a sale.

14. A dealer of shirts has white, green, yellow, blue, and pink shirts for sale. The sales indicate that when a person buys a shirt, the probability that he chooses a white, a green, a yellow, a blue, and a pink shirt are $\frac{1}{2}$, $\frac{1}{5}$, $\frac{1}{10}$, $\frac{3}{20}$, and $\frac{1}{20}$, respectively. Find the probability that he will choose:

    a) Either a blue or a green shirt;

    b) Not a white shirt;

    c) Neither a pink nor a yellow shirt.

15. Box 1 contains 5 defective and 8 nondefective items and Box 2 contains 6 defective and 11 nondefective items. An item is drawn at random from one or the other of the two boxes. Find the probability of drawing a defective item.

16. The ABC Company has two vacancies for an executive position. Mr. Cooper, Mr. Burns, and Mr. Smith are candidates for this position. The probability that Mr. Cooper will be selected is $\frac{5}{8}$, and the probability that Mr. Burns will be selected is $\frac{2}{8}$. What is the probability that:

    a) Mr. Cooper or Mr. Burns will be selected?

    b) Mr. Smith will be selected?

17. In a lot of 5,000 screws, 10% have minor defects, 5% have major defects, and 1% have both minor and major defects. If a screw is selected at random from such a lot, what is the probability that:

    a) It is nondefective?

    b) It has minor or major defects?

18. An automobile dealer sells sedans, hardtops, and station wagons. The probabilities that he will sell a sedan, a hardtop, and a station wagon during the next week are 0.3, 0.2, and 0.1, respectively. What is the probability that;

    a) He will not be able to make a sale at all?

    b) He will sell a sedan, or a hardtop or a station wagon?

    c) He will sell a sedan, a hardtop, and a station wagon?

19. A construction company is bidding for two contracts, A and B. The probability that the company will get contract A is 0.6, the probability that the company will get contract B is 0.3, and the probability that the company will get *both* contracts is 0.1. What is the probability that the company will get contract A or B?

20. A company makes color television sets. The marketing research department of thi company assigns the following probabilities to the various events.

Event	Probability
Selling less than 100,000 units	0.4
Selling 100,000–150,000 units	0.3
Selling 150,001–200,000 units	0.2
Selling more than 200,000 units	0.1
Total	1.0

What is the probability of:

a) Sales of at least 150,000?

b) Sales of more than 150,000?

c) More than 100,000 but less than 200,000?

21. Suppose a cartage company is located in a City A. The truck driver of the compan has to deliver goods in City B and in City C. The driver can take three routes, a, b, c from City A to City B, and four routes d, e, f, g, from City B to City C. If all th routes are equally likely, what is the probability that the driver will select routes a and e in going from City A to City C?

22. Let A and B be two events of a sample space S, such that $P(A)=0.4$, $P(B)=0.6$, an $P(A$ and $B)=0.2$. Find:

a) $P(A|B)$;   b) $P(B|A)$.

23. Let A and B be two events of a sample space S, such that $P(A|B)=\frac{1}{3}$, $P(B|A)=\frac{1}{2}$ and $P(A$ and $B)=\frac{1}{5}$. Find $P(A)$ and $P(B)$.

24. Let A and B be two events of a sample space S, such that $P(A|B)=0.7$, $P(A$ an $B)=0.2$, and $P($not $A)=0.3$. Find $P(A$ or $B)$.

25. An oil company is bidding on two tracts of land, one tract in region A and the othe in region B. The probability of winning the bids in regions A and B are 0.3 and 0.7 respectively. Based on past experience, the company believes that the probability tha the tract in region A has oil is 0.1, and that the tract in region B has oil is 0.3. Wha is the probability that the company will get more oil on account of these bids?

26. Three boxes called Box I, Box II, and Box III contain defective and nondefectiv items. The mixture of defective and nondefective items is as follows:

	Nondefective	Defective	Total
Box I	17	3	20
Box II	20	5	25
Box III	23	7	30
Total	60	15	75

An item is drawn at random; what is the probability that:

a) The item is defective?

b) The item came from Box I?

c) The item came from Box II and is defective?

27. A bakery assigns probabilities to the following events:

Event	Probability
Selling fewer than 1500 loaves per day	0.60
Selling 1500–2500	0.20
Selling 2501–3500	0.15
Selling more than 3500	0.05

Calculate the probability of:

a) Selling more than 2500 loaves;

b) Selling fewer than 3500 loaves;

c) Selling between 1500 and 3500 loaves.

d) Draw a sketch of the probabilities.

e) Describe the cumulative distribution.

28. In a particular university, 70% of the student body is male and 30% is female. 35% study business, and of this 35%, 40% are female and 60% are male. Calculate the probability that:

a) A student selected at random studies business;

b) A male student, selected randomly, studies business.

c) One male and one female selected randomly both study business.

29. In Exercise 28, what is the probability that:

a) A student studies business, given that the student is male?

b) A student studies business, given that the student is female?

c) A student is male, given that the student studies business?

30. In a large company 60% of the employees wear ties. Of this group, 70% hold management positions. What is the probability that

   a) A person is in management, given that he wears a tie?

   b) A person is not in management, given that he wears a tie?

31. Machine 1 makes 60% of a company's ball bearings and machine 2 makes the other 40%. Machine 1 makes 5% defective bearings, while machine 2 makes 2% defective bearings. What is the probability that a defective bearing came from machine 1?

32. 30% of Company A's executives and 50% of Company B's executives play golf. Of Company A's executives, 70% score in the 80's, while 40% of Company B's executives score in the 80's. What is the probability that an executive is from Company A, given that he scores in the 80's?

33. The percentages of a company's employees in accounting, sales, manufacturing, and "other departments" are 10, 40, 40, and 10, respectively. The respective absentee percentages are 8, 2, 3, and 6. Calculate the probabilities that an absent employee is from (a) accounting, (b) sales, or (c) manufacturing.

RANDOM VARIABLES
AND PROBABILITY
DISTRIBUTIONS 6

6.1. RANDOM VARIABLES

In Chapter 5 we considered a number of random experiments, such as picking an item from a box containing defective and nondefective items; selecting a director of a company as a representative for an international meeting; observing the sale of a particular item, etc. When an experiment is performed, the possible outcomes may be numerical quantities such as the number of units sold, or the number of defective items in the sample, or the number of cars sold weekly or monthly by a car dealer, or the life length of an electronic tube. On the other hand, the outcomes may be of a *qualitative* nature such as "defective or nondefective," "good or bad," etc. Often when the outcome of an experiment depends upon judgments of a qualitative nature, all we can do is count the items that "pass the test" or meet our criteria; but it is possible to grade (or rank) the items and even to assign some arbitrary values to the results.

No matter the nature of the experiment or its outcome, so long as that outcome is subject to variation, the numerical values counted, measured, or (arbitrarily) assigned to the outcomes are called variables. And such quantities whose values are determined by the outcome of a *random experiment* are called random variables.

> **Random Variable**
>
> A variable whose value is determined by the outcome of a random experiment is called a *random variable*.

A random variable may be *discrete* or *continuous*. If the random variable takes on the integer values such as $0, 1, 2, \ldots$, then it is called a *discrete random variable.* For example, the number of defective items in a sample, or the weekly or the monthly number of cars sold by a car dealer, or the number of telephone calls made in a given period of time—all these are discrete random variables. If the random variable takes on all values, within a certain interval, then the random variable is called a *continuous random variable.* Any variable involving measurements of height, weight, time, volume, etc., is essentially a *continuous* random variable. For example, the weekly or monthly volume of gasoline sold by a certain gas dealer is a continuous random variable, or the weekly consumption of milk by a family, or the average cost per unit of an item.

Example 1 Consider the experiment of picking one item, at random, from a box containing defective and nondefective items. Let D represent a defective item and N represent a nondefective item. Then the sample space for this experiment is:

$$S = \{D, N\}.$$

The random variable "number of defective items" in this sample space is a *discrete* random variable. Since an item could be nondefective or defective, the random variable "number of defective items" takes on the value 0 or 1.

Example 2 Consider the experiment of picking two items, at random, with replacement, one at a time, from a box containing defective and nondefective items. Let D represent a defective item and N represent a nondefective item. Then the sample space for this experiment is

$$S = \{NN, ND, DN, DD\},$$

where DD means that both items are defective. The random variable "number of defective items picked" is a discrete random variable and takes on the values 0, 1, or 2.

Example 3 (Refer to Example 26 of Chapter 5.) Suppose a supermarket manager is interested in studying the sales of a particular item. The record of the number of units sold of a particular item in the last 100 sales days is a *discrete random variable.* (See Table 6.1.)

Example 4 Suppose an automatic machine is filling coffee jars each with one pound of coffee. Due to some faults in the automatic process, the weight of a jar could vary from jar to jar. The weight of a jar is a *continuous* random variable since the possible values of this random variable could be 0.90, 0.95, 0.99, 1.00, 1.01, 1.05 pounds and since infinitely many more values could lie between any two values, say 0.95 and 1.05 pounds.

TABLE 6.1

Value of the random variable	Days (Frequency)
1	4
2	6
3	25
4	35
5	19
6	11

6.2. PROBABILITY DISTRIBUTIONS OF RANDOM VARIABLES

Frequency distributions were discussed in Chapter 2. Probability distributions of a random variable are closely related to frequency distributions. A frequency distribution is a table in which the measurements or observations of a sample are grouped into classes or intervals. A probability distribution of a random variable is a listing of the various values of a random variable with a corresponding probability associated with each value of the random variable. For instance, consider the example of the sale of a particular item by a retailer. The record of the number of units of a particular item sold in the last 100 selling days is as shown in Table 6.2.

TABLE 6.2

(1) Value of the random variable	(2) Days (Frequency)	(3) Relative frequency or probability
1	4	0.04
2	6	0.06
3	25	0.25
4	35	0.35
5	19	0.19
6	11	0.11
Total	100	1.00

Column (1) of Table 6.2 lists all possible values of the random variable (the number of units sold), column (2) lists the frequency of the number of units sold (the number of days on which a given number of sales occurred), and column

(3) lists the relative frequency (or the probability of occurrence) of each of the values 1, 2, 3, 4, 5, and 6. A table like Table 6.2, which gives a probability to every possible value of the random variable, is called a *probability distribution*. The graph of the probability distribution given in Table 6.2 is shown in Fig. 6.1.

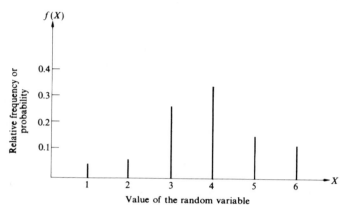

6.1   Probability distribution for Table 6.2.

Probability Distribution

A listing of the probabilities for every possible value of a random variable is called a probability distribution. Since the probabilities are assigned to every possible value of a random variable, the sum of all the corresponding probabilities must be equal to 1.

Example 5 Consider the experiment of picking an item from a box containing equal numbers of defective and nondefective items. The sample space for the experiment is:

$$S = \{D, N\}.$$

Let X represent the random variable "number of defective items"; then the possible values of X are 0, 1. Let $f(X)$ represent the probability of the random variable X; then

$$f(X) = P(\text{Of a random variable } X).$$

Thus,

$$f(0) = P(\text{An item is nondefective}) = P(X = 0) = \frac{1}{2}.$$

$$f(1) = P(\text{An item is defective}) = P(X = 1) = \frac{1}{2}.$$

We note that the values of X exhaust all possible values, and the sum

$$f(0) + f(1) = 1.$$

The possible values of the random variable X, and their corresponding probabilities, are shown in Table 6.3.

TABLE 6.3

Value of the random variable X	Probability $f(X)$
0	$\frac{1}{2}$
1	$\frac{1}{2}$
Total	1.0

The graph of the probability distribution given in Table 6.3 is shown in Fig. 6.2.

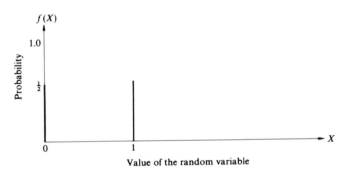

6.2   Probability distribution for Table 6.3.

Example 6 (Refer to Example 2.) Consider the experiment of picking two items at random with replacement, one at a time, from a box consisting of an equal number of defective and nondefective items. The sample space for this experiment is

$$S = \{ NN, ND, DN, DD \}$$

where DD means both items are defective. The random variable "number of defective items picked" takes on the values 0, 1, or 2. The possible values of the random variable X, and their corresponding probabilities, are shown in Table 6.4.

TABLE 6.4

Value of the random variable X	Probability f(X)
0	$\frac{1}{4}$
1	$\frac{1}{2}$
2	$\frac{1}{4}$
Total	1.0

The graph of the probability distribution given in Table 6.4 is shown in Fig. 6.3.

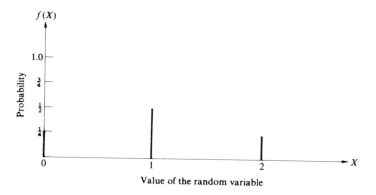

6.3   Probability distribution for Table 6.4.

Example 7 (Refer to Example 4.) Suppose an automatic machine is filling coffee jars each with one pound of coffee. Since the actual weight could vary from jar to jar due to some faults in an automatic process, the weight of a jar is a random variable. Table 6.5 gives the weight in pounds of 100 jars just filled by the machine.

Table 6.5 lists the observed values of the continuous random variable, corresponding frequencies, and their probabilities. Thus, Table 6.5 is a probability distribution of the continuous random variable "weight of a jar."

Suppose we want to know the probability of selecting a jar at random with a weight lying between any two values, say 0.95 and 1.05 pounds. The number of possible values of this random variable lying between 0.95 and 1.05 pounds is

TABLE 6.5

Value of the random variable X	Number of jars (Frequency)	Probability f(X)
0.90	1	0.01
0.95	7	0.07
0.99	25	0.25
1.00	32	0.32
1.01	30	0.30
1.05	5	0.05
Total	100	1.00

not finite. Therefore, the problem of assigning probabilities to continuous random variables cannot be treated in the same way as for discrete random variables. However, the concept of area and its properties proves powerful in assigning probabilities to continuous random variables. It was pointed out, in Section 2.4, that the total area under the graph of a frequency distribution such as a polygon or a histogram is proportional to the total number of observations. In other words, the total area under the graph of a relative frequency curve, called a frequency curve, is equal to 1. Now the relative frequency or the probability of a continuous random variable lying between any two values is given by that portion of the area (under the frequency curve) that lies *between* the two specified values.

The graph of a relative frequency distribution (or the probability distribution) of a continuous random variable may be represented as in Fig. 6.4.

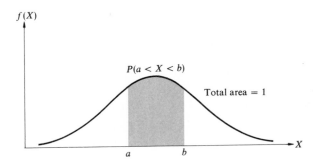

6.4   Probability distribution of a continuous random variable.

6.3. THE EXPECTED VALUE OF A RANDOM VARIABLE

The concept of "expectation" has always been associated with games of chance, lotteries, and bets. At one time or another you must have heard terms such as " fair bet," or "a fair game," or "the player's expected winnings," or "the expectation of winning a prize." For example, suppose an item is drawn from box containing an equal number of defective and nondefective items. Suppos you get $3 if the item is nondefective, and you lose $1 if the item is defective That is, for the single draw of an item, you would either receive $3 or lose $1 Thus, your payoffs are $3 and $-$1$, and their respective probabilities are $\frac{1}{2}, \frac{1}{2}$ The average amount you would make if you drew an item, with replacement over and over, is

$$\$3\left(\frac{1}{2}\right) - \$1\left(\frac{1}{2}\right) = \$1.$$

That is, in an infinite number of draws you expect to make an average of one dollar per draw. This long-run average of one dollar per draw is called the "expectation" or "expected value."

To consider another example, suppose a supermarket manager buys 4 dozen loaves of bread daily, at a cost of $2 per dozen. Suppose, further, th probabilities that he will be able to sell them for $3 a dozen and $2.50 a dozen respectively, are 0.70 and 0.30. What is his expected daily profit? A supermarke manager can sell 40 dozen loaves of bread for $120 and $100, respectively, wit probabilities 0.70 and 0.30. Thus,

$$\text{Expected value} = \$120(0.70) + \$100(0.30)$$

$$= \$84 + \$30 = \$114;$$

and his expected daily profit is

$$\$114 - \$80 = \$34.$$

The reader should by now understand the concept of expectation or expected value. We may now define this concept as follows:

Expected Value

Let X represent a discrete random variable; then the expected value of the probability distribution of X is the *sum* of the products of each possible value of X with its corresponding probability. That is, if we denote the expected value by $E(X)$, then

$$E(X) = \sum Xf(X),$$

where the summation extends over all possible values of X.

This number $E(X)$ is also called the *mean* of the probability distribution, or the *population mean*. The Greek letter μ (mu) is often used to denote the expected value of X or the population mean.

Example 8 (Refer to Example 6.) Suppose two items are drawn at random, with replacement, one at a time, from a box containing an equal number of defective and nondefective items. The probability distribution of the random variable "number of defective items drawn" is given in Table 6.4. What is the expected value of the random variable "number of defective items drawn"?

Solution: The possible values of the random variable "number of defective items drawn" and their corresponding probabilities are given in Table 6.4. The expected value of the number of defective items drawn may be calculated as shown in that table.

TABLE 6.6

(1) Value of the random variable X	(2) Probability $f(X)$	(3) Product $Xf(X)$
0	$\frac{1}{4}$	0
1	$\frac{1}{2}$	$\frac{1}{2}$
2	$\frac{1}{4}$	$\frac{1}{2}$
		$\sum Xf(X) = 1$

Column (1) lists all possible values of the random variable "number of defective items drawn," column (2) lists the corresponding probabilities of each of these values, and the sum of column (3) gives the sum of the products of each possible value of the random variable with a corresponding probability, and is called the *expected value.* Thus

$$E(X) = E(\text{Number of defective items drawn})$$

$$= \sum Xf(X)$$

$$= 1.$$

Example 9 Consider the probability distribution of the sale of a particular item by a retailer in the last 100 selling days, given in Table 6.2. Find the expected value of the number of units sold per day.

Solution: The possible values of the random variable "number of units sold per

day" and their corresponding probabilities are given in Table 6.2. The expected
value of the number of units sold per day is calculated below.

TABLE 6.7.

Value of the random variable X	Probability $f(X)$	$Xf(X)$
1	0.04	0.04
2	0.06	0.12
3	0.25	0.75
4	0.35	1.40
5	0.19	0.95
6	0.11	0.66
		$\sum Xf(X) = 3.92$

Thus,

$$E(X) = E(\text{number of units sold per day})$$

$$= \sum Xf(X)$$

$$= 3.92.$$

Example 10 From an urn containing three white and two black balls, a player
draws two balls at random, with replacement. He gets $100 for every white ball
he draws and loses $50 for every black ball. Assuming he is interested in
maximizing his expected profits, should he participate in such a game?

Solution: To answer this question we should calculate the expected value of this
game. Let the random variable X be the payoffs in dollars. Table 6.8 shows a
sample space of possible outcomes, corresponding probabilities, and the payoffs
in dollars.

Thus the expected value of this game is $\sum Xf(X) = \$80$. Since the expected
value of this game is positive, it means that the player *should* participate in this
game. The average amount the player would make over the long run is $80.

6.4. THE VARIANCE OF A RANDOM VARIABLE

In Chapter 4 we discussed the variance and the standard deviation of a sample
and of a frequency distribution, as measures of *dispersion*. In this section we
shall discuss the variance and the standard deviation as measures of dispersion
of a probability distribution of a random variable. The expected value (or the

TABLE 6.8

Outcomes	Value of a random variable X (Payoffs)	Probability $f(X)$	$Xf(X)$
Both balls white	$200	$\frac{3}{5} \times \frac{3}{5} = \frac{9}{25}$	$200 \times \frac{9}{25} = \$72$
First ball white and second ball black	$50	$\frac{3}{5} \times \frac{2}{5} = \frac{6}{25}$	$50 \times \frac{6}{25} = \$12$
First ball black and second ball white	$50	$\frac{2}{5} \times \frac{3}{5} = \frac{6}{25}$	$50 \times \frac{6}{25} = \$12$
Both balls black	$-\$100	$\frac{2}{5} \times \frac{2}{5} = \frac{4}{25}$	$-100 \times \frac{4}{25} = -\$16$
Total			$\sum Xf(X) = \$80$

mean) of a probability distribution of a random variable tells us about the center of the probability distribution. It gives us quick information about the long-run average when an experiment is performed over and over. But it does not give any information about the "spread" or "dispersion" of the values of a random variable, from one experiment to another. We now introduce the most commonly used measures of dispersion of a probability distribution of a random variable, called the variance and the standard deviation. The variance of a probability distribution of a random variable is measured almost the same way as in the case of a frequency distribution. The only difference is that we now average the squared deviations from their expected value, instead of from the sample mean or the mean of a frequency distribution.

Variance

Let X represent a random variable; then the variance of the probability distribution of a random variable X is the *average squared deviation* measured from the expected value of a probability distribution of X. That is, if we denote the variance by σ^2, then

$$\sigma^2 = \sum (X - \mu)^2 f(X),$$

where the summation extends over all possible values of X.

We recommend that the student review the procedures for calculating the variance of a frequency distribution before proceeding further in this section. We illustrate the calculation of the variance of a probability distribution in the following examples.

Example 11 (Refer to Example 9.) Consider the probability distribution of the sale of a particular item by a retailer in the last 100 selling days, given in Table 6.2. The expected value or the mean of this distribution, as calculated in Example 9, is $\mu = 3.92$. Table 6.9 gives the computation of σ^2.

TABLE 6.9

Value of the random variable X	Probability $f(X)$	$(X - \mu)$	$(X - \mu)^2 f(X)$
1	0.04	−2.92	0.3411
2	0.06	−1.92	0.2212
3	0.25	−0.92	0.2116
4	0.35	0.08	0.0022
5	0.19	1.08	0.2216
6	0.11	2.08	0.4759
Total			1.4736

Thus, $\sigma^2 = 1.4736$. Since the standard deviation is the positive square root of the variance, $\sigma = \sqrt{1.4736} = 1.2139$.

Example 12 (Refer to Example 10.) Consider the probability distribution of the payoffs of the player drawing two balls at random with replacement from an urn containing three white and two black balls, given in Table 6-8. The expected value or the mean of this distribution, as calculated in Example 10, is $\mu = \$80$. The following table gives the computation of the variance σ^2 and the standard deviation σ of this distribution.

Thus $\sigma^2 = 6192$. Since the standard deviation is the positive square root of the variance, the standard deviation is

$$\sigma = \sqrt{6192} = 78.69.$$

The expected value of this game is $80 and the standard deviation is $78.69.

TABLE 6.10

Value of a random variable X (Payoffs)	Probability $f(X)$	$(X-\mu)$	$(X-\mu)^2 f(X)$
200	$\dfrac{9}{25}$	$200-80=120$	$(120)^2 \times \dfrac{9}{25} = 576$
50	$\dfrac{6}{25}$	$50-80=-30$	$(-30)^2 \times \dfrac{6}{25} = 216$
50	$\dfrac{6}{25}$	$50-80=-30$	$(-30)^2 \times \dfrac{6}{25} = 216$
-100	$\dfrac{4}{25}$	$-100-80=-180$	$(-180)^2 \times \dfrac{4}{25} = 5184$
Total			6192

SUMMARY

A variable whose value is determined by the outcome of a random experiment is called a random variable. Random variables may be discrete or continuous. The probability distribution of a random variable is closely related to a frequency distribution. The frequency distribution is a table in which the measurements or observations of a sample are grouped into classes or intervals with a corresponding frequency in each class or interval. The probability distribution of a random variable is the listing of the various values of a random variable with a corresponding probability associated with each random variable. The probability of a continuous random variable lying between any two values is given by the area between the two specified values under the frequency curve or the probability curve of a continuous random variable. The cumulative probability of a random variable is the probability that a random variable is less than or equal to one of its possible values.

The expected value or the mean of the probability distribution of a random variable is the center of the probability distribution and is given by the sum of the products of each possible value of the random variable with its corresponding probability. The variance of the probability distribution of a random variable measures the spread or dispersion of the values of a random variable, and is given by the average squared deviation measured from the expected value of a probability distribution of a random variable.

Words to Remember

Random variable
Discrete random variable
Continuous random variable
Probability distribution
Probability as an area

Cumulative probability
Expected value
Variance
Standard deviation

Symbols to Remember

$f(X)$ Probability distribution of a random variable X
$E(X)$ Expected value of a random variable X
μ Population mean
σ^2 Population variance
σ Population standard deviation

EXERCISES

1. Classify the following random variables as discrete or continuous.

   a) The number of customers in a bank during a lunch hour on a particular day.

   b) The weekly or monthly volume of gasoline sold by a certain gas dealer.

   c) The number of items of a particular stock sold daily at a certain store.

   d) The weekly consumption of milk by a family.

2. Describe two phenomena in business or industry with which you are familiar, one with a discrete and one with a continuous random variable.

3. A box contains 3 defective and 7 nondefective items. Two items are drawn at random one at a time, with replacement.

   a) Describe the sample space.

   b) How many points are there in the sample space?

   c) Attach a probability to each point in the sample space.

4. A committee of two people is to be chosen from among six men and four women.

   a) Describe the sample space.

   b) How many points are there in the sample space?

   c) Assign a probability to each point.

5. In Exercise 3, find the expected value and the variance of the number of defectives drawn.

6. A box contains 4 defective and 6 nondefective items. Three items are drawn at random, one at a time, with replacement. Let the random variable X represent the number of defective items drawn.

   a) Find the probability distribution of X.

   b) Find the cumulative probability distribution of X.

7. Exercise 9 of Chapter 2 presented data on 1060 people who bought new cars in 1971 and the length of time they had owned their previous cars.

   a) Make a probability distribution for the length of time the customer had owned a previous car.

   b) Find the expected length of time the customer had owned a previous car.

8. The following table gives the probability distribution of the number of cars sold by an automobile dealer during a given month.

Number of cars sold	Probability
0	0.01
10	0.05
20	0.39
30	0.45
40	0.10
Total	1.00

a) Find the average number of cars sold during a given month.

b) Find the variance of the number of cars sold during a given month.

9. The table below lists the number of gallons of regular gasoline sold weekly by a service station (rounded to the nearest thousand gallons) with the respective probabilities.

Number of gallons of gasoline sold	Probability
2	0.13
4	0.27
6	0.32
8	0.21
10	0.07
Total	1.00

a) Find the mean weekly gasoline sales.

b) Find the variance of the mean weekly gasoline sales.

10. A man purchases a lottery ticket. He can win a first prize of $4,000 or a second prize of $3,000, with probabilities 0.005 and 0.008, respectively. What should be a fair price to pay for the ticket?

11. A construction company is bidding for a contract, on which it hopes to make a profit of $20,000. The cost of making a proposal is $2,000. If the probability that the company will get the contract is 0.6, what is the expected profit of the company on account of this contract?

12. A man walking into a car showroom will choose cars A, B, or C with probabilities 0.4, 0.3, 0.2, respectively. The cars cost $2,000, $2,500 and $3,000, respectively. What is the expectation of the amount the man will pay?

13. Exercise 20 of Chapter 5 gives the probability distribution of the sales of colored television sets for the next year for a certain company.

   a) What is the expected number of sales for the next year?

   b) What is the variance of sales for the next year?

14. Exercise 27 of Chapter 5 gives the probability distribution of the sales of loaves of bread on a given day by a certain bakery.

   a) What is the expected number of loaves of bread sold on a given day?

   b) What is the variance of the number of loaves sold on a given day?

   c) What is the expected number of loaves sold on two successive days?

   d) What is the variance of the number of loaves on two successive days?

15. A company is working on three independent projects A, B, and C. Expected revenues on projects A, B, and C are $100,000, $50,000, and $25,000, with variances 10,000, 5,000, and 3,000, (dollars)^2, respectively.

   a) What is the total expected revenue on the three projects?

   b) What is the total variance?

16. If $Y=3X-5$ and $E(X)=4$, $\text{Var}(X)=2$, what are the mean and variance of Y?

17. If a random variable X has mean 10 and variance 6, find $E(X^2+3X)$.

18. If the length X of a piece of pipe is a random variable, with expectation 10 and variance 5, what are the expectation and variance of the sum of three such pieces of pipe?

19. The following table gives the probability distribution of the number of customers arriving in a store in a five-minute period.

Number of customers	Probability
0	0.90
1	0.05
2	0.03
3	0.02
Total	1.00

Find the expected number of customers arriving in the next five-minute period.

Using properties 1, 2, and 3 of summation, given in Chapter 2, or otherwise, verify the following properties of the expected value.

20. Let a be a constant. Then the expected value of a is a; that is, $E(a)=a$.

21. Let X be a random variable and a be a constant. Then the expected value of aX is equal to the expected value of X multiplied by a; that is, $E(aX)=aE(X)$.

22. The expected value of a sum of random variables is equal to the sum of their individual expected values, that is, $E(X_1 + X_2 + \cdots + X_N) = E(X_1) + E(X_2) + \cdots + E(X_N)$.

23. Let X be a random variable with $E(X) = \mu$ and variance σ^2. Show that the variance σ^2 is given by the short-cut formula

$$\sigma^2 = E(X^2) - \mu^2.$$

(*Hint*: Use Exercises 20, 21, and 22.)

Using properties of the expected value, given in Exercises 20, 21, and 22, or otherwise, verify the following properties of the variance.

24. Let a be a constant; then the variance of a is zero.

25. Let a be a constant, and X be a random variable; and let the symbol Var(X) be the variance of X. Then

$$\text{Var}(aX) = a^2 \text{Var}(X).$$

In words, this means multiplying each random variable by the same factor a multiplies the variance of the random variable by a^2.

26. Let X and Y be two independent random variables, with $E(X) = \mu_1$, $E(Y) = \mu_2$, and variances $\text{Var}(X) = \sigma_1^2$ and $\text{Var}(Y) = \sigma_2^2$, respectively; then

$$\text{Var}(X \pm Y) = \sigma_1^2 + \sigma_2^2.$$

(*Hint*. Two random variables are said to be independent if $E(XY) = E(X)E(Y)$.)

TWO DISCRETE PROBABILITY DISTRIBUTIONS 7

7.1. INTRODUCTION

In this chapter two discrete probability distributions, the binomial distribution and the Poisson distribution, are discussed. Both of these distributions have wide applications in business and economics. Because of their importance and the wide areas of relevant applications in business and economics, we shall discuss these two probability distributions in detail, and study some of their general properties.

7.2. THE BINOMIAL DISTRIBUTION

We have considered a number of experiments in Chapters 5 and 6 involving one or more trials, such as picking one or more items from a box consisting of defective and nondefective items. In many applications of business and economics, we are interested in a trial which has only two possible outcomes, generally called *success* and *failure*; such trials are called "binomial trials." For example, picking an item from a box consisting of defective and nondefective items is a binomial trial; randomly selecting a family in a particular district and asking them whether they like the taste of Brand X coffee is a binomial trial. The experiment in which a sequence of binomial trials is performed is called a binomial experiment; randomly selecting 50 families in a particular district and asking them whether they like the taste of Brand X coffee is a binomial experiment. The binomial experiment must satisfy the following four conditions:

1. The number of trials is fixed.
2. Each trial has only two outcomes, either S (success) or F (failure).
3. The probability p of a success is the same for each trial.
4. The trials are independent.

We shall use below our earlier examples from Chapters 5 and 6, to derive and illustrate the *probability distribution* of the random variable in a binomial experiment, called the *binomial distribution*.

Example 1 (Refer to Example 5, Chapter 6.) Consider the experiment of picking one item at random from a box containing equal numbers of defective and nondefective items. Then the sample space for this binomial experiment is

$$S = \{D, N\},$$

where D represents the defective item and N represents the nondefective item. Let X represent the random variable "number of defective items"; then the possible values of X are 0, 1. Let p be the probability that an item is defective and $q = 1 - p$ be the probability that an item is nondefective. The probability distribution of the random variable X is given in Table 7.1.

TABLE 7.1 Binomial Distribution
$(n = 1)$

Value of the random variable X	Probability distribution $f(X)$
0	$f(0) = P(N) = q$
1	$f(1) = P(D) = p$
Total	$q + p = 1$

Example 2 (Refer to Example 6, Chapter 6.) Consider the experiment of picking two items at random, one at a time, with replacement, from a box consisting of defective and nondefective items. The sample space for this binomial experiment is

$$S = \{NN, ND, DN, DD\},$$

where DD means both items are defective. Let X represent the random variable "number of defective items picked"; then the possible values of X are 0, 1, or 2. Let p be the probability that an item is defective and $q = 1 - p$ be the probability that an item is nondefective. The binomial distribution of the random variable X is given in Table 7.2.

TABLE 7.2 Binomial Distribution $(n=2)$

Value of the random variable X	Probability distribution $f(X)$
0	$f(0)=P(NN)=P(N)P(N)=q^2$
1	$f(1)=P(ND)+P(DN)=P(N)P(D)$ $+P(D)P(N)=qp+pq=2qp$
2	$f(2)=P(DD)=P(D)P(D)=p^2$
Total	$q^2+2qp+p^2=(q+p)^2=1$

Example 3 Now consider the experiment of picking three items at random, one at a time, with replacement, from a box containing defective and nondefective items. The sample space for this experiment is

$$S=\{NNN, NND, NDN, DNN, NDD, DND, DDN, DDD\},$$

where NNN means all three items are nondefective. Let X represent the random variable "number of defective items picked"; then the possible values of X are 0, 1, 2, 3. Let q be the probability that an item is defective and $p=1-q$ be the probability that an item is nondefective. The binomial distribution of the random variable X is given in Table 7.3.

TABLE 7.3 Binomial Distribution $(n=3)$

Value of the random variable X	Probability distribution $f(X)$
0	$f(0)=P(NNN)=P(N)P(N)P(N)=q^3$
1	$f(1)=P(NND)+P(NDN)+P(DNN)$ $=q^2p+q^2p+q^2p=3q^2p$
2	$f(2)=P(NDD)+P(DND)+P(DDN)$ $=qp^2+qp^2+qp^2=3qp^2$
3	$f(3)=P(DDD)=P(D)P(D)P(D)=p^3$
Total	$q^3+3q^2p+3qp^2+p^3=(q+p)^3=1$

From Tables 7.1, 7.2, and 7.3, we note the following points:

1. There are $2^1 = 2$ possible outcomes when $n = 1$; $2^2 = 4$ possible outcomes when $n = 2$; $2^3 = 8$ possible outcomes when $n = 3$. It must be quite clear by now that there will be 2^n possible outcomes in a binomial experiment of n trials.

2. The coefficients of the probabilities for various values of the random variable in binomial experiments can be obtained by using binomial coefficients, as follows:

$$(q+p)^1 = q + p = \binom{1}{0}q + \binom{1}{1}p.$$

$$(q+p)^2 = q^2 + 2qp + p^2 = \binom{2}{0}q^2 + \binom{2}{1}qp + \binom{2}{2}p^2.$$

$$(q+p)^3 = q^3 + 3q^2p + 3qp^2 + p^3$$

$$= \binom{3}{0}q^3 + \binom{3}{1}q^2p + \binom{3}{2}qp^2 + \binom{3}{3}p^3.$$

Hence, the binomial probabilities for the various values of the random variable are given by the binomial expansion $(q+p)^n$ and the coefficient of the probability for X number of successes in n trials is $\binom{n}{X}$.

3. The exponent of p corresponds to the number of successes and the exponent of q corresponds to the number of failures; and the sum of the exponents of p and q is equal to n, the total number of trials.

Following the above discussion, we find that the probability of X successes and $(n - X)$ failures in n binomial trials is

$$\binom{n}{X} p^X q^{n-X}.$$

Binomial Distribution

The probability distribution of the random variable X, the number of successes in n binomial trials, is called the binomial distribution, and is given by the formula

$$f(X) = \binom{n}{X} p^X q^{n-X}$$

where

i) $X = 0, 1, 2, \ldots, n$;

ii) $p =$ the probability of success in a single trial;

iii) $q =$ the probability of failure in a single trial;

iv) $p + q = 1$.

Example 4 From a lot containing 20 items, five of which are defective, four items are drawn *with replacement*. What is the probability of:

a) Getting exactly one defective item?
b) Getting at least one defective item?

Solution: Let X be the number of defective items drawn; then the possible values of X are 0, 1, 2, 3, 4. Let $p = P$(defective item); then $p = \frac{5}{20} = \frac{1}{4}$. The random variable X is a binomial, since the items are drawn with replacement and the probability of drawing a defective item does not change from trial to trial.

a) Since $n = 4$, $X = 1$, $p = \frac{1}{4}$, and $q = \frac{3}{4}$, we have, using the formula for the binomial distribution,

$$P(X=1) = f(1) = \binom{4}{1}\left(\frac{1}{4}\right)^1\left(\frac{3}{4}\right)^3$$

$$= \frac{4!}{1!3!}\left(\frac{1}{4}\right)^1\left(\frac{3}{4}\right)^3$$

$$= 4\left(\frac{1}{4}\right)\left(\frac{27}{64}\right) = \frac{27}{64}.$$

b) The probability of at least one defective item is:

$$P(X \geqslant 1) = P(X=1) + P(X=2) + P(X=3) + P(X=4)$$

$$= 1 - P(X=0)$$

$$= 1 - \binom{4}{0}\left(\frac{1}{4}\right)^0\left(\frac{3}{4}\right)^4$$

$$= 1 - \left(\frac{3}{4}\right)^4$$

$$= 1 - \frac{81}{256} = \frac{175}{256}.$$

Example 5 A salesman for ABC Company makes 10 house calls a day. The probability that he will make a sale at a randomly selected house is 0.2. Find the probability of a salesman:

a) Making no sales in a day;
b) Making one or more sales in a day;
c) Making exactly 3 sales in a day.

Solution: Let X represent the number of sales the salesman makes in a day. Then the possible values of X are $0, 1, 2, \ldots, 10$. Since $n = 10$, $p = 0.2$, $q = 0.8$, using the binomial formula for the binomial distribution, we have

a) $f(X) = \binom{n}{X} p^X q^{n-X}$. Hence the probability that the salesman makes *no* sales

is:

$$P(X = 0) = f(0) = \binom{10}{0}(0.2)^0(0.8)^{10}$$

$$= (0.8)^{10}$$

$$= 0.1074.$$

b) The probability that the salesman makes one *or more* sales is:

$$P(X \geqslant 1) = f(1) + f(2) + \cdots + f(10)$$

$$= 1 - f(0)$$

$$= 1 - (0.8)^{10}$$

$$= 0.8926.$$

c) The probability that the salesman makes *exactly three* sales is:

$$P(X = 3) = f(3) = \binom{10}{3}(0.2)^3(0.8)^7$$

$$= 120(0.008)(0.8)^7$$

$$= 0.2013$$

Example 6 A company makes color TV's, of which 15 percent are defective. Fifteen color TV's are shipped to a dealer. If each color TV assembled is considered an independent trial, what is the probability that the shipment of 15 color TV's contains:

a) No defective color TV?

b) One or less defective color TV?

Solution: Let X represent the number of defective color TV's shipped to a dealer. Then the possible values of X are $0, 1, 2, \ldots, 15$. Let $p = P(\text{Defective color TV})$; then $p = 0.15$ and $q = 0.85$.

a) Since $n = 15$, $p = 0.15$, $q = 0.85$, we have, using the formula for the binomial distribution,

$$P(X = 0) = f(0) = \binom{15}{0}(0.15)^0(0.85)^{15}$$

$$= (0.85)^{15}$$

$$= 0.0873.$$

b) $P(\text{One or less}) = P(X \leqslant 1) = P(X = 0) + P(X = 1)$

$$= f(0) + f(1)$$

$$= \binom{15}{0}(0.15)^0(0.85)^{15} + \binom{15}{1}(0.15)^1(0.85)^{14}$$

$$= (0.85)^{15} + 15(0.15)^1(0.85)^{14}$$

$$= 0.0873 + 0.2312$$

$$= 0.3185.$$

7.3. THE MEAN AND VARIANCE OF THE BINOMIAL DISTRIBUTION

In Chapter 6 we discussed the mean and variance of a probability distribution. We shall use formulas given in Chapter 6, to find the mean and variance of the binomial distribution. We shall find the mean μ and variance σ^2 for $n = 1$, 2, and 3; and then we will be able to recognize the pattern for n number of trials. The approach employed in this section is similar to the approach employed in Section 7.2 for the derivation of the binomial distribution. Refer to Section 6.3, where formulas for the mean of a probability distribution were given; these allow us to find the mean of the binomial distribution for $n = 1$, 2, and 3, as follows:

Example 7 (Refer to Example 1.) Consider the experiment of picking one item at random from a box containing defective and nondefective items. The probability distribution of the random variable "number of defective items" is given in Table 7.1. The mean of this distribution is calculated in Table 7.4.

TABLE 7.4 Mean of the Binomial Distribution $(n=1)$

Value of the random variable, X	Probability distribution, $f(X)$	$Xf(X)$
0	q	0
1	p	p
Total		$\mu = E(X) = p$

Example 8 (Refer to Example 2.) Consider the experiment of picking two items at random, one at a time, with replacement, from a box containing defective and nondefective items. The probability distribution of the random variable "number of defective items picked" is given in Table 7.2. The mean of this distribution is calculated in Table 7.5.

TABLE 7.5 Mean of the Binomial Distribution $(n=2)$

Value of the random variable, X	Probability distribution, $f(X)$	$Xf(X)$
0	q^2	0
1	$2pq$	$2pq$
2	p^2	$2p^2$
Total		$\mu = E(X) = 2p(q+p) = 2p$

The probability distribution of the random variable "number of defective items picked" in the experiment of picking three items, at random, one at a time, with replacement, from a box containing defective and nondefective items, is given in Table 7.3. You may verify that the mean of this distribution is $3p$.

Following the above discussion, we find that the mean of the binomial distribution is p when $n=1$; $2p$ when $n=2$; $3p$ when $n=3$. It must be clear by now that the mean of the binomial distribution will be np for n number of trials.

Mean of the Binomial Distribution

If p is the probability of success and q is the probability of failure in a single trial, then the expected number of successes in n trials is

$$\mu = np.$$

Example 9 (Refer to Example 4.) From a lot containing 20 items, five of which are defective, four items are drawn with replacement. What is the mean of the probability distribution of the number of defective items drawn?

Solution: The probability distribution of the number of defective items drawn is binomial with $n=4$, $p=\frac{1}{4}$. Using the formula for the mean of the binomial distribution, we have

$$\mu = np = 4\left(\frac{1}{4}\right) = 1.$$

Using the formula for the variance of a probability distribution, given in Section 6.4, we may calculate the variance of the binomial distribution for $n=1$, 2, and 3, as follows.

Example 10 (Refer to Example 1.) Consider the experiment of picking one item, at random, from a box containing defective and nondefective items. The probability distribution of the random variable "number of defective items" is given in Table 7.1. The mean of this distribution, as calculated in Table 7.4, is p. The variance of this distribution is calculated in Table 7.6.

TABLE 7.6 Variance of the Binomial Distribution ($n=1$)

Value of the random variable X	Probability distribution $f(X)$	$X - \mu$	$(X-\mu)^2 f(X)$
0	q	$-p$	$p^2 q$
1	p	$1-p$	$(1-p)^2 p$
Total			$\sigma^2 = p^2 q + q^2 p = pq(p+q) = pq.$

Example 11 (Refer to Example 2.) Consider the experiment of picking two items at random, one at a time, with replacement, from a box containing defective and nondefective items. The probability distribution of the random variable "number of defective items picked" is given in Table 7.2. The mean of this distribution, as calculated in Table 7.5, is $2p$. The variance of this distribution is calculated in Table 7.7.

TABLE 7.7 Variance of the Binomial Distribution ($n=2$)

Value of the random variable X	Probability distribution $f(X)$	$X - \mu$	$(X-\mu)^2 f(X)$
0	q^2	$-2p$	$4p^2q^2$
1	$2pq$	$1-2p$	$(1-2p)^2 2pq$
2	p^2	$2-2p$	$4(1-p)^2 p^2$

Total

$$\sigma^2 = 4p^2q^2 + (1-2p)^2 2pq + 4(1-p)^2 p^2$$
$$= 4p^2q^2 + (1+4p^2 - 4p)2pq + 4q^2p^2$$
$$= 8p^2q^2 + 2pq + 8p^3q - 8p^2q$$
$$= 8p^2q^2 + 2pq + 8p^2q(p-1)$$
$$= 8p^2q^2 + 2pq - 8p^2q^2 = 2pq.$$

Using $q = 1-p$ and some algebraic simplifications, we find that $\sigma^2 = 2pq$.

The probability distribution of the random variable "number of defective items picked" in the experiment of picking three items at random, one at a time, with replacement, from a box containing defective and nondefective items, is given in Table 7.3. It has already been pointed out that the mean of this distribution is $3p$. You may verify that the variance of this distribution is $3pq$.

Following the above discussion, we find that the variance of the binomial distribution is pq when $n=1$; $2pq$ when $n=2$; $3pq$ when $n=3$. It must be clear by now that the variance of the binomial distribution will be npq for n number of trials.

Variance of the Binomial Distribution

If p is the probability of success and q is the probability of failure in a single trial, then the variance of the binomial distribution in n trials is

$$\sigma^2 = npq.$$

Example 12 (Refer to Example 4.) From a lot containing 20 items, five of which are defective, four items are drawn with replacement. What is the variance of the probability distribution of the number of defective items drawn?

Solution: The probability distribution of the number of defective items drawn is binomial, with $n=4$, $p = \frac{1}{4}$, and $q = \frac{3}{4}$. Using the formula for the variance of the binomial distribution, we have

$$\sigma^2 = npq = 4\left(\frac{1}{4}\right)\left(\frac{3}{4}\right) = \frac{3}{4}.$$

7.4. THE POISSON DISTRIBUTION

The Poisson distribution has applications in many experiments which yield an infinite number of possible integer values $0, 1, 2, \ldots$ in a continuous time interval or in a continuous region of space. A unit of time may be a minute, an hour, a day, a week, and a unit of space may be a length, an area, or a volume. For example, in waiting-time problems, the number of telephone calls per minute at a telephone switchboard, the number of airplane arrivals per hour at an airport, or the arrival of customers per hour at a supermarket are often considered Poisson random variables. In quality control, the number of defects or scratches in a sheet of glass or a piece of furniture is considered a Poisson random variable. The Poisson distribution is applied to the number of occurrences per unit of space or per unit of time, whereas the binomial distribution is applied to the number of occurrences in a given number of trials.

The Poisson random variable must satisfy the following three conditions:

i) The number of successes in two disjoint time intervals or regions of space are independent.

ii) The probability of a success for a small time interval or region of space is proportional to the length of the time interval or region of space.

iii) The probability of two or more successes in a small time interval or region of space is negligible.

The probability distribution of the Poisson random variable, called the Poisson distribution, is given as follows:

The Poisson Distribution

Let X be a Poisson random variable with possible values $0, 1, 2, \ldots$ Then the probability distribution of the random variable X, called the Poisson distribution, is given by the formula

$$f(X) = \frac{e^{-\lambda}\lambda^X}{X!},$$

where

i) $X = 0, 1, 2, \ldots$;

ii) $e = $ A constant with the value of 2.71828;

iii) $\lambda = $ Mean number of successes in the given time interval or region of space.

Example 13 If the number of telephone calls an operator receives from 9:00 to 9:05 follows a Poisson distribution with $\lambda = 2$, what is the probability that the operator will not receive a phone call in the same time interval tomorrow?

Solution: Let X represent the number of calls the operator receives from 9:00 to

9:05. Then the random variable X has the Poisson distribution with $\lambda = 2$. Using the formula for the Poisson distribution, we have

$$P(X=0)=f(0)=e^{-2}\frac{2^0}{0!}$$

$$=0.135 \qquad \text{(From Appendix, Table 4)}.$$

Example 14 Flaws in a large plate of glass occur, on the average, one per 20 square feet. Using the Poisson distribution, find the probability that a $3' \times 10'$ sheet will contain

a) No flaws;

b) At least one flaw.

Solution: Let X represent the number of flaws in a $3' \times 10'$ sheet. Then the random variable X has the Poisson distribution with mean $\lambda = (3 \times 10)/20 = 1.5$ flaws in a $3' \times 10'$ sheet.

a) The probability that a $3' \times 10'$ sheet contains no flaws is:

$$P(X=0)=f(0)=\frac{e^{-1.5}(1.5)^0}{0!}$$

$$=0.223 \qquad \text{(From Appendix, Table 4, } e^{-1.5}=0.223).$$

b) The probability that a $3' \times 10'$ sheet contains at least one flaw is:

$$P(X \geqslant 1)=1-P(X=0)=1-0.223$$

$$=0.777.$$

Example 15 Suppose the average number of customers per minute arriving at a supermarket is two. Using the Poisson distribution, find the probability that during one particular minute exactly three customers will arrive.

Solution: Let X represent the number of customers arriving per minute at a supermarket. Then the random variable X has the Poisson distribution with mean $\lambda = 2$. Using the formula for the Poisson distribution, we have

$$P(X=3)=f(3)=\frac{e^{-2}(2)^3}{3!}$$

$$=\frac{0.135(8)}{6} \qquad \text{(From Appendix Table 4, } e^{-2}=0.135)$$

$$=0.180.$$

The mean and variance of the Poisson distribution may easily be found using formulas given in Chapter 6. The reader must have noted already that the constant λ in the formula of the Poisson distribution represents the mean. Using the formula for the computation of the variance of a random variable, it can be shown that the variance of the Poisson distribution is also λ.

The Mean and Variance of the Poisson Distribution

If λ is the mean number of successes occurring in a given time interval or region of space in the Poisson distribution, then the mean of the Poisson distribution is equal to λ, and the variance is also equal to λ.

7.5. THE POISSON DISTRIBUTION AS AN APPROXIMATION TO THE BINOMIAL DISTRIBUTION

The graphs or histograms of the Poisson and binomial distributions are *approximately* of the same shape when n is large and p is small. Hence, the Poisson distribution may be used as an approximation to the binomial distribution when n is large and p is small. For example, suppose machine parts turned out by an automatic process are known to be 1% defective, on the average. Suppose a quality-control expert takes a sample of 50 parts and would like to know the probability of 0, 1, 2,..., 50 defective parts. We may calculate these probabilities either directly, using the binomial distribution, with $n = 50$, $p = \frac{1}{100}$, $q = 99/100$ or (since the number of trials is large and p is small) we may use the Poisson distribution as an *approximation* to the binomial distribution. Results of the computation of probabilities using the binomial distribution and the Poisson distribution are shown in Table 7.8.

TABLE 7.8

Value of the random variable X	Binomial	Poisson
0	0.6050	0.6065
1	0.3056	0.3033
2	0.0756	0.0758
3	0.0122	0.0126
4	0.0015	0.0016
5	0.0001	0.0002

Example 16 A company makes electric motors. The probability that an electric motor is defective is 0.01. What is the probability that a sample of 300 electric motors will contain exactly five defective motors?

Solution: Let X represent the number of defective motors in a sample of 300. Then the random variable X has the binomial distribution, with $n = 300$, $p = 0.01$, $q = 0.99$. Since the number of trials is large and p is small, we may use the Poisson distribution as an approximation to the binomial distribution. Thus X has the Poisson distribution with mean

$$\lambda = np = 300(0.01) = 3.$$

The probability of exactly five defective motors in a sample of 300 is:

$$P(X = 5) = f(5) = \frac{e^{-3}(3)^5}{5!}$$

$$= \frac{0.05(243)}{120} \quad \text{(From Appendix Table 4, } e^{-3} = 0.05\text{)}$$

$$= 0.10.$$

SUMMARY

In this chapter two discrete probability distributions, the binomial distribution and the Poisson distribution, are discussed.

The binomial distribution has many applications in business and economics. In many business situations we are interested in a trial which has only two possible outcomes, generally called success and failure. Such trials are called *binomial* trials. The probability distribution of the number of successes in n binomial trials is called the binomial distribution. The binomial distribution must satisfy the following four conditions:

1. The number of trials n is fixed.
2. Each trial has only two outcomes, either S (success) or F (failure).
3. The probability p of a success is the same for each trial.
4. The trials are independent.

The mean of the binomial distribution in n trials is

$$\mu = np$$

The variance of the binomial distribution in n trials is

$$\sigma^2 = npq.$$

The Poisson distribution has many applications in business and economics. For example, in waiting-time problems, the number of telephone calls per minute at a telephone switchboard, the number of airplane arrivals per hour at an airport, the arrival of customers per hour at a supermarket, may be considered Poisson random variables. The probability distribution of the Poisson random variable is called the Poisson distribution. This distribution must satisfy the following three conditions:

1. The number of successes in two disjoint time intervals (or regions of space) are independent.
2. The probability of a success for a small time interval (or region of space) is proportional to the length of the time interval (or region of space).
3. The probability of two or more successes in a small time interval (or region of space) is negligible.

If λ is the mean number of successes occurring in a given time interval or region of space in the Poisson distribution, then the mean of the Poisson distribution is

$$\mu = \lambda.$$

The variance of the Poisson distribution is

$$\sigma^2 = \lambda.$$

Words to Remember

Binomial trial   Poisson distribution
Binomial distribution   Poisson approximation to binomial

Symbols to Remember

n	Number of trials
X	Number of successes
p	Probability of a success in a single trial
q	Probability of a failure in a single trial
$f(X)$	Probability of X successes in n trials
$\mu = np$	Mean of the binomial distribution
$\sigma^2 = npq$	Variance of the binomial distribution
$\mu = \lambda$	Mean of the Poisson distribution
$\sigma^2 = \lambda$	Variance of the Poisson distribution

EXERCISES

1. Three light bulbs are drawn at random from a large consignment of light bulbs, of which 10% are defective. What is the probability of getting
   a) No defectives?   b) 1 defective?   c) 2 defectives?   d) 3 defectives?
2. Suppose 40% of the people in a particular city smoke cigarettes. A group of 10 people is chosen. What is the probability that more than 4 are smokers?

3. The reliability (probability of working properly) of a transistor is 0.9. In a radio with 10 transistors, what is the probability that a breakdown will occur?

4. A school-chair manufacturer must put some chairs for lefthanded people in each classroom. If he knows that 10% of the population is lefthanded, what is the probability that in a class of twenty people:

   a) None will be lefthanded?

   b) Two will be lefthanded?

   c) More than two will be lefthanded?

5. The probability that a production line produces defective parts is 0.1. Eight pieces are selected from the line.

   a) What is the probability of at least one defective part?

   b) What is the probability that all pieces are good?

   c) If 50 pieces were selected, what would the expected number of defectives be?

6. Find the parameters n and p of the binomial distribution for which the mean is 7 and the variance is $28/5$.

7. Find the mean and the variance of the binomial distribution with $p = \frac{2}{3}$ and $n = 15$.

8. Find the values of n and p for a binomial distribution for which the mean is 7 and the variance is $14/3$.

9. A production line produces good articles with probability 0.7, and defective articles with probability 0.3. Ten articles are selected.

   a) What is the probability of selecting eight good articles?

   b) What is the probability that there is an equal number of good and defective articles?

10. In a population of 150 individuals, 15% wear glasses. What is the probability of getting at most three individuals wearing glasses in a sample of 15 (sampling with replacement)?

11. Suppose a manufacturing plant of a certain company has five machines. The probability that any one of the machines will break down during a given week is 0.2.

    a) What is the probability that exactly two machines will break down during a given week?

    b) What is the probability that *none* of the machines will break down during a given week?

    c) What is the probability that *all* machines will break down in a given week?

12. A company advertises its products in a local newspaper. The probability that a person who reads the advertisement about these products will actually make a purchase is 0.3. What is the probability that exactly three persons will make a purchase out of five persons selected at random from a group of people who read the advertisement?

13. In Exercise 12, suppose 1000 persons read the advertisement during a given week, How many persons do you expect will make a purchase?

14. In Exercise 2, how large must the group be in order that the probability is at least 0.90 that 1 or more of them smoke?

15. If 20% of the population is unqualified for a particular job and if fifteen people are selected at random from the population, what is the probability that:
    a) All are qualified?
    b) None are qualified?
    c) One is unqualified?

16. Let $X_1, X_2, \ldots, X_n$ be n independent binomial trials such that $E(X_i) = p$, $i = 1, 2, \ldots, n$. Show that:

    a) $E\left(\sum_{i=1}^{n} X_i\right) = np$ b) $\text{Var}\left(\sum_{i=1}^{n} X_i\right) = npq$

    Compare the results to the mean and variance of the binomial distribution.

17. A taxicab company has found that, on the average, two accidents occur in a month. Assuming that the accidents follow a Poisson distribution, find the probability of having five or more accidents during a given month.

18. Calculate the probability of exactly two accidents at a busy intersection on a given day if on the average there are 0.5 accidents per day.

19. Customers arrive at a service desk at the rate of two every five minutes. If the arrival of customers is assumed to follow a Poisson distribution, find the probability that:
    a) None arrive in a five-minute period.
    b) More than four arrive in a ten-minute period.

20. What is the probability that there will be three incoming telephone calls at a switchboard during a particular two-minute interval if, on the average, there are 1.5 incoming calls in a two-minute span?

21. Accidents in a very large plant of a company follow a Poisson distribution. If, on the average, four accidents occur in a week, what is the probability that more than four accidents will occur during the next week?

22. The actuary of a life-insurance company has found that the probability of a person having a fatal accident is 0.0001. If the company holds 50,000 life-insurance policies, what is the probability that the company will pay more than three claims next year due to fatal accidents?

23. A typist of a company, on the average, makes 3 errors per page. What is the probability of her typing a page:
    a) With no error?
    b) With at least 2 errors?

24. Assume that the number of items of a certain kind purchased in a store during a week's time follows a Poisson distribution, with $\mu = 2$. How large a stock should the merchant have on hand to yield a probability of 0.95 that he will be able to supply the demand?

25. Flaws in a large plate of glass occur on the average of one per 10 square feet. What is the probability that a $6' \times 10'$ sheet will contain (a) no flaws? (b) at least one flaw?

THE NORMAL
DISTRIBUTION 8

8.1. INTRODUCTION

Recall that in Chapter 6 we discussed the probability distribution of a discrete random variable, as well as of a continuous random variable. In Chapter 7 we discussed two discrete probability distributions, the binomial distribution and the Poisson distribution. It was pointed out, in Section 6.2, that the problem of assigning probabilities to continuous random variables cannot be treated in the same way as discrete random variables. However, the concept of area and its properties prove powerful in assigning probabilities to continuous random variables. That is, the area under a continuous probability distribution represents probability. Since the total area under the graph of a probability distribution of a continuous random variable is equal to 1, the probability of a continuous random variable lying between any two values is given by the area under the probability distribution between the two specified values. In this chapter, we introduce a well-known continuous distribution called the normal distribution. The normal distribution occupies the central position in probability and statistics. The normal distribution is the most frequently used of all probability distributions. It is also known as the "Gaussian Distribution." Its graph is called the "Normal Curve." The normal distribution occurs in many industrial and quality-control experiments. The normal distribution is also important because of the fact that the distribution of sample means and many other statistics for large sample sizes is approximately normally distributed, even though the original population may not be normal. This point will be discussed in a later chapter. The normal distribution has convenient mathematical properties and it also serves as an approximation to other discrete and continuous distributions. A continuous random variable X is said to be normally distributed if its probability density function is given by the formula

$$f(X) = \frac{1}{\sigma\sqrt{2\pi}} e^{-(1/2)[(X-\mu)/\sigma]^2},$$

where

i) The random variable X assumes any value from minus infinity to plus infinity;

ii) μ is the mean of the normal distribution;

iii) σ is the standard deviation of the normal distribution;

iv) $e = 2.71828$;

v) $\pi = 3.14159$.

The normal distribution with a mean μ and a variance σ^2 may be denoted by the symbol $N(\mu, \sigma^2)$. Since $f(X)$ is a probability distribution, the total area under the curve $f(X)$ is equal to 1. The probability that a normally distributed random variable with a mean μ and a variance σ^2 lies between two specified values a and b is:

$$P(a < X < b) = \text{Area under the curve } f(X) \text{ between the}$$
$$\text{specified values } X = a \text{ and } X = b.$$

The graph of three normal distributions with the same standard deviation but different means is shown in Fig. 8.1; and the graph of three normal distributions with the same mean but different standard deviations is shown in Fig. 8.2.

It can be seen, from the graphs of the normal distributions shown in Figs. 8.1 and 8.2, that the normal distribution possesses the following properties:

Properties of the Normal Distribution

1. The normal distribution is symmetrical about the mean μ; that is, the graph of the normal distribution to the left of the mean μ is the same as to the right.

2. The mean is in the middle and divides the area in half.

3. The total area under the curve is equal to 1.

The normal distribution possesses many other interesting properties. The importance and usefulness of this distribution will become evident in later chapters on sampling, estimation, and testing of hypotheses.

8.2. THE STANDARD NORMAL DISTRIBUTION

It has been pointed out, in Section 8.1, that the normal distribution is continuous. Since it is continuous, the random variable can assume an *infinite number of values* in an interval. In order to compute the probability of a random

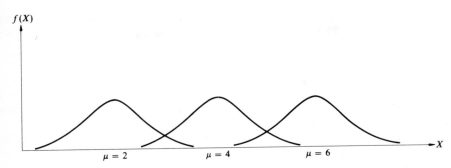

8.1   Normal distributions with $\mu = 2$, 4, 6 and same standard deviation.

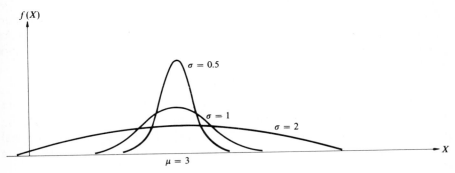

8.2   Normal distributions with $\mu = 3$ and $\sigma = 0.5$, 1, 2.

variable lying between two specified values, we need to know the area under the normal curve between two specified values. The formula for the normal distribution given in Section 8.1 involves two parameters, the mean μ and variance σ^2. Since μ and σ^2 could assume an infinite number of values, it is impossible to tabulate the areas under the curve for different values of μ and σ^2. However, we can overcome this difficulty by using the areas under the probability distribution called "the standard normal distribution."

A normal distribution with mean $\mu = 0$ and variance $\sigma^2 = 1$ is called the *standard normal distribution* (or the unit normal distribution) and is given by the formula:

$$f(Z) = \frac{1}{\sqrt{2\pi}} e^{-Z^2/2},$$

where the random variable Z assumes any value from minus infinity to plus infinity.

The normal distribution with mean 0 and variance 1 may be denoted by the symbol $N(0, 1)$. Since $f(Z)$ is a probability distribution, the total area under the curve $f(Z)$ is equal to 1. The probability that a normally distributed random variable with mean 0 and variance 1 lies between two specified values c and d is:

$$P(c < Z < d) = \text{Area under the curve } f(Z) \text{ between}$$
$$\text{two specified values } Z = c \text{ and } Z = d.$$

The table of areas (or probabilities) under the standard normal distribution is given in Appendix Table 5.

Any normal distribution with a mean μ and a variance σ^2 can be reduced to the standard normal distribution by using the transformation $Z = (X - \mu)/\sigma$. This transformation makes the mean of the normal distribution 0 and variance 1.

Standardization of the Normal Distribution

If a random variable X has the normal distribution with a mean μ and variance σ^2, then a random variable $Z = (X - \mu)/\sigma$ has the standard normal distribution with mean 0 and variance 1.

Example 1 If X is normally distributed with mean 3 and variance 4, what is the probability that X lies between 3 and 5?

Solution

$$Z_1 = \frac{X_1 - \mu}{\sigma} = \frac{3-3}{2} = 0; \qquad Z_2 = \frac{X_2 - \mu}{\sigma} = \frac{5-3}{2} = 1.$$

Thus,

$$P(3 < X < 5) = P(0 < Z < 1)$$

$$= \text{Area between } Z = 0 \text{ and } Z = 1,$$

From Appendix Table 5, the shaded area is 0.3413. Therefore, $P(3 < X < 5) = 0.3413$. (See Fig. 8.3.)

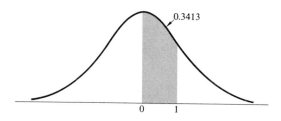

8.3

Example 2 Let X have a normal distribution with mean μ and variance σ^2. Find

a) $P(\mu-\sigma<X<\mu+\sigma)$;
b) $P(\mu-2\sigma<X<\mu+2\sigma)$;
c) $P(\mu-3\sigma<X<\mu+3\sigma)$.

Solution

a)
$$Z_1 = \frac{(\mu-\sigma)-\mu}{\sigma} = -1; \qquad Z_2 = \frac{(\mu+\sigma)-\mu}{\sigma} = 1.$$

Thus,

$$P(\mu-\sigma<X<\mu+\sigma) = P(-1<Z<1)$$
$$= \text{Area between } Z=-1 \text{ and } Z=1$$
$$= (\text{Area between } Z=-1 \text{ and } Z=0)$$
$$+ (\text{Area between } Z=0 \text{ and } Z=1).$$

By symmetry, the area between $Z=-1$ and $Z=0$ is the same as the area between $Z=0$ and $Z=1$. From Appendix Table 5, the area between $Z=0$ and $Z=1$ is 0.3413. Therefore,

$$P(\mu-\sigma<X<\mu+\sigma) = 0.3413+0.3413$$
$$= 0.6826.$$

This means that 68.26% of the area under the normal curve falls *within one standard deviation* of the mean (either way); see Fig. 8.4.

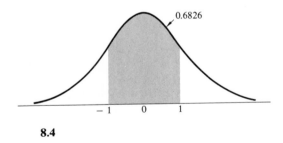

0.6826

8.4

b)
$$Z_1 = \frac{(\mu-2\sigma)-\mu}{\sigma} = -2; \qquad Z_2 = \frac{(\mu+2\sigma)-\mu}{\sigma} = 2.$$

Thus,
$$P(\mu-2\sigma<X<\mu+2\sigma) = P(-2<Z<2)$$
$$= (\text{Area between } Z=-2 \text{ and } Z=0)$$
$$+ (\text{Area between } Z=0 \text{ and } Z=2).$$

From Appendix Table 5, the area between $Z=0$ and $Z=2$ is 0.4773. Therefore,

$$P(\mu-2\sigma < X < \mu+2\sigma) = 0.4773 + 0.4773$$

$$= 0.9546.$$

This means that 95.46% of the area under the normal curve falls within two standard deviations of the mean (see Fig. 8.5).

c) $$Z_1 = \frac{(\mu-3\sigma)-\mu}{\sigma} = -3; \qquad Z_2 = \frac{(\mu+3\sigma)-\mu}{\sigma} = 3.$$

Thus,

$$P(\mu-3\sigma < X < \mu+3\sigma) = P(-3 < Z < 3)$$

$$= (\text{Area between } Z = -3 \text{ and } Z = 0)$$

$$+ (\text{Area between } Z = 0 \text{ and } Z = 3).$$

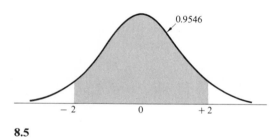

0.9546

8.5

From Appendix Table 5, the area between $Z=0$ and $Z=3$ is 0.4987. Therefore,

$$P(\mu-3\sigma < X < \mu+3\sigma) = 0.4987 + 0.4987$$

$$= 0.9974.$$

This means that 99.74% of the area under the normal curve falls within three standard deviations of the mean (see Fig. 8.6).

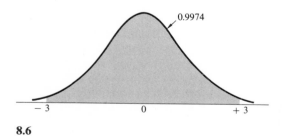

0.9974

8.6

Example 3 In a certain construction job, the average wage is $4.00 per hour and the standard deviation is $0.50. If the wages are assumed to follow a normal distribution, what percentage of the workers:

a) Receive wages more than $4.50 per hour?

b) Receive wages between $3.50 and $4.50 per hour?

Solution: Let X represent the hourly wage of a construction worker in dollars. Then X is normally distributed with mean $\mu = \$4.00$ and standard deviation $\sigma = \$0.50$.

a)
$$Z = \frac{X - \mu}{\sigma} = \frac{4.50 - 4.00}{0.50} = \frac{0.50}{0.50} = 1.$$

Thus,

$$P(X > 4.50) = P(Z > 1.0)$$

$$= \text{Area to the right of } Z = 1.0$$

$$= 0.5000 - 0.3413$$

$$= 0.1587.$$

This means that 15.87% of the construction workers receive wages more than $4.50 per hour.

b) $$Z_1 = \frac{3.50 - 4.00}{0.50} = \frac{-0.50}{0.50} = -1.0; \qquad Z_2 = \frac{4.50 - 4.00}{0.50} = \frac{.50}{.50} = 1.0.$$

Thus,

$$P(3.50 < X < 4.50) = P(-1 < Z < 1)$$

$$= \text{Area between } Z = -1 \text{ and } Z = 1$$

$$= (\text{Area between } Z = -1 \text{ and } Z = 0)$$

$$+ (\text{Area between } Z = 0 \text{ and } Z = 1).$$

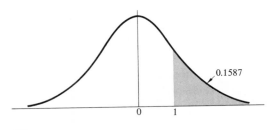

0.1587

8.7

From Appendix Table 5, the area between $Z=0$ and $Z=1$ is 0.3413. Therefore,

$$P(3.50 < X < 4.50) = 0.3413 + 0.3413$$

$$= 0.6826.$$

This means that 68.26% of the construction workers receive wages between $3.50 and $4.50 per hour (see Fig. 8.8).

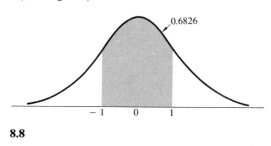

8.8

Example 4 The weights of one-pound coffee jars filled by an automatic machine are normally distributed with mean $\mu = 1.03$ pounds and standard deviation $\sigma = 0.02$ pounds. Find the probability that:

a)   The weight of any coffee jar is less than one pound;

b)   The weight is more than 1.06 pounds.

Solution: Let the random variable X represent the true weight in pounds of a coffee jar filled by an automatic machine. Then X is normally distributed, with mean $\mu = 1.03$ and standard deviation $\sigma = 0.02$.

a)
$$Z = \frac{X - \mu}{\sigma} = \frac{1.00 - 1.03}{0.02} = -1.5.$$

Thus,

$$P(X < 1) = P(Z < -1.5)$$

$$= \text{Area to the left of } Z = -1.5$$

$$= 0.5000 - 0.4332$$

$$= 0.0668.$$

Therefore, $P(X < 1) = 0.0668$. This means that the probability that the weight of any coffee jar is less than one pound is 0.0668. (see Fig. 8.9.)

b)
$$Z = \frac{X - \mu}{\sigma} = \frac{1.06 - 1.03}{0.02} = 1.5.$$

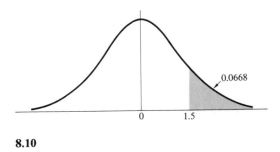

0.0668

−1.5   0

8.9

Thus,

$$P(X > 1.06) = P(Z > 1.5)$$

$$= \text{Area to the right of } Z = 1.5$$

$$= 0.5000 - 0.4332$$

$$= 0.0668$$

Therefore, $P(X > 1.06) = 0.0668$ (Fig. 8.10). This means that the probability that the weight of any coffee jar is more than 1.06 pounds is 0.0668.

0.0668

0   1.5

8.10

Example 5 The lifetimes of a certain make of battery are normally distributed. If 33% of the batteries have a lifetime under 45 hours and 10% of the batteries have lifetimes over 65 hours, find the mean and standard deviation of the distribution.

Solution: Let the random variable X represent the lifetime of a certain make of battery. Let μ and σ be the mean and standard deviation of the normal distribution.

Since the area lying to the left of $X = 45$ is 0.33, the area between $Z = 0$ and $Z_1 = (\mu - 45)/\sigma$ is $0.5000 - 0.3300 = 0.1700$. From Appendix Table 5, the value of

Z_1 corresponding to this area is 0.44. Therefore

$$\frac{45-\mu}{\sigma} = -0.44.$$

Since the area to the right of $X=65$ is 0.10, the area between $Z=0$ and $Z_2=(65-\mu)/\sigma$ is $0.50-0.10=0.40$. The value of Z_2 corresponding to this area is 1.28. Therefore

$$\frac{65-\mu}{\sigma} = 1.28.$$

Solving the above two equations, we get

$$\mu = 50.12, \qquad \sigma = 11.63.$$

Example 6 (Refer to Example 3.) In a certain field of construction work, the wages are assumed to follow a normal distribution, with an average wage of $4.00 per hour and a standard deviation of $0.50. If 20 workers received wages of more than $4.50 per hour, what is the total number of construction workers?

Solution: Let the random variable X represent the hourly wage in dollars of a construction worker. Then X is normally distributed with mean $\mu = \$4.00$ and standard deviation $\sigma = \$0.50$. Hence,

$$Z = \frac{X-\mu}{\sigma} = \frac{4.50-4.00}{0.50} = \frac{0.50}{0.50} = 1.0.$$

Thus,

$$P(X > 4.50) = P(Z > 1.0)$$

$$= \text{Area to the right of } Z = 1$$

$$= 0.5000 - 0.3413$$

$$= 0.1587.$$

Therefore, the total number of construction workers N is given by

$$\frac{1,587}{10,000} N = 20$$

or

$$N = \frac{200,000}{1,587} = 126.$$

That is, the total number of construction workers is 126.

8.3. THE NORMAL APPROXIMATION TO THE BINOMIAL DISTRIBUTION

In Section 7.2, we saw that the probability of X successes in n independent binomial trials is given by the formula

$$f(X) = \binom{n}{X} p^X q^{n-X},$$

where

i) $X = 0, 1, 2, \ldots, n$;

ii) p is the probability of success on a single trial;

iii) $q = 1 - p$ is the probability of failure on a single trial.

The computation of binomial probabilities becomes laborious and difficult as the number of trials n becomes large. In Section 7.5, we noted that, when the number of trials n is large and p is near 0 or 1, we can use the Poisson distribution to approximate the binomial distribution. The binomial probabilities can also be approximated by a normal distribution if the number of trials n is large. The accuracy of the normal approximation to the binomial distribution also depends on p. The approximation is more rapid if n is large and p and q are close to $\frac{1}{2}$. If p is close to 0 or 1, the normal approximation to the binomial distribution becomes poorer for any given n. The normal approximation can also be applied to many other probability distributions. We shall explain the normal approximation to the binomial distribution with the help of the following example.

Example 7 From a lot containing an *equal number* of defective and nondefective items, ten items are drawn at random, one at a time, with replacement. Find the probability of drawing 4, or 5, or 6 defective items by using:

a) The binomial distribution.

b) The normal approximation to the binomial distribution.

Solution

a) Let the random variable X represent the number of defective items drawn in 10 draws. Then X has the binomial distribution with $n = 10$, $p = \frac{1}{2}$, $q = \frac{1}{2}$. Hence,

$$f(X) = \binom{10}{X} \left(\frac{1}{2}\right)^X \left(\frac{1}{2}\right)^{10-X} \qquad (X = 0, 1, \ldots, 10).$$

Therefore,

$f(0) = 0.0009$	$f(3) = 0.1172$	$f(6) = 0.2051$	$f(9) = 0.0087$
$f(1) = 0.0087$	$f(4) = 0.2051$	$f(7) = 0.1172$	$f(10) = 0.0009$
$f(2) = 0.0440$	$f(5) = 0.2461$	$f(8) = 0.0440$	

Thus,

$$P(4 \leqslant X \leqslant 6) = f(4) + f(5) + f(6)$$

$$= 0.2051 + 0.2461 + 0.2051$$

$$= 0.6563.$$

The graph of the probability distribution for the number of defectives drawn in 10 draws from a lot containing an equal number of defective and nondefective items is shown in Fig. 8.11. In that figure, we superimpose a normal distribution with mean

$$\mu = np = 10\left(\frac{1}{2}\right) = 5$$

and

$$\sigma = \sqrt{npq} = \sqrt{10\left(\frac{1}{2}\right)\left(\frac{1}{2}\right)} = \sqrt{\frac{5}{2}} = 1.581.$$

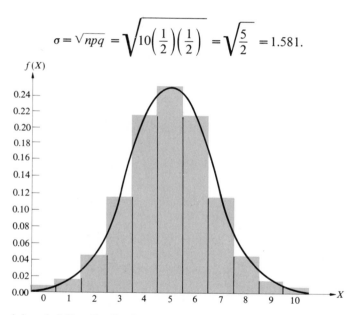

8.11 Binomial probability distribution and a normal curve for the number of defectives.

Since the binomial distribution is discrete and we want to use the normal approximation as if the data were continuous, we make a correction for the continuity by subtracting $\frac{1}{2}$ from the lower value and adding $\frac{1}{2}$ to the upper value. Thus,

$$Z_1 = \frac{3.5 - 5.0}{1.581} = -0.95, \qquad Z_2 = \frac{6.5 - 5.0}{1.581} = 0.95.$$

Therefore,

$$P(3.5 < X < 6.5) = P(-0.95 < Z < 0.95)$$
$$= (\text{Area between } Z = -0.95 \text{ and } Z = 0)$$
$$+ (\text{Area between } Z = 0 \text{ and } Z = 0.95)$$
$$= 0.3289 + 0.3289$$
$$= 0.6578.$$

(See Fig. 8.12.) The exact probability obtained in part (a) is 0.6563.

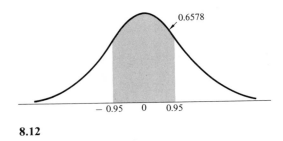

0.6578

-0.95 0 0.95

8.12

The normal approximation to the binomial distribution may be stated as follows.

Normal Approximation to Binomial

Let the random variable X be the number of successes in n independent binomial trials. Let p be the probability of success and $q = 1 - p$ be the probability of failure on a single trial; then the standardized binomial random variable

$$Z = \frac{X - np}{\sqrt{npq}}$$

is approximately normally distributed, with mean 0 and variance 1, if the number of trials n is sufficiently large.

Example 8 If 3% of the light bulbs made by a certain manufacturer are defective, what is the probability that, out of 600 light bulbs selected at random, more than 20 are defective?

Solution: Let the random variable X represent the number of defective bulbs in a sample of 600. Then X has the binomial distribution, with mean

$$\mu = np = 600\left(\frac{3}{100}\right) = 18,$$

and standard deviation

$$\sigma = \sqrt{npq} = \sqrt{600\left(\frac{3}{100}\right)\left(\frac{97}{100}\right)} = 4.18.$$

We require the probability that the number of defective bulbs is more than 20. Using the normal approximation to the binomial, we have

$$Z = \frac{20.5 - 18}{4.18} = \frac{2.50}{4.18} = 0.60.$$

Thus,

$$P(X > 20) = P(Z > 0.60)$$

$$= \text{Area to the right of } Z = 0.60$$

$$= 0.5000 - 0.2258$$

$$= 0.2742.$$

That is, the approximate probability that a sample of 600 light bulbs contains more than 20 defectives is 0.2742 (Fig. 8.13).

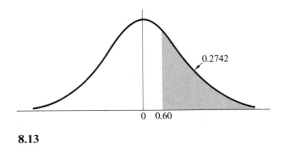

8.13

SUMMARY

Whereas the binomial probability distribution is the most frequently used of discrete probability distributions, the normal distribution is the most frequently used of all *continuous* probability distributions. The normal distribution occurs in many industrial and quality-control experiments. The normal distribution possesses many convenient mathematical properties and can serve as an approximation to the binomial distribution and many other probability distributions. The normal distribution is also important because of the fact that the distribution of sample means and many other statistics for large sample sizes *is* normal, even though the original population may not be normal. This point will be discussed in detail in a later chapter.

Words to Remember

Normal distribution	Unit normal
Normal curve	Normal approximation to binomial
Standard normal	Continuity correction

Symbols to Remember

$N(\mu, \sigma^2)$ Normal distribution with mean μ and variance σ^2.

$N(0,1)$ Normal distribution with mean 0 and variance 1, or the *standard normal distribution* (or the *unit* normal distribution)

Z Standard normal variable

EXERCISES

1. Find the area under the standard normal distribution in each of the following cases.
   a) Between $Z=0$ and $Z=0.7$;
   b) Between $Z=-0.5$ and $Z=0$;
   c) Between $Z=-0.6$ and $Z=2.2$;
   d) Between $Z=0.4$ and $Z=1.4$;
   e) To the right of $Z=-1.1$;
   f) To the left of $Z=-1.2$.

2. Find the values of Z in the following cases where the area refers to that under the standard normal distribution.
   a) Area to the left of Z is 0.4032;
   b) Area to the left of Z is 0.7580;
   c) Area between -1.2 and Z is 0.1269;
   d) Area between -1.2 and Z is 0.6429.

3. If a random variable Z is normally distributed, with mean 0 and variance 1, find:
   a) $P(Z \geqslant 1.96)$;
   b) $P(-3 \leqslant Z \leqslant 3)$;
   c) $P(Z \leqslant -1)$.

4. If a random variable X is normally distributed, with $\mu=2$ and $\sigma=1$, find:
   a) $P(X>4)$;
   b) $P(0<X<2)$.

5. If a random variable X is normally distributed with $\mu = 3$ and $\sigma = 6$, find a number X_0 such that:

   a) $P(X > X_0) = 0.05$.

   b) $P(X > -X_0) = 0.95$.

6. The diameters of ball bearings produced by a certain firm are normally distributed with a mean of 0.5 inches and a standard deviation of 0.05 inches. Ball bearings with a diameter between 0.45 and 0.55 inches are accepted, and those with diameter outside these limits will be termed defective and be rejected. Find the percentage of defective ball bearings manufactured by the company.

7. In a certain construction job, the average wage is $4.00 per hour and the standard deviation is $0.50. If the wages are assumed to follow a normal distribution, what percentage of the workers receive wages between $2.50 and $3.00 per hour?

8. It is known that the life of a certain make of transistor is normally distributed with a mean of 30 months. If a purchaser requires at least 95 percent of them to have lives exceeding 20 months, what is the largest value the standard deviation can have and still keep the purchaser satisfied?

9. The mean life length of type A light bulbs is 1500 hours, with standard deviation 50 hours. If the life length is assumed to be normally distributed, what is the probability that:

   a) A bulb will fail before 1400 hours?

   b) A bulb will last longer than 1550 hours?

   c) A bulb will last between 1450 and 1550 hours?

10. In a normal distribution, 30 percent of the items are under 40 and 10 percent are over 65. Find the mean μ and standard deviation σ of the distribution.

11. The marks in a certain business-statistics examination are normally distributed, with a mean of 70 and a standard deviation of 10. The top five percent of the students are to receive grade "A." What is the minimum mark a student must get in order to obtain a grade of "A"?

12. A normal distribution with mean 75 and standard deviation 10 has 300 items between 65 and 95. How many items are there in the whole distribution?

13. The weight of oranges from a large shipment averages 10.0 ounces, with a standard deviation of 1.5 ounces. If these weights are normally distributed, what percent of all these oranges would be expected to weigh between 7.9 ounces and 12.4 ounces?

14. A supermarket manager finds that the rate of arrival of customers in the supermarket is 1750 customers an hour. Past experience reveals that 55 percent of the customers entering the supermarket make a purchase. Using a normal approximation to the binomial distribution, find the probability that more than 1000 customers will make a purchase during the next hour.

15. Suppose the revenue of a company in a year is normally distributed with mean $100 million and standard deviation $5 million.

   a) What is the probability that the revenue will exceed $100 million?

   b) What is the probability that the revenue will be less than $100 million?

c) What is the probability that the revenue will be between $100 million and $110 million?

16. If X and Y are two normal random variables with means 10 and 15, and standard deviations 2 and 4, respectively, find:

a) $P(X + Y > 25)$;

b) $P(10 < X + Y < 30)$;

c) $P(X + Y < 5)$.

17. In a certain city, the weekly wages of men are normally distributed with mean 100 and standard deviation 10, while women's wages are normally distributed (independently of the men's) with mean 80 and standard deviation 7. For a man and a woman chosen randomly, find the probability that the sum of their wages is:

a) Greater than 200;

b) Between 160 and 210;

c) Less than 145.

18. The lengths of two pipe parts (in inches) are random variables X and Y, which are independently normally distributed, with means 3 and 7, and variances 16 and 25, respectively. Find the probability that the sum of the lengths of two parts chosen at random is:

a) Less than eight inches;

b) More than eleven inches;

c) Between eight and eleven inches.

19. A firm manufactures a product which requires machining and finishing. Suppose machining and finishing operations are statistically independent normal random variables, with means 15 minutes and 12 minutes, and standard deviations 3 and 2 minutes, respectively. What is the probability that:

a) The total time to machine and finish a product will exceed 30 minutes?

b) The total time to machine and finish a product will take less than 30 minutes?

c) The total time to machine and finish a product will take between 25 minutes and 30 minutes?

20. Suppose 60 percent of the employees of a certain company favor a person A in an election for president of a local union. A sample of 100 employees is taken; what is the probability that between 60 and 70 employees in the sample favor person A?

21. It is known that 5 percent of the light bulbs from a certain factory are defective. A sample of 200 light bulbs is taken. What is the probability of finding:

a) Exactly 5 defectives?

b) More than 10 defectives?

c) Between 0 and 5 (inclusive) defectives?

22. A new drug made by a pharmaceutical company is effective on 60 percent of the patients. What is the probability that, of 12 patients, three or fewer will show no effect?

23. If the probability that a tire is defective in its first 100 hours of use is 0.05, use the normal approximation to the binomial to calculate the probability that more than 460 out of a total of 10,000 will be defective in their first 100 hours of use.

24. If the probability that a man is qualified for a particular job is 0.8, find the probability that, in a group of 50 men, there are between 35 and 45 qualified men.

SAMPLING 9

9.1. SOURCES OF BUSINESS DATA

The businessman working with data must decide, early in the game, whether he wants to simply describe the data that he has (in which case the discussions of Chapters 2,3, and 4 are sufficient for his purposes), or whether he wants to *generalize* from his data to a *larger population*. If he desires to do the latter, which is the basis of inferential statistics, then he will want to obtain his sample data from reliable sources. In this chapter we will discuss the meaning of sampling, the methods of sampling, and their major advantages and disadvantages. Before proceeding, however, consider again some of the examples of Chapter 1.

In Example 4 of Chapter 1, the measurements of the 200 parts in the sample are not of interest in themselves. The quality-control expert wants to use them to *generalize* about the whole day's run. Suppose that there are several machines turning out the parts, and he took his sample all from the same machine. If the machine happens to make more defective parts than the others, then there will be a greater probability of rejecting the whole day's run than there should be. We would say that the sample *does not represent* (or *is not representative of*) the population. In Example 6 of Chapter 1, the wages of a sample of the skilled technicians are not of interest in themselves. If they adequately represent the population, then the company *can* generalize from them to the population; if they do *not* represent the population, no generalization can be made, and a large part of their usefulness has been lost.

Basically, sampling may be defined as the *selection of part of a population* in such a way that the sample is *representative* of the population, so that judgments about the population can be made on the basis of the sample. Thus, before starting to sample, the observer must be able to *define his population* clearly. In

Example 2 of Chapter 1, for instance, he should know whether he wants to generalize his results to the whole country, or to (say) New York State. An oil company, wanting to measure average daily sales, might want to separate its study into two populations, namely city sales and country sales, or weekday sales and weekend sales.

Once the population has been defined, what are the sources of data and information available to the sampler? We will make a broad division here between data that are available inside a company, called *internal*, and data that are available from outside sources, called *external*. Examples of internal data are inventory, payroll, sales, accounting records, purchase orders, etc., and data generated in experiments (e.g., testing the quality of products). Examples of external data are unemployment records, gross national product, total tax receipts, and various governmental agency figures. Some problems rely only on internal data, some only on external data, and some require a mixture of the two.

By far the largest collector of business data is the Federal government. Data are collected on Census figures, employment and unemployment, distribution of manufacturing industries, international trade, domestic trade rates of growth, rates of inflation, etc. A major American source of business data is the *Survey of Current Business*, published monthly. This contains general business indicators relating to wholesale and retail prices, employment, salaries, hours of work, finance, international trade, and many other factors. Other important sources of business data include: (1) *Business Conditions Digest*, (2) *Federal Reserve Bulletin*, and (3) *Monthly Labor Review*.

Sampling plays a large part in business statistics; and the techniques of sampling are concerned with the practical methods, and the consequences, of various types of sampling techniques. The problem of actually choosing a sample is a practical one, and hence the technique of sampling requires attention to *practical* details; we shall not concern ourselves with some of the more *theoretical* details of sampling.

9.2. REASONS FOR SAMPLING

We have already pointed out that we wish to sample in order to obtain information about a population. Now, in a few instances, when the population has a *finite number* of members or elements, we can examine *all* the members of the population; this complete examination of all the members of a finite population is called a *census*. We might ask why we should sample, and thus obtain only partial information, when a census could provide us with complete information. A number of reasons are given below

1. The first reason for sampling is that a population may be too large for every element to be measured. When the population is very large (such as the number of toothpaste users in the U.S.) or is continuous (such as the amount of

iron in a particular region), we have little choice but to sample. Other populations may be discrete, and not too large, but their geographical spread is too great to measure the whole population. Suppose, for instance, that we wish to measure the number of tractors used on farms in Ontario; the number of farms is not too great, but the sheer size of the province makes measuring the population impractical. Size of the population alone does not necessarily dictate (or preclude) sampling; after all, in the U.S. census, the whole population is measured, presumably because the resources are adequate to do so.

2. A sample, using only part of the population, and hence requiring *less work*, is usually *cheaper* to obtain. Some types of information, such as that collected by interviews, can be extremely expensive. For example, imagine what it would cost to take a census of television viewers in New York State. (When discussing costs, one must be careful not to base the argument for sampling on *cost alone*, since one can always reduce the cost of a sample, but at the expense of the quality of the data.)

3. For some populations, there is less chance of making errors if a sample rather than the whole population is observed. People assigned to do the large amounts of repetitive, tedious, clerical work that would be required to compile *all* data (the population) suffer a loss in efficiency over time because of fatigue and boredom; the possibility for error thus introduced more than offsets the uncertainty inherent in sampling. Consider, for example, a trucking company's going through its past records to find the average weight of a delivery. To measure the whole population, the clerk must go through every one of a large number of records, and record the weight of the delivery. If instead, a sample of records is taken, the amount of work (in this case tedious work) can be reduced. This has the added advantage that the uncertainty due to sampling can be calculated, whereas the uncertainty due to subjective factors like fatigue and boredom is unknown and variable.

4. For some questions, preliminary answers are needed quickly. In such cases the whole population might have been measured but a sample of the results is used for "quick" or "first" results. This is particularly useful in situations where it may take several years to process the population data completely. An example of such a situation is the census, which is taken every 10 years. If we want to know how many school-age children there are in a particular area (for sales purposes), we may not want to wait for the whole population to be enumerated. A *sample* of the results will be sufficient.

5. Sampling is imperative when the testing procedure is destructive; that is, if the sample is destroyed by testing, then the experimenter *cannot* test the population but must choose a sample. (Consider, for instance, the case where ammunition is being tested (for its range, for instance) or where the life length of light bulbs is being measured.)

6. For those businesses concerned with products whose testing involves ethical considerations (such as drugs and pharmaceuticals), it may be unlawful

or unethical to consider testing a whole population. Consider, for example, the development of a new drug for the relief of hypertension (high blood pressure). It is out of the question to consider testing the drug on the whole population of hypertensive patients, and the testing must be done on a sample.

Before continuing let us, for practical purposes, make a distinction between two types of populations.

1. We will call a population that consists of *distinct elements* (such as people, light bulbs, machine parts, toothpaste users, television buyers) a *discrete population*.
2. We will call a population that consists of a bulk or liquid product, such as a tank of gasoline, a truckload of cement, the iron in a particular region (i.e., any population which is *not* discrete) a *continuous population*.

It it easier to limit our discussion of sampling to discrete populations. For this reason, we shall assume that populations mentioned, unless otherwise stated, *will* be discrete. (Comments on continuous populations will be made later.)

Example 1 The melting points of iron in region A (Example 3 of Chapter 1) represent a *continuous* population, while the number of parts made in a day's run (Example 4 of Chapter 1) represents a *discrete* population.

9.3. METHODS OF SELECTING A SAMPLE

There is no "best" method of obtaining a sample. A sampling method used to estimate workers' wages in Chicago may not be the one to use to estimate voters' preferences in California. Samples must be obtained in a way that reflects the particular characteristics of the population being sampled. Basically, there are two broad classifications of methods of sampling:

    i) Probability sampling, and

    ii) Judgment sampling.

In probability sampling, each element of the population has a known chance of being included in the sample. Only probability sampling allows us to calculate sampling errors, and therefore permits us to judge the goodness of our estimates. The major probability sampling methods are:

1. Simple random sampling, where the sample is selected purely on a random basis.
2. Stratified random sampling, where the population is divided into homogeneous groups called *strata*. The strata are then sampled independently.
3. Systematic sampling, where, using a randomly chosen starting point, every kth member of the population is selected.
4. Cluster sampling, where natural population groups (or clusters) are sampled.

In contrast to probability samples, judgment samples depend to some extent on the personal feelings of the investigators. For example, a new product being tested for its market impact might be sold in only one or two cities, considered, by the investigator, to be "typical" or representative of the whole population. Although important decisions are sometimes made on the basis of judgment samples, it is not possible to use *statistical theory* to estimate the goodness of the results. Estimates of the goodness of such results are very much a matter of personal conviction.

We begin the discussion of sampling by defining a *simple random sample* (sometimes called a probability sample). Whether the population is discrete or continuous, we intuitively require that the sample *represent* the population. That is, we would like the sample to be a microcosmic representation of the population and truly represent its main characteristics. We might call such a sample "good." If we chose a sample to reflect our prejudices (i.e., picking only those values from the population that we liked, and ignoring the rest), we would have an untrue and unrepresentative picture or idea of the population. If, for instance, in sampling the quality of manufactured machine parts, we threw away those members of the sample which we didn't like, calculated the average quality of the remainder, and used it to describe the quality of the population, then the average sample quality would be too high (since all the pieces of poor quality would have been rejected). Thus, we would obtain a wrong idea of the average quality of the population. If a sampling procedure is such that it will lead to a false or distorted picture of the population, it is said to be *biased*. It is of the utmost importance to use sampling techniques which are *not* biased. Generally, the experimenter will not know whether his sampling technique is biased or not. The basis for probability sampling is the idea of the simple random sample (or probability sample), defined as follows.

Simple random sample or probability sample

A sample will be called a *simple random sample* (or probability sample) if every member of the population has an *equal chance* of being included in the sample. Alternatively, a sample of size n from a finite population will be called a simple random sample if all samples of size n have an equal chance of being chosen.

The word *random* should not be confused with *haphazard*. Randomness is a definite statistical concept; and a simple random sample is obtained by a well-defined procedure, which will be discussed below. Before proceeding, it is illustrative to consider some examples of *biased* sampling.

Example 2 Consider an investigator who is assigned the task of sampling households to find the average daily consumption of milk. Assume that, according to a sound (i.e., statistically based) plan, the investigator should proceed to a number of residences and record the daily milk consumption of the household. Assume, also, that since families with children are expected to drink more milk, the number of households to be visited is apportioned between childless families and families with children. Suppose, now, that the investigator, to save himself from having to go back to households where no answer is obtained, whenever he receives a "No answer," goes next door and records the response of *that* household instead. Since most of the families with no children will have both members out at work, whereas most families with children will have someone at home, the responses will become heavily weighted towards the families with children, and the resulting estimate of milk consumption will probably be too high.

Example 3 In order to produce good veal, ranchers take young cattle soon after birth and confine them in extremely restrictive quarters, where they can barely move. An animal protection group, in order to arouse public sympathy for its views, published the results of a survey where about 80% of 200 farmers expressed their opinion that such a practice (of confining the animals) was cruel and should be changed. What the group did not say is that the sample of 200 farmers included almost no *veal* raisers and was therefore extremely biased.

Many other examples of sampling bias can be found in Darrell Huff's book *How to Lie with Statistics* (W.W. Norton and Co., 1954).

9.4. THE TECHNIQUE OF SIMPLE RANDOM SAMPLING

We assume that we have a population with a finite number of members, which we can number from 1 to N. We wish to choose a sample of size n in such a way that all samples of size n have an equal chance of being chosen.

Example 4 Suppose that the population has four elements A, B, C, and D, and we wish to choose a sample of size two. Here $N = 4$, $n = 2$. There are $\binom{4}{2} = 6$ possible samples. They are AB, AC, AD, BC, BD, and CD. If the sample is chosen randomly, then each possible sample has a 1-in-6 chance of being selected.

A simple way of selecting the sample in the above example would be to put the letters A, B, C, and D on 4 separate slips of paper, put them into a box, shake well, and draw out two slips. Such a method is valid, but becomes tedious if repeated often, and is cumbersome for large populations. The use of *random number tables* is a more sophisticated way of drawing a random sample. The use of these tables is illustrated in the following example.

Example 5 Consider a town with 1473 families, from which we wish to select a sample of 100 to ask their opinion on a new street-lighting project. The list of families can be obtained easily enough from the city government. Suppose that the families are listed (probably, although not necessarily) in alphabetical order. We number them from 1 to 1473. Now turn to Appendix Table 2, and choose any four adjacent columns of digits, and read down. The first number chosen corresponds to the first family in the sample, the second number to the second family, etc. If a number is bigger than 1473, or has already been selected, ignore it and go on to the next one. Continue this process until 100 families have been selected. For illustrative purposes, suppose that we had started in the upper lefthand corner of the table; then the first few numbers would have been 921, 113, 669, 273, 567.

Example 6 An oil company wants to estimate the average daily sales of motor oil for all of its 750 service stations, by selecting a sample of 50 of them. The stations are listed in some order (it doesn't matter in what order) and numbered 1, 2, 3, etc. Using Appendix Table 2, one starts anywhere with three adjacent columns of figures and reads down. As in Example 5, ignore numbers too large or numbers repeated. Stop when fifty numbers have been selected.

Example 7 Consider a manufacturing process where a large number of machines produce similar products whose length we wish to measure. We wish to choose a sample of machines, and from each machine in the sample we wish further to choose five manufactured parts, and measure their lengths. Machines were selected first by associating a number with each, and selecting a sample of numbers using random-number tables. Then, for each machine (the parts are made sequentially, first, second, third, etc.), five parts are selected, to be measured by again using the random-number tables.

The above examples illustrate the technique of drawing a simple random sample from a population, using random numbers. A sample chosen by this technique is called a simple random sample (or a probability sample). The random-number table of Appendix Table 2 is necessarily brief in this text. For repeated use or for use with larger populations, it would be advisable to obtain a set of more extended tables. To be highly recommended is the Rand Corporation's *A Million Random Digits with 100,000 Normal Deviates* (published by the Free Press of Glencoe, New York, 1955).

Whether we use a random-number table or some other method of selecting a sample, it is possible to obtain a sample which looks decidedly nonrandom. It is possible, though not likely, in Example 5 to select 100 families *all of whom* favor a new street-lighting project. The results might not appear to have come from a randomly selected sample; yet if the ideas discussed above had been incorporated, we would have to say that the selection *was* random.

9.5. STRATIFIED RANDOM SAMPLING

In stratified random sampling, the population of N members is subdivided into subpopulations of sizes N_1, N_2, etc., where $N = N_1 + N_2 + \dots$ The subpopulations are called strata. A sample is drawn from each stratum using the technique of simple random sampling. The reasons for stratified random sampling are as follows:

i) If information is required for certain subdivisions of a population, it is reasonable to treat each subdivision as a "population" in its own right

Example 8 A company wants to know the buying habits of the population of Cleveland. It would also be convenient to know the buying habits, separately, of those earning at least $12,000 per year and those earning under $12,000 per year These two groups would constitute the two *strata*.

ii) It might be *administratively convenient* to stratify a population.

Example 9 If a sampling program is to be run from the regional marketing offices of a large company, then the area covered by each office could constitute a stratum.

iii) If a population is extremely heterogeneous, the variation (imprecision) of the results could probably be reduced by stratifying. This could be done by making each stratum relatively homogeneous with respect to some particular criterion.

Example 10 Suppose we are interested in the distance from an individual's home to the nearest hospital. Such distances are relatively small for city-dwellers but large for country-dwellers. Two reasonable strata would then be (a) city dwellers and (b) country-dwellers.

iv) The problems of sampling (their nature, size, etc.) may differ in different parts of a population. If we divide a population into strata, we can sample each stratum according to its own needs.

Example 11 If we want to sample the annual sales of a state's retail outlets, then it might be convenient to categorize the outlets according to size, for example small, medium and large, because most of the large ones will be in or near cities while many of the small ones will be in the country. They will therefore reflect different buying patterns and should probably be looked at separately.

We assume here that the strata into which a population will be divided are self-evident, depending on the problem. Some suggestions are:

a) *Income*: Less than $5,000, $5,001–$10,000, $10,001–$15,000, over $15,000.
b) *Sex*: Male, female
c) *Sales*: Less than 50,000, 50,000–150,000, over 150,000
d) *Politics*: Republican, Democrat, others
e) *Companies*: Large, medium, small

f) *Type of work*: Professional, skilled, unskilled, others.

g) *Families*: No children, 1 or 2 children, more than two.

After stratification we may sample from each stratum, using simple random sampling. We may take the sample size proportional to the stratum size, in which case we have *proportional* stratified sampling; or we may make the sample size independent of the stratum size, which we call *nonproportional* stratified sampling (the word random may be dropped when it is understood).

Example 12 Consider the case where the population consists of 10,000 retail outlets, to be sampled for their annual sales. Let us suppose that there are four strata, as follows:

Strata (Annual sales $000,000)	Number of outlets	Number in sample (Proportional sampling)
Over 10	500	5
5–10	1000	10
1–5	1500	15
Less than 1	7000	70
Total	10,000	100

We see that the ratio of sample size to stratum size is constant (in this case 1/100). We might argue, however, that the bigger outlets sell so much more than the smaller ones that the sample should be relatively greater for the bigger outlets. Further, if the small stores are more rural and therefore farther apart, the cost of sampling from them may be more than it is worth. Thus, nonproportional sampling might be justified in this case on economic grounds.

9.6. SYSTEMATIC SAMPLING

In some situations, simple random sampling may be possible but awkward to apply. For instance, suppose that we have 5,000 receipts from which we wish to draw a sample of size 100. In order to draw a simple random sample, we would have to number the receipts from 1 to 5,000, and 100 numbers would have to be selected using random-number tables.

Another method, called *systematic* sampling, consists in randomly selecting a starting point between, say, 1 and 50, which we will call k, and then including in the sample the kth receipt, the $(k+50)$th receipt, the $(k+100)$th receipt, and so on. For instance, if $k = 7$, then the sample would include receipts numbered 7, 57, 107, 157, etc.

Systematic sampling has the advantage that the sample is often easier to draw, especially if the population is organized in an orderly way. Its major disadvantage is that hidden periodicities might be present in the data. Thus in selecting light bulbs for testing purposes, if we select every tenth, and yet every fifth one has a defect, our results would be biased.

9.7. CLUSTER SAMPLING

In order to obtain either a simple random sample or a stratified random sample, it is necessary to be able to list the population. If there are machine parts to be sampled, they can be numbered as they are produced $1, 2, \ldots, N$. If the population is all the gas stations in Ohio or the farmers in Ontario or the retail outlets in New York, a list of the population can be obtained.

For some problems, such a list is (i) very expensive to make (e.g., listing all the voters in the US.) or (ii) difficult to obtain (such as the number of people who have more than $50,000 worth of life insurance) or (iii) *impossible* to obtain (such as the number of people who use Brand X margarine). Finally suppose that a project involved sampling over a large geographical area. Using simple or even stratified sampling might necessitate the need to travel extensively, thus incurring great costs.

For all of the above described situations, a viable alternative to simple or stratified sampling is *cluster sampling*. We illustrate the idea of cluster sampling by considering the problem of choosing a sample of families in the U.S. We might first break the country into precincts (or election districts) and choose a number of these. Within each of these we might then choose one or more blocks of houses, and interview every family in those blocks. (It is true that many precincts *will not appear* in the sample, but they *had a chance* of doing so, and that is all that is required in order for the sample to be random.)

The major disadvantage of cluster sampling is that the information will be less precise than through simple random sampling (for the same sample size), but the cost of cluster sampling may be such that it is the only method that can reasonably be considered. (It is interesting to note that the famous Gallup polls by Dr. George Gallup generally use the technique of cluster sampling.)

9.8. SAMPLING FROM A CONTINUOUS POPULATION

All of the discussion of this chapter has presupposed that the population is discrete and that, at least theoretically, the members of the population *could* be counted. For sampling from continuous populations, no hard and fast rules exist, and experience and common sense must be relied on heavily. Nevertheless, the concepts of simple random sampling should be borne in mind. Thus, when sampling a bulk material such as cement or sand, or oil in a tank, the amount sampled must be large enough to obtain a measured response, yet not too large for (say) laboratory analysis. To ensure that the samples are truly representative, one might mix the substance being sampled after each amount is removed, or obtain amounts from different places. Sometimes if the population is of a reasonable size, it can be divided into a number of smaller parts, and analysis of these parts can be performed by simple random sampling. In summary, though, for continuous populations, one must rely heavily on techniques that are intuitively reasonable.

SUMMARY

In this chapter we discuss the techniques for choosing a sample from a population in such a way that the sample *represents the population*. Reasons why we sample instead of measuring the whole population are given, and the distinction between discrete and continuous populations is discussed. The sometimes abstract notion of randomness is explained in terms of a simple random sample (not to be confused with a *haphazard* sample); and some examples of nonrandom (or *biased*) samples help to illustrate this notion. Various techniques are then discussed, including simple random sampling, stratified random sampling, cluster sampling, systematic sampling, and sampling from a continuous population.

Words to Remember

Discrete population
Continuous population
Biased sample
Representative
Random sample

Stratified random sampling
Proportional sampling
Cluster sampling
Systematic sampling
Random numbers

EXERCISES

1. Consider a manufacturing plant with two assembly lines. We wish to sample the parts on each line to obtain some idea of their relative quality, and then we wish to say something about overall quality. Suggest and discuss a sampling technique which could be used.

2. Classify the following populations as discrete or continuous:
    i) A truckload of cement;
   ii) The entire United States population, considered for census purposes;
  iii) The amount of gasoline in a large oil company's storage tank;
   iv) All the transistors in a T.V. set;

    v) Your personal gasoline purchase receipts for last year;

   vi) The number of tractors in Nebraska;

  vii) The yield of corn per acre in Iowa;

 viii) The number of people with degrees in statistics;

   ix) The number of people (in a country) earning more than $15,000 per year;

    x) The average daily rainfall in Dallas last year.

3. Suggest and describe some sampling techniques to find out how many Americans have degrees in business. Discuss the relative merits (statistical and economic) of each technique.

4. Records in a small town give the name, address, age, and income for each resident. These records may be easily sorted into any order; and additional information may be obtained by inquiry from each person chosen. What methods of sampling would you suggest for estimating:

    i) The proportion of men and women in the town?

   ii) The number of people earning over $5,000/year?

  iii) The number of Democrats?

   iv) The number of people living more than one mile from the city center?

    v) The number of people having more than five days of sickness per year.

5. Consider the following quality-control problem. A large batch of parts must be rejected if the proportion defective is greater than $\frac{1}{6}$. The quality-control engineer selects 12 parts from each batch, and rejects the batch if three or more of the twelve are defective. Simulate his selection procedure by throwing a single die twelve times, and denoting a "six" as a defective (i.e., count the number of "sixes" in the twelve throws). Repeat this procedure about 100 times and comment on the results; that is, discuss the quality-control engineer's selection procedure.

ESTIMATION 10

10.1. INTRODUCTION

We stated in Chapter 1 that, in studying almost all business problems, the true relationships between the variables being studied are generally obscured by lack of complete information and by variability. The examples of Chapter 1 are discussed in terms of this variability. In Chapters 2, 3, and 4, we saw how to organize a mass of data and how to obtain measures of central *location* and *dispersion*. In Chapter 6, we discussed what a random variable is, and how to obtain its expected value and its variance.

We hope that it is now clear to the student that the approach of statistics to any business problem is to look at some variable (for instance, gasoline sales, wages, the number of workers in a plant, the number of days of sick leave, the melting point of iron, the sales of product X), and regard this variable as a *random variable* the form of whose distribution is assumed to be known. It might have a normal, or a binomial, or a Poisson distribution. The distributions are described in terms of their parameters; but the actual values of these parameters are unknown because the whole population has not been (or cannot be) measured. The best that can be done is to take a sample and perform the measurements on it, and then calculate the *sample statistics*. If the student is not clear about these ideas, it would be wise for him to reread Chapter 1, Sections 1.1 and 1.2, and Chapters 2, 3, 5, and 6.

This chapter is concerned with the problem of estimating the *unknown* population parameters by means of the *known* sample statistics.

Example 1 Reconsider Example 2 of Chapter 1, concerning the company that tests an increased toothpaste flavor-level by giving samples to 10,000 families and asking for their reaction. Suppose that a family can answer either (a) "Yes," meaning that it likes the new flavoring, or (b) "No," meaning that it doesn't like the new flavoring. If each family's response is independent of all the other families' responses, then the distribution of the number of families answering "Yes" will be binomial. (This follows from the discussion of the binomial distribution in Chapter 7.) The company might now ask "What is the expected value of this binomial distribution?" We know from Chapter 7 that the expected value of a binomial distribution is np, where $n = 10,000$ and p is the probability that an individual family answers "Yes." However, since p is unknown, the expected value is unknown. The question then becomes "How can we estimate p and, hence, np?"

Example 2 In Example 3 of Chapter 1, we may assume that iron-ore melting-point measurements have normal distributions. Those measurements from region A have one normal distribution with mean μ_A and variance σ_A^2, while those from region B have another normal distribution with mean μ_B and variance σ_B^2. Of course μ_A, σ_A^2, μ_B, and σ_B^2 are unknown. They can be estimated, however, from the data given in Example 3 of Chapter 1.

Example 3 In Example 4 of Chapter 1, each of the 200 parts examined by the quality-control expert has an unknown probability p of being defective. The number of defective parts is therefore a binomial random variable. We need to estimate p in order to estimate the mean np and the variance npq.

We restate something now which we have tried to emphasize throughout the text. We will never know the parameters unless we measure the *whole* population. Using the *sample* figures will necessarily mean that our knowledge of the parameters will be incomplete, and that any statement about the parameters will contain an element of *uncertainty*. We continue our discussion with a study of some important statistics, and attempt to see the relationship between them and the *population* parameters.

10.2. SAMPLING DISTRIBUTIONS

We shall introduce the ideas of this section, which are perhaps the most fundamental ideas of statistical inference, by looking at one particular statistic, the sample mean $\overline{X}$. We shall be concerned with what a sampling distribution is, how it arises, and how it is related to the population from which the sample comes. The sample mean or just the mean, $\overline{X}$, is discussed in Chapter 3. It is particularly important and plays a major role in the discussion of estimation. Thus we start this chapter by looking at some properties of $\overline{X}$.

Suppose that a firm wants to estimate the mean hourly wages of (say) skilled workers in Seattle. A carefully chosen random sample of sixteen observations yields the following results: $6.80, 9.40, 7.90, 12.10, 10.84, 7.74, 8.84, 9.60, 8.30, 11.60, 10.48, 9.00, 10.20, 12.50, 8.00, 9.20. The sample mean for these figures is their sum, divided by 16, or

$$\overline{X} = \frac{\$152.50}{16} = \$9.53.$$

Thus we could say that $\overline{X}$, our estimate of the mean hourly wages of skilled Seattle workers, is $9.53. We must recognize, however, that if we had taken a different set of sixteen observations, we would probably have obtained a different estimate. Indeed, if we had repeated the sampling again and again, we would have obtained a sequence of different values for $\overline{X}$, such as $9.83, 9.07, 10.09, ...

If we assume that the differences between these figures are due to chance variation and not due to any real differences between the samples, then studying the different values of $\overline{X}$, we can obtain some idea of the distribution of $\overline{X}$. This distribution will be called the *sampling distribution of $\overline{X}$*. The distribution of $\overline{X}$ will tell us how close to or far from the true population mean μ we can expect $\overline{X}$ to be. Now we realize that, in fact, we have only one value of $\overline{X}$ with which to estimate μ. What must be clearly understood is that the single value of $\overline{X}$ is but one of *many possible values* which might have been chosen. The distribution of the totality of the possible values of $\overline{X}$ is called the *sampling distribution of the mean $\overline{X}$*, and the actual values of these many $\overline{X}$'s would scatter about the true value of μ according to some pattern. In order to be able to use $\overline{X}$ as an estimator of μ, we need to know some of the important characteristics of the distribution of $\overline{X}$. Specifically, we need to know how the parameters of the distribution are related to the parameters of the original sampled distribution.

Suppose that we are dealing with a population which has n elements (for instance, the number of skilled workers in Seattle), and we wish to estimate μ, the mean weekly wages of the population. The unknown mean of the population is μ and the variance (sometimes known and sometimes unknown) is σ^2. Then the mean of the distribution of the mean (or the mean of the distribution of $\overline{X}$), denoted by $\mu_{\overline{X}}$ is $\mu_{\overline{X}} = \mu$ and the variance of the distribution, denoted by $\sigma_{\overline{X}}^2$ is

$$\sigma_{\overline{X}}^2 = \frac{\sigma^2}{N}\left(\frac{n-N}{n-1}\right).$$

If n is thought of as being very large compared to N, as is usually the case, then the variance of the distribution of $\overline{X}$ is $\sigma_{\overline{X}}^2 = \sigma^2/N$. In short, we say that the expected value of the mean, $\mu_{\overline{X}}$, is equal to μ and that the variance of $\overline{X}$, $\sigma_{\overline{X}}^2$, is equal to σ^2/N. Both of these results are extremely powerful and important, and allow us to evaluate the quality of our estimator $\overline{X}$. Another way of stating it is to say that, if from a population with mean μ and variance σ^2, samples of size N are taken, then the population of the means of these samples has mean μ and variance σ^2/N.

Thus the variance of the means is smaller (σ^2/N is smaller than σ^2) than the variance of the original values. This means that the means are *less variable* than the original values. This somewhat obvious but important property is extremely useful in estimating parameters.

Sampling Distribution of Sample Means

The sampling distribution of the means of N observations, taken from a population with mean μ and variance σ^2, has mean μ and variance σ^2/N.

We illustrate the above discussion for a simple situation where n, the population size, is 3 and N, the sample size, is 2. If the population values are 3, 4, and 5, then

$$\mu = \frac{3+4+5}{3} = 4,$$

and

$$\sigma^2 = \frac{(3-4)^2 + (4-4)^2 + (5-4)^2}{3} = \frac{2}{3}.$$

There are $\binom{3}{2} = 3$ possible samples of size 2; their means are

$$\overline{X}_1 = \frac{3+4}{2} = 3.5, \qquad \overline{X}_2 = \frac{3+5}{2} = 4, \qquad \overline{X}_3 = \frac{4+5}{2} = 4.5.$$

So $\mu_{\overline{X}}$, the mean of the means, is

$$\mu_{\overline{X}} = \frac{3.5+4.0+4.5}{3} = 4.0.$$

This shows that $\mu_{\overline{X}} = \mu$. Similarly, the variance of the means is

$$\sigma_{\overline{X}}^2 = \frac{(3.5-4)^2 + (4.0-4)^2 + (4.5-4)^2}{3}$$

$$= \frac{1}{6} = \frac{2/3}{4} = \frac{\sigma^2}{4}.$$

This shows that

$$\sigma_{\overline{X}}^2 = \frac{\sigma^2}{N}\left(\frac{n-N}{n-1}\right).$$

If a population, in addition to having μ and σ^2 as its mean and variance, is normal, then we may state the following result.

> **Distribution of Sample Means from a Normal Population**
>
> The sampling distribution of the means of N observations, taken from a normal population with mean μ and variance σ^2, is normal, with mean μ and variance σ^2/N.

The distribution of sample means from a normal population for $N=9$ and $N=25$ is illustrated in Fig. 10.1.

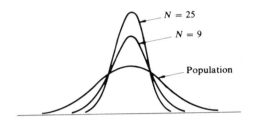

10.1   Distribution of means of samples of sizes 9 and 25 from a normal population.

Example 4   (Refer to Exercise 6, Chapter 8.) In a certain construction job, wages are assumed to follow a normal distribution, with mean $4.50 per hour and standard deviation $0.50. What is the probability that the average wage of four workers is greater than $5.00?

Solution: If X represents the wage of an individual worker, then X is normally distributed with $\mu=\$4.50$ and $\sigma=\$0.50$. We are interested in the distribution of the means of samples of size four. We know that the distribution of the means of samples of size four is normal, with

$$\mu_{\bar{X}} = \mu = 4.50,$$

$$\sigma_{\bar{X}}^2 = \frac{\sigma^2}{N} = \frac{0.25}{4} = 0.0625,$$

$$\sigma_{\bar{X}} = \sqrt{0.0625} = 0.25.$$

So

$$Z = \frac{\bar{X} - \mu}{\sigma_{\bar{X}}} = \frac{5.00 - 4.50}{0.25} = 2.00,$$

and $P(\overline{X} > 5.00) = P(Z > 2.00) = 0.0227$. Thus the probability that the average wage of four workers is greater than $5.00 is 0.0227.

It will be necessary to consider the distribution of the difference of the means of two normal populations. For this we shall use the result of Exercise 26 of Chapter 6, namely, that the variance of two independent variables is equal to the sum of their respective variances.

Distribution of the Difference of Two Normal Means

If, from a normal population 1, with mean μ_1 and variance σ_1^2, $\overline{X}_1$ is the mean of a sample of size N_1, and similarly from a normal population 2, with mean μ_2 and variance σ_2^2, $\overline{X}_2$ is the mean of a sample of size N_2, then $\overline{X}_1 - \overline{X}_2$ is normally distributed with mean $\mu_1 - \mu_2$ and variance

$$\frac{\sigma_1^2}{N_1} + \frac{\sigma_2^2}{N_2}.$$

Example 5 In one industry, the workers' wages are normally distributed, with mean $5.00 and variance 0.50. In a second industry, the workers' wages are normally distributed with mean $4.00 and variance 0.25. From the first industry, the mean wage of a sample of size 20 is calculated, while from the second industry, the mean wage of a sample of size 10 is calculated. What is the probability that the *difference* of these two means is less than $0.75?

Solution: We are interested in $\overline{X}_1 - \overline{X}_2$. This is normal, with mean

$$\mu_1 - \mu_2 = 5.00 - 4.00 = 1.00,$$

and variance

$$\frac{\sigma_1^2}{N_1} + \frac{\sigma_2^2}{N_2} = \frac{0.50}{20} + \frac{0.25}{10} = 0.05.$$

So

$$Z = \frac{(\overline{X}_1 - \overline{X}_2) - (\mu_1 - \mu_2)}{\sqrt{\dfrac{\sigma_1^2}{N_1} + \dfrac{\sigma_2^2}{N_2}}} = \frac{0.75 - 1.00}{0.2236}$$

$$= -1.12.$$

Now $P(\overline{X}_1 - \overline{X}_2 < 0.75) = P(Z < -1.12) = 0.1314$. Thus the probability that the difference in the mean wages of workers from the two industries is less than $0.75 is 0.1314.

Finally in this section we present a remarkable characteristic of the distribution of the sample mean.

The Central Limit Theorem

If repeated samples of size N are drawn from a population with mean μ and variance σ^2, then for N large, the distribution of $\overline{X}$, the sample mean, is approximately normal, with mean μ and variance σ^2/N, and this approximation becomes better as N becomes larger.

Notice that this result says nothing about the distribution of the *original population*. It justifies the use of the normal distribution in a wide range of problems where the sample size is large enough. It is difficult to say what is meant by "large," but except in the case of unusual populations, N greater than 30 will usually suffice.

10.3. POINT AND INTERVAL ESTIMATION

A large number of problems concerned with business statistics consist of taking a sample from a population and using it to estimate a parameter of the population. There are two kinds of estimation with which we shall be concerned. The first of these is called *point estimation* where, from the values of the sample, a single number is obtained as an estimate of the parameter. Thus, in estimating p, the proportion of voters who will vote for Mr. Smith, we may use x/n, where x is the number who will vote for Mr. Smith in a sample of size n.

Example 6 The toothpaste manufacturer in Example 1 may estimate that 37% of the population likes the new flavoring. The estimate $\hat{p}$ of p, the proportion liking the new flavoring, is $\hat{p} = 0.37$. Here $\hat{p}$ is a *point estimator*.

(When we are talking of the statistic in general, we will call it an *estimator*. If the estimator takes on a specific value we call it an *estimate*.) The major disadvantage of point estimation is that it is unlikely for the estimate to coincide with the *population* parameter. Even in the small example in Section 10.2, if we use the sample mean to estimate the population mean, there are two chances in three that they will be different. A more desirable approach to estimation is to calculate an *interval* which has a preassigned probability of containing the unknown parameter. We then know that it is highly unlikely for the parameter to be *outside* this interval. Estimation based on the calculation of such intervals is called *interval estimation*; the intervals themselves are called confidence intervals; and the probability with which the interval includes the parameter is called the *degree of confidence* or *confidence coefficient*.

10.4. ESTIMATION OF THE POPULATION MEAN

Suppose that a firm wishes to estimate the mean hourly wage of skilled workers in Seattle, and obtains a sample of forty observations, from which $\bar{X}$ is calculated to be \$9.50. If only a point estimate of μ is desired, then the problem is answered. If we require an interval estimate of μ, then we can consider the statistic

$$Z = \frac{\bar{X} - \mu}{\sigma/\sqrt{N}}.$$

If the original population is normal, then Z will have a standard normal distribution. If the original distribution is not normal, then by the central-limit theorem (Section 10.2), Z will be approximately normally distributed. From Chapter 8, we know that the probability that Z lies between -1.96 and 1.96 is 0.95. Thus

$$P(-1.96 < Z < 1.96) = 0.95$$

or

$$P\left(\bar{X} - 1.96\frac{\sigma}{\sqrt{N}} < \mu < \bar{X} + 1.96\frac{\sigma}{\sqrt{N}}\right) = 0.95.$$

The values $\bar{X} - 1.96(\sigma/\sqrt{N})$ and $\bar{X} + 1.96(\sigma/\sqrt{N})$ are said to form an interval which includes μ with a probability 0.95. The interval itself is called a *95% confidence interval*. Of course, if a different sample had been taken, we would have a different value for $\bar{X}$ and hence a different interval. We know, however, that if we calculated a large number of these intervals, 95% of them would contain μ. We thus say that we are 95% confident that the said interval will contain μ. If we define $Z_{\alpha/2}$ such that the probability that Z lies between $-Z_{\alpha/2}$ and $Z_{\alpha/2}$ is $(1 - \alpha)$, then a $100(1 - \alpha)\%$ confidence interval is

$$\left(\bar{X} - Z_{\alpha/2}\frac{\sigma}{\sqrt{N}}, \quad \bar{X} + Z_{\alpha/2}\frac{\sigma}{\sqrt{N}}\right).$$

For instance, for 90% or 99% confidence intervals, $Z_{\alpha/2}$ is 1.65 and 2.58, respectively.

Notice that, in order to calculate the confidence interval, we must know σ, the population standard deviation. For sample sizes greater than 30, we may, in general, safely replace σ by

$$s = \sqrt{\frac{\sum (X_i - \bar{X})^2}{N - 1}}.$$

The case where σ is unknown and N is small needs a modification which will be discussed below.

Confidence Interval

If $X_1, X_2, \ldots, X_N$ is a sample of size N from a normal population with known variance σ^2 and unknown mean μ, then the interval between $\overline{X} - 1.96\sigma/\sqrt{N}$ and $\overline{X} + 1.96\sigma/\sqrt{N}$ is called a 95% *confidence interval* for μ. The endpoints of the interval are called the confidence limits and 0.95 is called the *confidence coefficient*. A $100(1-\alpha)\%$ confidence interval for μ is between $\overline{X} - Z_{\alpha/2}\sigma/\sqrt{N}$ and $\overline{X} + Z_{\alpha/2}\sigma/\sqrt{N}$. If the population is nonnormal but $N > 30$, the central-limit theorem assures us that the results still hold. If σ is unknown but $N > 30$, with s replacing σ, the results will be approximately valid.

Example 7 In order to estimate the average amount that families of four people spend on groceries each week at the PQR supermarket, a sample of 15 people was taken. The sample mean was calculated as $27.85, while the standard deviation σ was assumed known and equal to 1.50 (obtained from other similar studies). The data could safely be assumed to be normal. Calculate a 99% confidence interval for the average weekly amount spent on groceries at the PQR supermarket by families of four in the population.

Solution

i) $\overline{X} = 27.85$, $\sigma = 1.50$, $N = 15$;

ii) $2.58\dfrac{\sigma}{\sqrt{N}} = \dfrac{(2.58)(1.50)}{\sqrt{15}} = 1.00$;

iii) $\overline{X} - 2.58\dfrac{\sigma}{\sqrt{N}} = 27.85 - 1.00 = 26.85$,

$\overline{X} + 2.58\dfrac{\sigma}{\sqrt{N}} = 27.85 + 1.00 = 28.85$;

iv) The 99% confidence interval is between $26.85 and $28.85. This is interpreted as saying that, although the average amount spent is unknown, we are 99% confident that it lies in this interval.

If the standard deviation σ is unknown and N is small, then the preceding work is not valid. If the original distribution has approximately the shape of the normal distribution, then the statistic is:

$$\frac{\overline{X} - \mu}{s/\sqrt{N}},$$

whose sampling distribution is called the Student t-distribution (or simply, the t-distribution). The name Student is a pseudonym for the discoverer of this density, W. S. Gosset. Until his discovery in 1908, it was common practice to treat $(\overline{X} - \mu)\sqrt{N}/s$ as having an approximate standard normal distribution. We denote the statistic by the letter t_{N-1}, thus:

$$t_{N-1} = \frac{\overline{X} - \mu}{s/\sqrt{N}}.$$

The mathematical form for the Student's t-distribution can be shown to depend on $N - 1$, where N is the sample size. The quantity $N - 1$ is usually referred to as the degrees of freedom, and is abbreviated d.f..

Student's t-Distribution

If $X_1, X_2, \ldots, X_N$ is a sample of size N from a normal population with unknown mean μ and unknown variance σ^2, and if $\overline{X}$ and s^2 are the sample mean and sample variance, respectively, then

$$t_{N-1} = \frac{\overline{X} - \mu}{s/\sqrt{N}}$$

has a Student's t-distribution with $N - 1$ degrees of freedom. If the original distribution is only approximately normal, the above result is still valid.

The t-distribution is similar to the normal. It is bell-shaped, symmetrical, and has zero mean. The actual shape depends on the degrees of freedom $N - 1$, but for N greater than 30, the t-distribution and the normal are practically indistinguishable. The graphs of the Student's t-distribution for degrees of freedom 5, 15, and 30 are shown in Fig. 10.2.

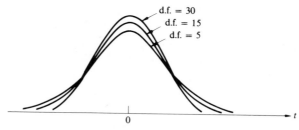

10.2   Student's t-distribution for degrees of freedom 5, 15, and 30.

When dealing with the normal distribution, we saw that 95% of the area (and hence of the probability) lay between -1.96 and 1.96. The corresponding values for the t-distribution are denoted by $-t_{0.025}$ and $t_{0.025}$, as shown in Fig. 10.3.

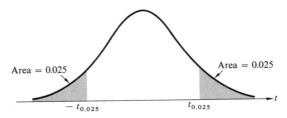

10.3 Showing $t_{0.025}$ and $-t_{0.025}$ on a typical t-distribution.

Since these values change with varying degrees of freedom, they must be looked up in a set of tables; $t_{0.025}$ and other values of t are given in Appendix Table 6.

Repeating the reasoning of Section 10.4, we can see that, based on the t-distribution, we can write

$$P(-t_{0.025} < t_{N-1} < t_{0.025}) = 0.95,$$

or

$$P\left(-t_{0.025} < \frac{\bar{X} - \mu}{s/\sqrt{N}} < t_{0.025}\right) = 0.95,$$

which can be rearranged to give

$$P\left(\bar{X} - t_{0.025}\frac{s}{\sqrt{N}} < \mu < \bar{X} + t_{0.025}\frac{s}{\sqrt{N}}\right) = 0.95.$$

This interval

$$\left(\bar{X} - t_{0.025}\frac{s}{\sqrt{N}}, \quad \bar{X} + t_{0.025}\frac{s}{\sqrt{N}}\right)$$

is the 95% confidence interval for μ.

More generally, for a $100(1-\alpha)\%$ confidence interval, we define $t_{\alpha/2}$ so that the probability that t lies between $-t_{\alpha/2}$ and $t_{\alpha/2}$ is $1-\alpha$; then a $100(1-\alpha)\%$ confidence interval for μ is

$$\left(\bar{X} - t_{\alpha/2}\frac{s}{\sqrt{N}}, \quad \bar{X} + t_{\alpha/2}\frac{s}{\sqrt{N}}\right).$$

The actual values of $t_{\alpha/2}$, which vary for α and for $(N-1)$, can be found in Appendix Table 6. We summarize as follows.

Confidence Interval

If $X_1, X_2, \ldots, X_N$ is a sample of size N from a normal population with unknown mean μ and unknown variance σ^2, then the interval between $\overline{X} - t_{0.025} s/\sqrt{N}$ and $\overline{X} + t_{0.025} s/\sqrt{N}$ is called a 95% confidence interval for μ. A $100(1-\alpha)\%$ confidence interval for μ is between $\overline{X} - t_{\alpha/2} s/\sqrt{N}$ and $\overline{X} + t_{\alpha/2} s/\sqrt{N}$. The results are still valid if the original distribution is approximately normal.

Example 8 Reconsider Example 7, and suppose that σ is not known (perhaps data from other similar studies were not available). From the sample of size 15, with mean \$27.85, the sample standard deviation was calculated to be \$2.00. Calculate 95% and 99% confidence intervals for μ, the population weekly average grocery bill of a family of four at the PQR supermarket.

Solution

i) $\overline{X} = 27.85$, $s = 2.00$, $N = 15$, d.f. $= N - 1 = 14$.

ii) For a 95% confidence interval, $t_{0.025} = 2.145$; for a 99% confidence interval, $t_{0.005} = 2.977$.

iii) $\dfrac{s}{\sqrt{N}} t_{0.025} = \dfrac{(2.145)(2.00)}{\sqrt{15}} = 1.11,$

$\dfrac{s}{\sqrt{N}} t_{0.005} = \dfrac{(2.977)(2.00)}{\sqrt{15}} = 1.54.$

iv) $\overline{X} \pm \dfrac{s}{\sqrt{N}} t_{0.025} = 27.85 \pm 1.11 = 26.74$ and 28.96;

$\overline{X} \pm \dfrac{s}{\sqrt{N}} t_{0.005} = 27.85 \pm 1.54 = 26.31$ and 29.39.

v) The 95% confidence interval is between \$26.74 and \$28.96, while the 99% confidence interval is between \$26.31 and \$29.39.

10.5. ESTIMATION OF THE POPULATION VARIANCE AND STANDARD DEVIATION

The last section centered on the problem of estimating the mean of a population. We now concern ourselves with estimating the variance and standard deviation of a normal population. (The requirement that the population be normal, or very nearly so, is much more important here than in the last section, since the estimates of the *mean* are not so sensitive to departures from normality as are the estimates of the variance and standard deviation.). The point estimate of the variance σ^2 is:

$$s^2 = \frac{1}{N-1} \sum_{i=1}^{N} \left(X_i - \overline{X} \right)^2,$$

and of the standard deviation σ is

$$s = \sqrt{\frac{1}{N-1} \sum_{i=1}^{N} \left(X_i - \overline{X} \right)^2}.$$

Now the quantity $(N-1)s^2/\sigma^2$ is said to have a chi-square distribution with $N-1$ degrees of freedom. We denote the statistic using the Greek letter X (chi). Thus

$$\chi_{N-1}^2 = \frac{(N-1)s^2}{\sigma^2}.$$

The chi-square distribution has an asymmetrical shape. A typical distribution is shown in Fig. 10.4, although the actual shape depends on the quantity $N-1$, called the degrees of freedom.

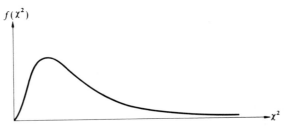

$f(\chi^2)$

10.4   The shape of a typical chi-square distribution.

When dealing with the *t*-distribution, we saw that 95% of the area lay between the two points $-t_{0.025}$ and $t_{0.025}$. For the chi-square distribution, we may say that there correspond two values called $\chi_{0.025}^2$ and $\chi_{0.975}^2$, as shown in Fig. 10.5, such that 95% of the area of the distribution lies between them.

Since the values of $\chi_{0.025}^2$ and $\chi_{0.975}^2$ change for different degrees of freedom, they must be looked up in a table. The actual values are shown in Appendix Table 7. We can repeat the reasoning used in Sections 10.4 and 10.5 and state that the probability that a chi-square statistic lies between $\chi^2_{0.025}$ and $\chi^2_{0.975}$ is 0.95, or

$$P(\chi^2_{0.025} < \chi^2_{N-1} < \chi^2_{0.975}) = 0.95.$$

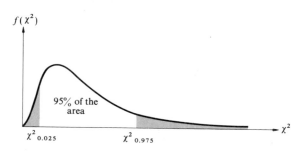

10.5

This may be manipulated to give

$$P\left(\frac{(N-1)s^2}{\chi^2_{0.975}} < \sigma^2 < \frac{(N-1)s^2}{\chi^2_{0.025}}\right) = 0.95.$$

This interval,

$$((N-1)s^2/\chi^2_{0.975}, \quad (N-1)s^2/\chi^2_{0.025}),$$

is a 95% confidence interval for σ^2. More generally, for a $100(1-\alpha)\%$ confidence interval, we define $\chi^2_{\alpha/2}$ and $\chi^2_{(1-\alpha)/2}$ so that the probability that χ^2_{N-1} lies between them is $1-\alpha$; then a $100(1-\alpha)\%$ confidence interval for σ^2 is

$$((N-1)s^2/\chi^2_{(1-\alpha)/2}, (N-1)s^2/\chi^2_{\alpha/2}).$$

The actual values of $\chi^2_{\alpha/2}$ and $\chi^2_{(1-\alpha)/2}$, which vary for α and $N-1$, can be found in Appendix Table 7. We summarize as follows.

Confidence Interval for the Variance of a Normal Population

If $X_1, X_2, \ldots, X_N$ is a sample of size N from a normal population with unknown mean and variance, then the interval between

$$(N-1)s^2/\chi^2_{1-\alpha/2} \quad \text{and} \quad (N-1)s^2/\chi^2_{\alpha/2}$$

is called a $100(1-\alpha)\%$ confidence interval for σ^2.

Example 9 Calculate a 95% confidence interval for the variance of the melting point of iron from region A in Example 3 of Chapter 1, if the values are assumed to be normally distributed.

Solution

i) $\Sigma X_i^2 = 22{,}702{,}503$, $N = 10$, $\overline{X} = 1506.7$,

$$s^2 = \frac{22{,}702{,}503 - (10)(1506.7)^2}{9} = 117.12.$$

ii) $N = 10$, d.f. $= N - 1 = 9$; thus from Appendix Table 7,

$\chi^2_{0.025} = 2.70$, $\chi^2_{0.975} = 19.02$.

iii) $\dfrac{(N-1)s^2}{\chi^2_{0.975}} = \dfrac{(9)(117.12)}{19.02} = 55.42,$

$\dfrac{(N-1)s^2}{\chi^2_{0.025}} = \dfrac{(9)(117.12)}{2.70} = 390.40.$

iv) A 95% confidence interval for σ^2 is between 55.42 and 390.40. This is interpreted as follows: The true value of σ^2 is unknown, but we are 95% confident that it lies between 55.42 and 390.40.

The calculation of a $100(1-\alpha)\%$ confidence interval for σ, the standard deviation, may be derived quite simply from the $100(1-\alpha)\%$ confidence interval for σ^2. The confidence interval for σ^2 is between

$$(N-1)s^2/\chi^2_{(1-\alpha)/2} \quad \text{and} \quad (N-1)s^2/\chi^2_{\alpha/2}.$$

The confidence interval for σ is found by taking the *square root* of these two values.

Confidence Interval for the Standard Deviation of a Normal Population

If $X_1, X_2, \ldots, X_N$ is a sample of size N from a normal population with unknown mean and variance, then the interval between

$$\sqrt{(N-1)s^2/\chi^2_{(1-\alpha)/2}} \quad \text{and} \quad \sqrt{(N-1)s^2/\chi^2_{\alpha/2}}$$

is called a $100(1-\alpha)\%$ confidence interval for σ.

Example 10 Find a 95% confidence interval for σ for the melting point of iron ore from region A, as given in Example 3 of Chapter 1.

Solution

i) From Example 9,

$$\frac{(N-1)s^2}{\chi^2_{0.975}} = 55.42, \qquad \frac{(N-1)s^2}{\chi^2_{0.025}} = 390.40.$$

ii)

$$\sqrt{\frac{(N-1)s^2}{\chi^2_{0.975}}} = \sqrt{55.42} = 7.44,$$

$$\sqrt{\frac{(N-1)s^2}{\chi^2_{0.025}}} = \sqrt{390.40} = 19.76.$$

iii) The 95% confidence interval for σ is between 7.44 and 19.76. Note that the point estimate of σ is $s = \sqrt{117.12} = 10.82$.

10.6. ESTIMATION OF A PROPORTION

Reconsider Example 2 of Chapter 1, where a toothpaste manufacturing company wishes to know the proportion of people in the population who would use the newly flavored brand of toothpaste. We may call this proportion p and its estimate $\hat{p}$. They take a sample of 10,000 families and ask whether they would use the new brand or not. Based on the number of favorable and unfavorable responses, they should be able to estimate the proportion of the population that prefers the new flavoring. As a start, if 7,500 of the 10,000 families give favorable responses, then a point estimate of p, the proportion of the population considered favorable, is

$$\hat{p} = \frac{7,500}{10,000} \qquad \text{or} \qquad \hat{p} = 0.75.$$

Consider now how to calculate a *confidence interval* for p.

The calculation of a proportion is the number of successes divided by n. But from Chapter 7, the number of successes follows a binomial distribution. Thus we assume that in estimating a proportion, the four conditions for a binomial experiment are satisfied (at least approximately). These conditions may be stated as follows:

1. The number of trials is fixed.

2. Each trial has two possible outcomes, success and failure.

3. The probability p of success is the same for each trial.

4. The trials are independent.

The calculation of confidence intervals for p does present some problems, in that they are expressed in terms of p, the unknown proportion. To overcome this difficulty, in the case where n is small, graphical procedures are sufficient for most purposes. The upper and lower limits for 95% and 99% confidence intervals can be read from Appendix Tables 8(a) and 8(b).

Example 11 A sample of 100 randomly chosen people are asked if they favor Brand X toothpaste and 43 answer "Yes." Find a 95% confidence interval for the proportion of the population that favors Brand X toothpaste.

Solution: The number of successes is $x = 43$ while $n = 100$, thus $\hat{p} = x/n = 0.43$. Turning to Appendix Table 8(a), we read the value of $\hat{p} = 0.43$ on the bottom horizontal scale. We go up vertically until we reach the two contour lines with the number 100 on them. These numbers on the contour lines correspond to n, the sample size. The confidence limits are read off on the lefthand scale. In this case, they are 0.33 and 0.53. Thus a 95% confidence interval for p is between 0.33 and 0.53. If the value of $\hat{p}$ should be greater than 0.5, read $\hat{p}$ on the upper horizontal scale, and read the upper and lower confidence limits on the righthand scale.

When n is large and p is not too close to zero or one, we may use the central-limit theorem, which says that $\hat{p} = x/n$ is approximately normal with mean p and variance $p(1-p)/n$. The $100(1-\alpha)\%$ confidence interval for p, from Section 10.4, is

$$\left(\hat{p} - Z_{\alpha/2}\sqrt{p(1-p)/n} \ , \ \ \hat{p} + Z_{\alpha/2}\sqrt{p(1-p)/n} \ \right).$$

Unfortunately, this interval also depends on p which is, of course, unknown. However, if we approximate p by $\hat{p}$, then an approximate $100(1-\alpha)\%$ confidence interval for p will be

$$\left(\hat{p} - Z_{\alpha/2}\sqrt{\hat{p}(1-\hat{p})/n} \ , \ \ \hat{p} + Z_{\alpha/2}\sqrt{\hat{p}(1-\hat{p})/n} \ \right).$$

Confidence Interval for a Proportion

If x is the number of successes in a sample of n trials, $\hat{p}$ is the point estimate of p, and the four conditions for a binomial experiment are satisfied, then a $100(1-\alpha)\%$ confidence interval for p, the population proportion of successes, is between

$$\hat{p} - Z_{\alpha/2}\sqrt{\frac{\hat{p}(1-\hat{p})}{n}} \quad \text{and} \quad \hat{p} + Z_{\alpha/2}\sqrt{\frac{\hat{p}(1-\hat{p})}{n}} \ .$$

Example 12 In Example 11, where 43 of 100 people favored Brand X tooth-paste, use the normal approximation to calculate a 95% confidence interval.

Solution

i) $n = 100$, $x = 43$, $\hat{p} = 0.43$;

ii) $\sqrt{\dfrac{\hat{p}(1-\hat{p})}{n}} = \sqrt{\dfrac{(0.43)(0.57)}{100}} = 0.0495$;

iii) $\hat{p} - 1.96\sqrt{\dfrac{\hat{p}(1-\hat{p})}{n}} = 0.43 - (1.96)(0.0495) = 0.33$,

$\hat{p} + 1.96\sqrt{\dfrac{\hat{p}(1-\hat{p})}{n}} = 0.43 + (1.96)(0.0495) = 0.53$;

iv) A 95% confidence interval for p, using the normal approximation, is between 0.33 and 0.53.

10.7. SAMPLE SIZE ESTIMATION

In planning an experiment designed to estimate the population mean, the sample size N required is often of great importance. Below we give a method for estimating N, the size of a sample from a normal population, when (a) the variance σ^2 is known, and (b) σ^2 is unknown.

If σ^2 is known, then a 95% confidence interval is given by

$$\bar{X} - 1.96\frac{\sigma}{\sqrt{N}} < \mu < \bar{X} + 1.96\frac{\sigma}{\sqrt{N}}.$$

The length of this interval is $(2)(1.96)\sigma/\sqrt{N}$. Suppose that we require the length of the interval to be less than some value E; then

$$E \geqslant (2)(1.96)\frac{\sigma}{\sqrt{N}}$$

or

$$N \geqslant \frac{(3.92)^2\sigma^2}{E}.$$

If σ^2 is unknown and has to be estimated by s^2, then the length of the confidence interval is $(2)(t_{0.025})s/\sqrt{N}$. If the length of the interval should not exceed E, then

$$E \geqslant (2)(t_{0.025}) \frac{s}{\sqrt{N}}$$

or

$$N \geqslant \frac{(4)(t_{0.025}^2)s^2}{E^2}.$$

Since $t_{0.025}$ depends on the value of N, a direct solution for N is not possible. However, the value of N can be found by a simple iterative procedure, illustrated as follows.

Example 13 We wish to estimate the amount of money that a family of four spends weekly at the PQR supermarket. It is assumed that the amount spent follows a normal distribution, and an earlier study yielded the estimate of the standard deviation as $s = \$2.00$. We would like to use a 95% confidence interval and we would prefer the length of the interval to be no longer than 1.75. What is the minimum sample size required?

Solution: $s^2 = (2.00)^2 = 4$, $E = 1.75$;

$$N \geqslant \frac{(4)(t_{0.25}^2)s^2}{E^2} = \frac{(4)(4)t_{0.025}^2}{3.0625} = 5.22 t_{0.025}^2.$$

1. Try $t_{0.025} = 2.262$, $5.22\ t_{0.025}^2 = 26.71$, but $N = 10$;
2. Try $t_{0.025} = 2.086$, $5.22\ t_{0.025}^2 = 22.71$, but $N = 21$;
3. Try $t_{0.025} = 2.080$, $5.22\ t_{0.025}^2 = 22.58$, but $N = 22$;
4. Try $t_{0.025} = 2.074$, $5.22\ t_{0.025}^2 = 22.45$, but $N = 23$.

Thus for $N = 22$ or 23, the equality will come closest to being satisfied. To be on the safe side, choose $N = 23$.

SUMMARY

This chapter is concerned with estimating the unknown parameters of a population by taking a sample from that population and using the information in the sample to calculate an interval, called a confidence interval, such that one has a prescribed level of confidence that the interval contains the unknown parameter. Preliminary to this discussion, we discuss some of the properties of the statistic $\overline{X}$. The following are then discussed from the point of view of interval estimation:

a) the estimation of the mean of a normal population when the variance is known,

b) the estimation of the mean of a normal population when the variance is unknown,

c) the estimation of the variance and standard deviation of a normal population,

d) the estimation of the difference of two normal means,

e) the estimation of a proportion.

Words to Remember

Distribution of the sample mean t-distribution
Point estimation χ^2-distribution
Interval estimation   Degrees of freedom
Confidence interval   Normal approximation
Proportion

Symbols to Remember

N	Sample size
μ	Population mean
σ^2	Population variance
σ	Population standard deviation
$\overline{X}$	Sample mean
s^2	Sample variance
s	Sample standard deviation
σ^2/N	Variance of the distribution of $\overline{X}$
$\sigma/\sqrt{N}$	Standard deviation of the distribution of $\overline{X}$
p	Population proportion
$\hat{p}$	Sample proportion
t_{N-1}	Student's t-variable, with $(N-1)$ degrees of freedom
χ^2_{N-1}	Chi-square variable, with $(N-1)$ degrees of freedom

EXERCISES

1. If the population consists of the elements 3, 4, 5, and 6, show that $\mu_{\overline{X}} = \mu$ and

$$\sigma_{\overline{X}}^2 = \frac{\sigma^2}{N}\left(\frac{4-N}{3}\right)$$

for (a) $N=2$, and (b) $N=3$.

2. If a population consists of the elements X_1, X_2, and X_3, show, for samples of size 2, that $\mu_{\overline{X}} = \mu$ and $\sigma_{\overline{X}}^2 = \sigma^2/4$.

3. If $n=5$, $N=3$, show that $\mu_{\bar{X}}=\mu$ and that $\sigma_{\bar{X}}^2=\sigma^2/6$.

4. If $n=5$ and $N=2$, show that $\mu_{\bar{X}}=\mu$ and that $\sigma_{\bar{X}}^2=3\sigma^2/8$.

5. Verify that the mean and standard deviation of the finite population 1, 2, 3, 4, and 5 are $\mu=3$ and $\sigma=\sqrt{2}$.

6. Find the mean and variance of the distribution of the mean of samples of size 2 taken from the population in Exercise 5.

7. The wages at plants A and B are normally distributed, with means \$2.00 and \$5.00, and variances 1 and 2, respectively. Samples of 10 and 20 people are taken from the two plants (respectively), and the average wage is computed. Find the distribution, mean, and variance for:

   a) The sum of the two sample means;

   b) The difference of the two sample means.

8. The weights of sacks of flour are normally distributed with mean 100 lbs. and variance 64 lbs. What is the probability that the average weight of N sacks will exceed 110 lbs., if $N=1$, 9, 16, 36?

9. In a savings bank, the average account is \$160.00, with a standard deviation of \$20.00. What is the probability that the average of a group of 400 randomly chosen accounts exceeds \$165.00?

10. If the blood pressure of company executives is normally distributed with mean 130 and standard deviation 10, what is the probability that the mean of a sample of 10 executives will be greater than 125?

11. A bus is designed to carry a maximum load of 6000 lbs., and has a capacity of 30 people. The bus company wants to know the probability that it will be overloaded if people's weights are normally distributed with mean 180 lbs., and standard deviation 25 lbs.

12. The sample 1.1, 0.7, 2.3, 1.7, and 1.0 is drawn from a normal population with mean μ and variance 10. Find 95% and 99% confidence intervals for μ.

13. In Exercise 12, calculate 95% and 99% confidence intervals for μ if σ^2 is unknown.

14. A tailoring company knows that men's heights are normally distributed with mean μ. A sample of size 25 yields a mean of 65 inches and a standard deviation of 10 inches. Calculate a 90% confidence interval for μ.

15. In Exercise 24 of Chapter 3 are given the sizes of 300 pairs of shoes sold during the week at a certain shoe store. Assuming the shoe sizes are normally distributed, find a 95% confidence interval for the mean shoe size sold at the shoe store.

16. In Exercise 23 of Chapter 3 are given the life lengths of 200 light bulbs of type A made by a certain manufacturer. Assuming the life lengths of light bulbs are normally distributed, find a 99% confidence interval for the mean life of a light bulb produced by the manufacturer.

17. In Exercise 22 of Chapter 3 are given the wages of 500 employees in a certain factory. Assuming the wages are normally distributed, find a 95% confidence interval for the mean wage paid at the factory.

18. Consider two normal distributions with population means μ_1 and μ_2 and variances σ_1^2 and σ_2^2. From each a sample is drawn of sizes N_1 and N_2, respectively, and the sample means $\bar{X}_1$ and $\bar{X}_2$ are calculated. Using the fact (from Section 10.2) that $\bar{X}_1 - \bar{X}_2$ is normal, and that

$$Z = \frac{(\bar{X}_1 - \bar{X}_2) - (\mu_1 - \mu_2)}{\sqrt{\dfrac{\sigma_1^2}{N_1} + \dfrac{\sigma_2^2}{N_2}}}$$

has mean zero and variance one, calculate 95% and 99% confidence intervals for the difference of the two population means $\mu_1 - \mu_2$.

19. In order to compare the amount that families of four spend each week on groceries at two supermarkets, two samples of size 15 and 12, respectively, are taken. The sample means are $27.85 and $22.47. The standard deviations were known (from other similar studies) to be $1.50 and $1.80, and the data were assumed to be normally distributed. Construct a 99% confidence interval for the difference of the population means of the amounts spent at the two supermarkets.

20. In one restaurant the average amount spent for breakfast is normal, with mean 75 cents and standard deviation 10 cents, while the average amount in a second restaurant is normal with mean 83 cents and standard deviation 15 cents. If random samples of size 50 are taken from the first restaurant and 80 from the second, find the probability that the difference between the two means is less than 5 cents in absolute value.

21. A random sample of 40 cans of peaches has a mean weight of 65 ounces and a standard deviation of 3 ounces. Calculate a 99% confidence interval for the mean weight of the population of cans of peaches.

22. The standard deviation of lifetimes of 50 electric bulbs was calculated to be 50 hours. Find 95% and 99% confidence intervals for the standard deviation of all such light bulbs.

23. The standard deviation of the voltage of a sample of 30 batteries from a large consignment is found to be 0.5 volts. Find a 95% confidence interval for the variance of the voltage of the batteries.

24. The variance of the lifetimes of 20 electric bulbs was computed to be 2500 (hours)^2. Find the (a) 95%, (b) 99% confidence interval for the variance of all such electric bulbs.

25. The standard deviation of the voltage of a sample of 25 batteries from a large consignment is found to be 0.5 volts. Find the (a) 90%, (b) 98% confidence limits on the variance of the voltage of such batteries.

26. In Exercise 12, find a 90% confidence interval for σ, using the sample of five observations given there.

27. In Exercise 14, give a 90% confidence interval for the standard deviation of men's heights.

28. From a population with a proportion of p smokers, a sample of 100 yields 25 smokers. Calculate a 95% confidence interval for p by two methods.

29. If there is a proportion p of the population which is colorblind and a sample of 50 yields 5 colorblind people, calculate a 95% confidence interval for p. (Use the graphical method.)

30. If a sample of 200 executives shows that 130 are over 45 years of age, give a 99% confidence interval for the proportion of the population (of executives) that are over 45.

31. Of 100 job applicants who take a written test, 75 pass it; give a 95% confidence interval for the proportion of the *population* of applicants who could pass the same test.

32. If in Exercise 18, σ_1^2 and σ_2^2 are unknown but if s_1^2 and s_2^2 may be substituted for them, we may say that

$$\frac{(\bar{X}_1 - \bar{X}_2) - (\mu_1 - \mu_2)}{\sqrt{\dfrac{s_1^2}{N_1} + \dfrac{s_2^2}{N_2}}}$$

has approximately a t-distribution with $N_1 + N_2 - 2$ degrees of freedom if N_1 and N_2 are large (>30). Calculate 95% and 99% confidence intervals for $\mu_1 - \mu_2$.

33. Reconsider Exercise 19 concerning the amount spent on groceries at two supermarkets, and suppose that σ_1 and σ_2 are not known. From the sample of size 15 with $\bar{X}_1 = \$27.85$, the sample standard deviation is calculated to be \$2.00; while from the sample of size 12 with $\bar{X}_2 = \$22.47$, the sample standard deviation is calculated to be \$2.50. Calculate a 95% confidence interval for the unknown difference of the mean amounts spent by the populations shopping at the two supermarkets.

34. Consider two binomial distributions with parameters p_1 and p_2. From each, samples of size n_1 and n_2 are drawn, yielding $\hat{p}_1 = x_1/n_1$ and $\hat{p}_2 = x_2/n_2$. Using the fact that for n_1 and n_2 reasonably large,

$$Z = \frac{(\hat{p}_1 - \hat{p}_2) - (p_1 - p_2)}{\sqrt{\dfrac{\hat{p}_1(1 - \hat{p}_1)}{n_1} + \dfrac{\hat{p}_2(1 - \hat{p}_2)}{n_2}}}$$

is approximately normal, with zero mean and unit variance, calculate 95% and 99% confidence intervals for $p_1 - p_2$.

35. In City A, 140 people out of 200 favor strong antipollution legislation, while in City B, 192 out of 200 favor similar legislation. Calculate a 90% confidence interval for the difference of the proportions in the two cities favoring strong antipollution legislation.

36. A sample of 100 nails manufactured by a company has 10 defectives. A sample of 150 similar nails manufactured by another company has 12 defectives. Find the (a) 90%, (b) 99% confidence interval for the difference in proportions of defective nails manufactured by the two companies.

37. A box contains an unknown number of defective ball bearings. A sample of 75 bearings showed that 20% were defective. Find a 95% confidence interval for the actual percentage of defective bearings in the box.

38. From a large consignment of apples, a sample of 200 showed that 30 were bad. Find a 99% confidence interval for the actual proportion of bad apples.

39. In City A, 130 of 180 people shopped at the XYZ chain store, while in City B, 240 out of 300 shopped at XYZ. Give a 95% confidence interval for the difference in the proportions of the residents of the two cities who shop at XYZ.

HYPOTHESIS
TESTING 11

11.1. INTRODUCTION

In Chapter 10, we discussed the problems of estimating population means, variances, standard deviations, and proportions. In this chapter we discuss statistical problems of a different type. In general, we shall be dealing with a random variable whose *distribution* is known (e.g., it might be normal or binomial) but whose *parameters* are unknown. Unlike estimation problems, where we must estimate the value of the unknown parameter, here we want to decide whether the value of the parameter is equal to some preassumed or prescribed value. Problems of this nature are called problems in *hypothesis testing*.

Example 1 If the population is next week's gasoline sales, we might hypothesize that the sales will be greater than 100,000 gallons.

Example 2 If a population consists of light bulbs whose life length we wish to measure, we might hypothesize that the average life length of the light bulbs is greater than 1000 hours.

Example 3 Reconsider Example 2 of Chapter 1 and suppose that the toothpaste manufacturer knows that it will be profitable to market the new flavor of toothpaste if 60% (or more) of the population favors it. It will be unprofitable to market the new flavor of toothpaste if less than 60% favors it. It will then be necessary to take a sample of (say) 10,000 families, count the number who favor the new brand, and use this information to see whether the proportion of the population that is favorable is 60%.

In each of these examples, we will never know the answer for sure. For instance, if in Example 3, we obtain 6100 families who favor the new brand, can we say that 61% of the population favors the brand? Perhaps, due to chance variations or fluctuations, only 59% of the population favors the brands but we just happened to obtain 6100 (out of 10,000) favorable responses.

Example 4 Consider a machine that packages potatoes. Each bag is supposed to contain 20 lbs. of potatoes, but due to obvious variation, some bags are heavier and some are lighter than 20 lbs. The weights of packages are known to have a normal distribution with a standard deviation of 2 lbs. The company operating the machine wants to make every effort to control the weight of the bags at the 20-lb. mark, because if the weight is greater than 20 lbs., produce is being given away unnecessarily, while if the weight is less than 20 lbs., there may be punitive action from government agencies. In order to keep a check on the weight, 25 bags are taken each four hours and weighed. If the mean weight falls between 19 and 21 lbs., the machine will be adjudged in control, while if the mean is less than 19 or greater than 21, it will be adjudged out of control, shut down, and readjusted.

In statistical terms we are testing the hypothesis that the mean weight μ of all bags being filled equals 20, versus an alternative hypothesis that μ is not equal to 20. In fact we would say: Reject the hypothesis that $\mu = 20$ if the mean weight of the 25 bags, $\overline{X}$, is less than 19 or greater than 21. We shall call the main hypothesis the *null hypothesis* and denote it by H_0. The alternative hypothesis will be denoted by H_A.

Null and Alternative Hypotheses

The main hypothesis that we wish to test is called the null hypothesis and denoted by H_0. Any other hypotheses are called alternative hypotheses and are denoted by H_A.

Example 5 In Example 4, the main hypothesis is that $\mu = 20$, while the alternative hypothesis is that $\mu \neq 20$. We may write:

$$H_0: \mu = 20, \qquad H_A: \mu \neq 20.$$

Now in Example 4, the machine is either in control or out of control, but since we are going to make a decision about the machine, based on a random variable $\overline{X}$, we might make one of two possible errors. We might say the machine is out of control when in fact it is not, or we might say it is in control when in fact $\mu \neq 20$. Such errors might arise because of sampling errors associated with weighing the 25 bags. For instance, even though, on the average, the mean weight is 20, a bag with too many large potatoes might be overweight. In general if a null hypothesis is true and we reject it (conclude it is false), we call the error a Type I error, whereas if the null hypothesis is false and we accept it (conclude it is true), we call the error a Type II error. The probability of making a Type I error is denoted by α (alpha), and called the significance level of the test. The probability of making a Type II error is denoted by β (beta).

Errors

A Type I error is made when H_0 is wrongly rejected. A Type II error is made when H_0 is wrongly accepted. The probabilities of Types I and II error are called α and β respectively. α is called the significance level of the test.

These types of errors can be summarized as in the accompanying box chart.

True situation	Decision	
	Accept H_0	Reject H_0
H_0 true	Correct decision	Type I error
H_0 false	Type II error	Correct decision

We would, of course, like to make correct decisions all the time. However, since we are dealing with random variables whose outcomes are unpredictable, this cannot be done, and we must learn to live with Type I and Type II errors.

Let us now see, for Example 4, the probabilities of making Type I errors and Type II errors. If either of these probabilities is too large, we will probably consider our decision-making procedure "poor" and seek to redesign it. We recall, from Chapter 10, that if we take a sample from a normal population, then the distribution of $\overline{X}$ will also be normal. Now the probability of a Type I error is α, which is the probability of getting a mean of less than 19 or greater than 21 when the sample is drawn from a population with mean $\mu = 20$ and standard deviation $\sigma = 2$. This probability is shown shaded in Fig. 11.1.

Figure 11.1 represents the distribution of $\overline{X}$, which is normal, with mean $\mu = 20$ and standard deviation $\sigma/\sqrt{N} = 2/\sqrt{25} = 0.4$. In terms of the standard normal distribution, 21 is two and a half standard deviations above the mean, while 19 is two and a half below the mean. Now, from Appendix Table 5, the probability of being above or below $2\frac{1}{2}$ standard deviations from the mean is

$$(2)(0.0062) = 0.0124.$$

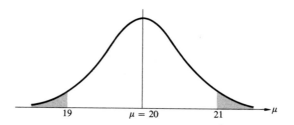

11.1   Distribution of $\overline{X}$ when $\mu = 20$.

Therefore, the probability that $\overline{X}$ will be less than 19 or greater than 21, if the true mean μ is 20, is 0.0124. Thus, $\alpha = 0.0124$. In practical terms this means that if the machine is in control, there is a probability of 0.0124 (about 12 in a thousand) of concluding that it is out of control.

In order to calculate β, the probability of a Type II error, we must assume that H_0 is false ($\mu \neq 20$). For a minute, suppose that something has happened to the machine and that the mean μ is now 21. The probability of making a Type II error is the probability of not detecting this new mean. This is the probability that $\overline{X}$ falls between 19 and 21 when in fact $\mu = 21$. This is shown shaded in Fig. 11.2, representing the distribution of $\overline{X}$, which is normal, with mean 21 and standard deviation 0.4. In terms of the standard normal distribution, 21 corresponds to zero standard deviations from the mean while 19 corresponds to 5 standard deviations. From Appendix Table 5, the probability is 0.5. Therefore the probability that $\overline{X}$ will fall between 19 and 21, if in fact $\mu = 21$, is 0.5. Thus $\beta = 0.5$. In practical terms this means that if the process has shifted its mean to 21 we will not detect it once in two times.

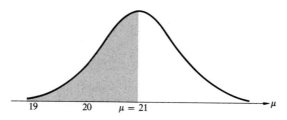

11.2   Distribution of $\overline{X}$ when $\mu = 21$.

Our computations for β assumed that μ had shifted to a specific new value, 21. There are, in fact, an infinite number of possible values for μ if H_A is true, and for each of them, there is a probability β. We calculate β in the same way, except that the underlying distribution has a different value of μ each time. We can then plot values of μ against corresponding values of β. This has been done below for nine values of μ.

Value of μ	Probability of Type II error (β)
18.0	0.0062
18.5	0.1056
19.0	0.5000
19.5	0.8944
20.0	—
20.5	0.8944
21.0	0.5000
21.5	0.1056
22.0	0.0062

No value is given for $\mu = 20$, because in that case, since H_0 is true, no Type II error can occur. These values are plotted in Fig. 11.3. This curve shows that, as the true value of μ gets farther away from 20, there is less chance of making a Type II error. For values of μ close to 20, the chances of detecting it are small, while for μ far away from 20, the chances of detecting it are large. Such a procedure surely represents a "common-sense requirement" of a testing procedure, and any procedure behaving otherwise would have little to justify itself.

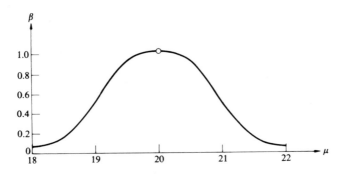

11.3   Comparing μ and β.

We notice, in the preceding discussion, that under the null hypothesis, μ takes on a specific value ($\mu = 20$), whereas under the alternative hypothesis, μ could take on infinitely many values. The reason for this is that then the value of α could be calculated. In effect, the null hypothesis says that the machine has not changed. In general, the null hypothesis is the hypothesis of no change. As we have seen, it is a simple matter to calculate α for a specific null hypothesis. The matter of calculating β depends on the value of μ.

Notice also, in Example 4, that the alternative hypothesis was $H_A : \mu \neq 20$.

This is called a two-sided test, because it considers, under the null hypothesis, cases where μ is greater than 20 and cases where μ is less than 20. If we had been concerned only with overfilling the bags (i.e., we wanted to detect a change only if μ increased), then our alternative hypothesis would have been $H_A: \mu > 20$. This is called a one-sided test.

The decision to accept the null hypothesis or the alternative hypothesis is made on the basis of a statistic computed from the sample. Such a statistic is called a *test statistic*. The values of the test statistic for which the null hypothesis is rejected make up the *critical region*. The values of the test statistic for which the null hypothesis is accepted may be called the *acceptance region*.

Test Statistic

The statistic on which the testing procedure is based is called the *test statistic*.

Example 6 In Example 4, the test statistic is $\overline{X}$, the mean weight of 25 bags. The *critical region* is composed of values of $\overline{X}$ greater than 21 or less than 19. The *acceptance region* is between 19 and 21.

The usual procedure followed in hypothesis testing is to formulate the null and alternative hypotheses, choose a level of α (often, for pragmatic reasons, $\alpha = 0.01$, 0.05, or 0.10), select a *test statistic*, and calculate the *critical region*. In Example 4, let us choose $\alpha = 0.05$ and calculate the critical region. We wish to keep the probability of a Type I error to 0.05. This means that since $\overline{X}$ has a normal distribution, with mean $\mu = 20$ and standard deviation 0.4, we will reject the null hypothesis if $\overline{X}$ falls more than 1.96 standard deviations from the mean μ. But 1.96 standard deviations is $(1.96)(0.4) = 0.784$. Thus we will reject H_0 if $\overline{X}$ is less than $20 - 0.784 = 19.216$ or greater than 20.784. So our conclusion is:

$$\text{If } \overline{X} < 19.216 \quad \text{or} \quad \overline{X} > 20.784, \qquad \text{reject } H_0;$$

$$\text{If } 19.216 < \overline{X} < 20.784, \qquad \text{accept } H_0.$$

Another, more usual procedure is to calculate $Z = (\overline{X} - 20)/0.4$ and reject H_0 if $Z < -1.96$ or $Z > 1.96$.

11.2. TESTING MEANS

We now apply the general remarks discussed in the first section to some specific problems concerning means. Consider that the underlying population is normal with known variance σ^2. We obtain N measurements $X_1, X_2, \ldots, X_N$, and wish to test hypotheses about the population mean μ. Our null hypothesis is $H_0: \mu = \mu_0$. For now, we consider just the two-sided test with alternative $H_A: \mu \neq \mu_0$. When

H_0 is true, $\overline{X}$ has a normal distribution with mean μ_0 and standard deviation σ. If the significance level is 0.05, then we reject H_0 if $\overline{X}$ is more than 1.96 standard deviations from μ_0, or alternatively, as illustrated at the end of the last section, if $Z = (\overline{X} - \mu_0)/(\sigma/\sqrt{N})$ is less than -1.96 or greater than 1.96. For other significance levels, for instance 0.01 or 0.10, one would use ± 2.58 and ± 1.65 (respectively) to define the critical region for Z. For the one-sided test $H_0 : \mu = \mu_0$, $H_A : \mu > \mu_0$, $\alpha = 0.05$, we reject H_0 if $Z = (\overline{X} - \mu_0)/(\sigma/\sqrt{N})$ is greater than 1.65, and for the other one-sided test reject H_0 if $Z < -1.65$.

If the distribution from which the sample is taken is nonnormal, then the central-limit theorem assures us that $\overline{X}$ is approximately normal for large enough sample sizes. In practice, $N > 30$ should suffice. In many cases the variance σ^2 is not known and must be estimated by s^2. If the sample size is large ($N > 30$) we may replace σ by s and proceed as discussed above, using the normal tables. The case where $N < 30$ and σ^2 is unknown is discussed subsequently.

Test of Hypothesis for the Mean of a Normal Population (Known Variance).

Null hypothesis $H_0 : \mu = \mu_0$;

Test statistic $Z = \dfrac{\overline{X} - \mu_0}{\sigma/\sqrt{N}}$.

Alternative hypothesis	Reject H_0 at the 0.05 significance level if:
$\mu \neq \mu_0$	$Z > 1.96$ or $Z < -1.96$
$\mu > \mu_0$	$Z > 1.65$
$\mu < \mu_0$	$Z < -1.65$

Figure 11.4(a) shows the distribution of $\overline{X}$ (normal with mean μ_0 and standard deviation $\sigma/\sqrt{N}$). Any values of $\overline{X}$ more than 1.96 standard deviations from the mean will reject the null hypothesis. Figure 11.4(b) shows the distribution of Z, the standardized form of $\overline{X}$. Any value of Z greater than 1.96 or less than -1.96 will reject H_0.

Example 7 A manufacturing company is thinking of moving its plant from City X to City Y. The wages of skilled technicians in City X are $10.00/hour. It wishes to test the hypothesis that the average wages of skilled technicians in City Y are the same as in City X, versus an alternative hypothesis (put forward by the plant manager) that the average wage is lower. A random sample of 100 skilled technicians is taken, and their average wage is found to be $\bar{X}=9.80/\text{hour}$, with standard deviation $s=0.50$. Test the hypothesis at the 5% level of significance.

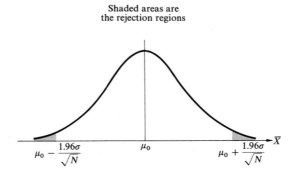

Shaded areas are
the rejection regions

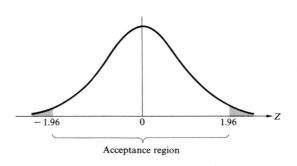

Acceptance region

11.4

Solution: The null and alternative hypotheses are $H_0: \mu = 10.00$, $H_A: \mu < 10.00$. We reject the null hypothesis if $Z < -1.65$ (for a 5% level of significance). Note that although σ is estimated from the sample, N is large enough to use Z.

$$Z = \frac{\bar{X} - \mu_0}{s/\sqrt{N}} = \frac{9.80 - 10.00}{0.50/10} = -4.00.$$

We reject H_0 and conclude that the average wage in City Y is lower than $10.00/\text{hour}$.

In cases where the population is normal with unknown standard deviation, and where N is less than 30, the above procedure cannot be used. We saw, in Chapter 10, that the distribution of

$$t = \frac{\bar{X} - \mu_0}{s/\sqrt{N}}$$

is called the Student's t-distribution with $(N-1)$ degrees of freedom. The procedure for $N < 30$ is similar to that using the Z statistic, except that instead of using the normal tables of Appendix Table 5, we use the t-table in Appendix Table 6. Thus, to test $H_0 : \mu = \mu_0$ against $H_A : \mu \neq \mu_0$ at the 5% level of significance, we reject the null hypothesis if $t = (\bar{X} - \mu_0)/(s/\sqrt{N})$ is greater than $t_{0.025}$ or less than $-t_{0.025}$, where $t_{0.025}$ is read from Appendix Table 6, with $(N-1)$ degrees of freedom. For testing $H_0 : \mu = \mu_0$ against $H_A : \mu > \mu_0$, we reject the null hypothesis if $t > t_{0.05}$ (at the 5% level of significance). Of course the appropriate t-values for other levels of significance can be found in Appendix Table 6.

Test of Hypothesis for the Mean of a Normal Population (Unknown Variance)

Null hypothesis $H_0 : \mu = \mu_0$;

Test statistic $t = \dfrac{\bar{X} - \mu_0}{s/\sqrt{N}}$.

Alternative hypothesis	Reject H_0 at the 0.05 level of significance if:
$\mu \neq \mu_0$	$t > t_{0.025}$ or $t < -t_{0.025}$
$\mu > \mu_0$	$t > t_{0.05}$
$\mu < \mu_0$	$t < -t_{0.05}$

Example 8 The average time required to perform a certain industrial task is known (from extensive past experience) to be 12.5 minutes. Ten new employees are hired and trained. During a testing period, their times for completion of the task were as follows: 9.3, 12.1, 15.7, 10.3, 12.2, 14.8, 15.1, 13.2, 15.9, 14.5. If these times are assumed to come from a normal distribution, test the hypothesis that these employees are no different from the average, using a significance level of 5%.

Solution

i) $H_0 : \mu = 12.5$, $H_A : \mu \neq 12.5$.

ii) $\sum X_i^2 = 1818.67$, $\sum X_i = 133.1$, $\bar{X} = 13.31$,
    $s^2 = 5.23$, $s = 2.28$, $N = 10$, $N - 1 = 9$.

iii) $t_{0.025} = 2.26$.

iv) $t = \dfrac{\overline{X} - \mu_0}{s/\sqrt{N}} = \dfrac{13.31 - 12.5}{2.28/\sqrt{10}} = 1.13$.

v) We reject H_0 if t is greater than 2.26 or less than -2.26. Since neither of these is true, we accept H_0, and conclude that the data do not indicate that the average time for these ten workers to perform their task is any different from the average time for other workers.

11.3. COMPARISION OF CONFIDENCE INTERVALS AND HYPOTHESIS TESTING

Perhaps it has become apparent by this time that there is a connection between the acceptance region of hypothesis testing and confidence intervals. For instance, to test

$$H_0 : \mu = \mu_0, \qquad H_A : \mu \neq \mu_0$$

at the 5% level of significance, with σ known, we know (see Fig. 11.4(a)) that the acceptance region can be expressed as:

$$\mu_0 - 1.96 \frac{\sigma}{\sqrt{N}} < \overline{X} < \mu_0 + 1.96 \frac{\sigma}{\sqrt{N}},$$

which can be rearranged as

$$-1.96 < \frac{\overline{X} - \mu_0}{\sigma/\sqrt{N}} < 1.96.$$

When H_0 is true, $(\overline{X} - \mu_0)/(\sigma/\sqrt{N}) = Z$, the standard normal distribution.

The 95% confidence interval for μ can be expressed as:

$$\overline{X} - 1.96 \frac{\sigma}{\sqrt{N}} < \mu < \overline{X} + 1.96 \frac{\sigma}{\sqrt{N}};$$

which can be rearranged as

$$-1.96 < \frac{\overline{X} - \mu}{\sigma/\sqrt{N}} < 1.96,$$

where $(\overline{X} - \mu)/(\sigma/\sqrt{N})$ again has a standard normal distribution. This shows that the acceptance region for testing $H_0 : \mu = \mu_0$ against $H_A : \mu \neq \mu_0$, when σ^2 is known, using a significance level of 5%, is the same as a 95% confidence interval for μ. More generally, the acceptance region for a significance level α is the same as the confidence interval with coefficient $(1 - \alpha)$.

In other words, accepting H_0 at an α level of significance is equivalent to μ_0 being contained in a confidence interval with coefficient $(1 - \alpha)$.

Hypothesis testing	Confidence interval
Accepting H_0	μ_0 contained in the confidence interval
Rejecting H_0	μ_0 outside the confidence interval

Similar equivalences hold for all the hypothesis-testing situations presented in the remainder of this chapter. For this reason, we will present subsequent results concerning hypothesis testing in an abbreviated form, since the main ideas have already been discussed in Chapter 10.

11.4. TESTING DIFFERENCES BETWEEN MEANS

There are many business problems where we must decide whether there is a difference between two population means. Suppose that we have two normally distributed populations, called population 1 and population 2, with respective means μ_1 and μ_2 and respective variances σ_1^2 and σ_2^2. We wish to test that $\mu_1 = \mu_2$. In practice, this might be equivalent to testing that the average profits of two investment plans are equal, or that the efficacy of two drugs is equal, or that the abilities of two groups of workers are equal. From the two populations, we take samples of sizes N_1 and N_2, respectively, and compute the sample means $\overline{X}_1$ and $\overline{X}_2$. If, for a start, we assume that σ_1^2 and σ_2^2 are known, then we know that

$$Z = \frac{(\overline{X}_1 - \overline{X}_2) - (\mu_1 - \mu_2)}{\sqrt{\dfrac{\sigma_1^2}{N_1} + \dfrac{\sigma_2^2}{N_2}}}$$

has a standard normal distribution. In order to test $H_0 : \mu_1 = \mu_2$ (which is equivalent to testing $H_0 : \mu_1 - \mu_2 = 0$) against $H_A : \mu_1 - \mu_2 \neq 0$, then, using the reasoning discussed in Section 11.2, we would reject H_0, for a 0.05 level of significance, if Z is greater than 1.96 or less than -1.96. Note that the formula for Z is:

$$Z = \frac{\bar{X}_1 - \bar{X}_2}{\sqrt{\dfrac{\sigma_1^2}{N_1} + \dfrac{\sigma_2^2}{N_2}}},$$

because $\mu_1 - \mu_2 = 0$ when H_0 is true. The one-sided tests are conducted in the same way as in Section 11.2, except, of course, that Z is calculated as above. Other levels of significance, for instance, 0.10 or 0.01, would lead to different criteria for rejection. If the two populations are not exactly normal, or if the variances are unknown and are replaced by their estimates s_1^2 and s_2^2 and both sample sizes N_1 and N_2 are greater than 30, the above discussion will still be approximately true.

Testing the Means of Two Normal Populations

Null hypothesis $H_0: \mu_1 = \mu_2$;

Test statistic $Z = \dfrac{\bar{X}_1 - \bar{X}_2}{\sqrt{\dfrac{\sigma_1^2}{N_1} + \dfrac{\sigma_2^2}{N_2}}}$

Alternative hypothesis	Reject H_0 at the 0.05 significance level if:
$\mu_1 \neq \mu_2$	$Z > 1.96$ or $Z < -1.96$
$\mu_1 > \mu_2$	$Z > 1.65$
$\mu_1 < \mu_2$	$Z < -1.65$

Example 9 In order to measure the efficacy of a training program in improving the ability of workers to perform an intricate job, two groups of workers were selected. One group was trained and one group was not trained. They each performed the same job, and the time in minutes taken to complete it was measured. There were 12 workers in the trained group and 15 in the untrained group, and their respective mean times were 6.8 minutes and 9.3 minutes. The respective variances, from past records, were known to be 9 minutes and 12 minutes. Use a 5% level of significance to test that there is no difference between the two groups; that is, that the training has no effect on the worker's speed in doing the job.

Solution

i) $H_0: \mu_T = \mu_{UT}$, $H_A: \mu_T \neq \mu_{UT}$.

ii) $\bar{X}_1 = 6.8$, $\bar{X}_2 = 9.3$, $N_1 = 12$, $N_2 = 15$, $\sigma_1^2 = 9$, $\sigma_2^2 = 12$.

iii)
$$Z = \frac{(\bar{X}_1 - \bar{X}_2)}{\sqrt{\dfrac{\sigma_1^2}{N_1} + \dfrac{\sigma_2^2}{N_2}}} = \frac{(6.8 - 9.3)}{\sqrt{\dfrac{9}{12} + \dfrac{12}{15}}} = -2.02.$$

iv) Since $Z < -1.96$, we reject H_0 and conclude that there is indeed a difference between the two groups. Training does appear to have an effect upon a worker's ability to perform his job.

If the populations are normal (or approximately normal) with unknown variances, and the samples are small, then we can perform the usual tests of hypothesis using the t statistic. In order to use this we must assume that the two variances are equal ($\sigma_1^2 = \sigma_2^2$). The appropriate test statistic is

$$t = \frac{\bar{X}_1 - \bar{X}_2}{s\sqrt{\dfrac{1}{N_1} + \dfrac{1}{N_2}}}$$

with

$$s = \sqrt{\frac{(N_1 - 1)s_1^2 + (N_2 - 1)s_2^2}{N_1 + N_2 - 2}}.$$

The value of t has a t-distribution with $N_1 + N_2 - 2$ degrees of freedom.

Example 10 Reconsider Example 9 and suppose that no past records were available; thus the variances of the trained and untrained groups are unknown, although assumed equal, and have to be estimated from the samples. Suppose that $s_1^2 = 10.3$ and $s_2^2 = 15.7$. Use a 5% level of significance to test that there is no difference between the trained and untrained groups.

Solution

i) $H_0: \mu_T = \mu_{UT}$, $H_A: \mu_T \neq \mu_{UT}$.

ii) $\bar{X}_1 = 6.8$, $\bar{X}_2 = 9.3$, $s_1^2 = 10.3$, $s_2^2 = 15.7$, $N_1 = 12$, $N_2 = 15$.

iii)
$$s = \sqrt{\frac{(N_1 - 1)s_1^2 + (N_2 - 1)s_2^2}{N_1 + N_2 - 2}}$$

$$= \sqrt{\frac{(11)(10.3) + (14)(15.7)}{25}}$$

$$= 3.65.$$

iv) $t_{0.025} = 2.06$, with 25 degrees of freedom.

v)
$$t = \frac{\left(\overline{X}_1 - \overline{X}_2\right)}{s\sqrt{\dfrac{1}{N_1} + \dfrac{1}{N_2}}} = \frac{(6.8 - 9.3)}{(3.65)\sqrt{\dfrac{1}{12} + \dfrac{1}{15}}} = -1.77.$$

vi) Since $t \not< -2.06$ we accept H_0. There is no evidence that training has an effect on the worker's ability to perform the task.

11.5. TESTING VARIANCES

If we desire to test σ^2 or σ of a normal population, on the basis of a random sample $X_1, X_2, \ldots, X_N$, we know, from Chapter 10, that $(N-1)s^2/\sigma^2$ has a χ^2-distribution with $(N-1)$ degrees of freedom. To test H_0: $\sigma^2 = \sigma_0^2$ against H_A: $\sigma^2 \neq \sigma_0^2$, using the discussion in Section 11.3 concerning confidence intervals and testing procedures, we reject H_0 if $(N-1)s^2/\sigma_0^2$ is greater than $\chi^2_{0.975}$ or less than $\chi^2_{0.025}$ (see Fig. 10.5), at the 5% level of significance. To test H_0: $\sigma^2 = \sigma_0^2$ against H_A: $\sigma^2 > \sigma_0^2$, we reject H_0 if $(N-1)s^2/\sigma_0^2$ is greater than $\chi^2_{0.95}$. The other one-sided alternative is handled similarly. Other levels of significance can, of course, be used. The values of $\chi^2_{0.025}$, $\chi^2_{0.975}$, $\chi^2_{0.95}$, etc., are found in Appendix Table 7.

Testing the Variance of a Normal Population

Null hypothesis H_0: $\sigma^2 = \sigma_0^2$;

Test statistic $\chi^2 = (N-1)s^2/\sigma_0^2$.

Alternative hypotheses	Reject H_0 at the 0.05 significance level if:
$\sigma^2 \neq \sigma_0^2$	$\chi^2 < \chi^2_{0.025}$ or $\chi^2 > \chi^2_{0.975}$
$\sigma^2 > \sigma_0^2$	$\chi^2 > \chi^2_{0.95}$
$\sigma^2 < \sigma_0^2$	$\chi^2 < \chi^2_{0.05}$

Example 11 A laboratory uses a standard method for determining the impurity level in its product. This method is precise (that is, it has a small variance) but it is expensive. A newer, cheaper method is suggested, but it is believed to be less precise (that is, it is believed to have a larger variance). The laboratory supervisor wishes to test that the new method is as precise as the standard, versus an alternative hypothesis that the new method is *less* precise than the standard. From past records, the variance of the standard method is $\sigma^2 = 7\%$. Twenty observations are obtained with the new method and are believed to be normally distributed. If the value of s^2 calculated from the sample is 8%, test the hypothesis at the 5% level of significance.

Solution

i) H_0: $\sigma^2 = 7$, H_A: $\sigma^2 > 7$.

ii) $s^2 = 8$, $N = 20$, $N - 1 = 19$.

iii) $\chi^2_{0.95} = 30.14$.

iv)
$$\chi^2 = \frac{(N-1)s^2}{\sigma_0^2} = \frac{(19)(8)}{7} = 21.71.$$

v) We reject H_0 if $(N-1)s^2/\sigma_0^2 > \chi^2_{0.95}$. Since $\chi^2 = 21.71$ is not greater than $\chi^2_{0.95} = 30.14$, we accept H_0, and conclude that the new method is *no less precise* than the standard method.

11.6. TESTING A PROPORTION

The first paragraph of Section 10.6 is an appropriate introduction to this section. We assume that we wish to test hypotheses concerning p, the population proportion. If the conditions for a binomial experiment are satisfied, then a point estimate of p is X/n, where n is the number of trials and X is the number of successes. We deal only with the case where n is large enough for the normal approximation to be valid. To test H_0: $p = p_0$ against H_A: $p \neq p_0$, the test statistic

$$Z = \frac{(X/n) - p_0}{\sqrt{\dfrac{p_0(1-p_0)}{n}}}$$

has approximately a standard normal distribution, so that we reject H_0 at the 5% level of significance if Z is less than -1.96 or greater than 1.96. To test H_0: $p = p_0$ against H_A: $p > p_0$, reject H_0 if the test statistic Z is greater than 1.65. Similarly for the other one-sided case.

Testing a Proportion

Null hypothesis $H_0: p = p_0;$

Test statistic $Z = \dfrac{(X/n) - p_0}{\sqrt{\dfrac{p_0(1 - p_0)}{n}}}.$

Alternative hypothesis	Reject H_0 at the 0.05 significance level if:
$p \neq p_0$	$Z > 1.96$ or $Z < -1.96$
$p > p_0$	$Z > 1.65$
$p < p_0$	$Z < -1.65$

Example 12 From past sales records, it is known that 30% of the population buys Brand X toothpaste. A new advertising campaign is completed, and to test its effectiveness, 1000 people are asked whether they buy Brand X toothpaste now. If 334 answer yes, does this indicate that the advertising campaign has been successful? Use a 5% level of significance to test that the advertising campaign has increased Brand X's share of the market.

Solution

i) $H_0: p = 0.30,$ $H_A: p > 0.30.$

ii) $n = 1000, X = 334, X/n = 0.334.$

iii) $\sqrt{p_0(1 - p_0)/n} = \sqrt{(0.3)(0.7)/1000} = 0.01449.$

iv) $Z = \dfrac{(X/n) - p_0}{\sqrt{p_0(1 - p_0)/n}} = \dfrac{0.334 - 0.3}{0.01449} = 2.35.$

v) Since $Z > 1.65$, reject H_0 and conclude that the advertising campaign has been successful.

If we have two populations with proportions p_1 and p_2, from which we take samples of sizes n_1 and n_2, obtaining X_1 and X_2 successes, respectively, then we know that:

$$Z = \frac{\left(\dfrac{X_1}{n_1} - \dfrac{X_2}{n_2}\right) - (p_1 - p_2)}{\sqrt{\dfrac{p_1(1-p_1)}{n_1} + \dfrac{p_2(1-p_2)}{n_2}}}$$

has approximately a standard normal distribution for n_1 and n_2 large. To test H_0: $p_1 = p_2$ against H_A: $p_1 \neq p_2$ proceed as follows. Since we are hypothesizing that $p_1 = p_2$, then we may call our variable just p ($= p_1 = p_2$), and estimate p by pooling the data from both samples.

The total number of successes is $X_1 + X_2$ and the total number of observations is $n_1 + n_2$; thus the estimate of p is

$$\hat{p} = \frac{X_1 + X_2}{n_1 + n_2}.$$

The test statistic is

$$Z = \frac{\dfrac{X_1}{n_1} - \dfrac{X_2}{n_2}}{\sqrt{\hat{p}(1-\hat{p})\left(\dfrac{1}{n_1} + \dfrac{1}{n_2}\right)}}$$

At the 0.05 level of significance, reject H_0 if Z is greater than 1.96 or less than -1.96.

Example 13 To compare the effects of two sales offices on the market share of product X, samples 1 and 2, of sizes 1000 and 1500, are taken in the sales areas of the two offices. The number of people buying product X is 420 and 615, respectively. Use a 1% level of significance to see if the market shares are different for the two offices.

Solution

Calling p_1 and p_2 the market shares of product X in the two sales regions, we have:

i) H_0: $p_1 = p_2$, H_A: $p_1 \neq p_2$;

ii) $n_1 = 1000$, $n_2 = 1500$, $X_1 = 420$, $X_2 = 615$,

$$\frac{X_1}{n_1} = 0.42, \qquad \frac{X_2}{n_2} = 0.41, \qquad \hat{p} = \frac{1035}{2500} = 0.414;$$

iii) $\sqrt{\hat{p}\,(1-\hat{p}\,)\left(\dfrac{1}{n_1}+\dfrac{1}{n_2}\right)} = \sqrt{(0.414)(0.586)\left(\dfrac{1}{1000}+\dfrac{1}{1500}\right)} = 0.0201;$

iv) $\dfrac{\dfrac{X_1}{n_1}-\dfrac{X_2}{n_2}}{\sqrt{\hat{p}\,(1-\hat{p}\,)\left(\dfrac{1}{n_1}+\dfrac{1}{n_2}\right)}} = \dfrac{0.42-0.41}{0.0201} = 0.50;$

v) Since the test statistic is between -2.58 and 2.58, we accept H_0 and conclude that there is no evidence that the market shares are different in the two areas.

SUMMARY

We have been concerned with testing hypotheses concerning the values of population parameters, based on *samples* of observations taken from the populations. We introduce the chapter by discussing the nature of the errors that could be made, namely Type I and Type II errors, and their respective probabilities α (called the significance level) and β. We note the relationship between hypothesis testing and the calculation of confidence intervals. Because of this relationship, and because most of the relevant ideas have already been discussed in Chapter 10, most of the results are developed in an abbreviated way. The following tests of hypothesis are discussed:

a) The mean of a normal population with known variance,
b) The mean of a normal population with unknown variance,
c) The variance and standard deviation of a normal population,
d) The difference of two normal means,
e) Proportions,
f) The difference of two proportions.

Words to Remember

Null hypothesis   Critical region
Alternative hypothesis   Acceptance region
Test statistic   Significance level
Type I error   One-sided test
Type II error   Two-sided test

Symbols to Remember

α Probability of Type I error

β Probability of Type II error

Z Test statistic for normal means (large samples)

t Test statistic for normal means (small samples)

χ^2 Test statistic for normal variance

EXERCISES

1. Reconsider Example 4, and calculate the level of significance if the decision rule is as follows:

$$\text{Reject } H_0 \text{ if } \bar{X} < 20; \quad \text{Accept } H_0 \text{ if } \bar{X} \geqslant 20.$$

2. Suppose that the height of American males is normally distributed with mean 71 inches and standard deviation 5 inches. A sample of 100 male university students has $\bar{X} = 72.5$ inches. Use a 5% level of significance to test whether male university students are taller than the average American male.

3. The average daily sales per store for a large chain of retail stores is $10,000, with a standard deviation of $75, and is normally distributed. A particular store has its sales recorded for ten days, for which the average sales is $9,500. Use a 1% level of significance to test whether this store has lower than average sales.

4. If the grades of students in a business-course examination are known from experience to be approximately normally distributed, with mean 60 and standard deviation 5, and a new class of 30 students obtains an average of 64, can we say the new class is better than average?

5. In Exercise 2, calculate β if, in fact, the mean height of male university students is 72 inches.

6. In Exercise 3, calculate β if, in fact, the mean sales for that particular store is $9,750.

7. In Exercise 4, calculate β if, in fact, the mean score for the class is 65.

8. Consider students entering a typing course who have an average speed of 50 words per minute and a standard deviation of 4 words. At the end of the course, the speeds for ten students are 57, 62, 48, 51, 63, 55, 44, 46, 59, 50. Use a 5% one-sided test to see whether the course has been of value to them. (Assume that the speeds are normally distributed and that the standard deviation remains constant.)

9. In Exercise 8, find β when, in fact, the mean typing speed of these students at the end of the course is 55 words, and again if it is 60 words per minute. Plot a graph of β versus mean typing speed.

10. Do Exercise 8 without assuming that σ is known (that is, calculate and use s).

11. A drug company claims that its medication takes effect within ten minutes of administration. The following are the times that twelve patients required before

feeling the effects of the medication: 9.1, 13.2, 8.7, 8.1, 4.2, 9.5, 10.7, 15.3, 12.1, 14.0, 9.0, 12.7. If these times are assumed to be approximately normally distributed, test the company's claim with $\alpha = 0.01$.

12. The standard deviation of the lifetime of light bulbs produced by a company in the past was 125 hours. Some change was made in the machinery and a sample of 10 light bulbs showed a standard deviation of 90 hours. Investigate the significance of the apparent decrease in variability, using a 5 percent level of significance.

13. The mean lifetime of a sample of 81 transistors produced by a company is computed to be 1535 days with a standard deviation of 60 days. If μ is the mean life of all transistors produced by this company, test the null hypothesis $\mu = 1550$ days against $\mu \neq 1550$ days at the 0.05 level of significance.

14. Suppose an automatic filling machine fills coffee jars each with one pound of coffee. If the actual weight of coffee in a jar is greater than one pound, it will be a loss to the company, and if it is less than one pound, the company could lose potential customers. To check whether the filling machine is set up correctly, the weight of each of 50 jars filled by the filling machine is determined. The mean and standard deviation of these weights is found to be 16.5 ounces and 0.4 ounces, respectively. Test the company's claim that $\mu = 16$ ounces at the 5-percent level of significance.

15. The wages of skilled workers in a particular industry are known to be normally distributed with mean \$4.00/ hour. A certain large company claims to offer higher than average wages. The wages of fifteen randomly chosen workers are: 3.57, 4.23, 4.04, 4.71, 3.96, 3.88, 4.09, 4.27, 4.30, 4.50, 4.55, 3.50, 3.75, 4.45, 4.20. Test the company's claim, using a one-sided test with $\alpha = 0.05$.

16. In Exercise 8, test with $\alpha = 0.01$ that the standard deviation is in fact 4.

17. In Exercise 11, test that the standard deviation is 1.5 against the alternative that it is greater than 1.5. Use $\alpha = 0.05$.

18. In Exercise 15, test the hypothesis that the standard deviation of skilled workers is \$0.30. Use a 5% two-sided test.

19. A manufacturer of rope knows, from data accumulated over many years, that the rope's breaking strength has a standard deviation of 4 pounds. A change is made in the manufacturing procedure, and then a sample of 25 lengths of rope is tested, from which s is calculated to be 4.5. Has changing the procedure increased the variability of the rope's breaking strength? (Use $\alpha = 0.05$ with a one-sided test.)

20. A machinist makes parts whose diameter variance is known to be 0.01. A new machinist takes over the job and, to test his performance, the shop manager measures the diameters of a sample of twenty parts, obtaining the following results: 1.3, 1.7, 1.4, 1.2, 1.3, 1.4, 1.1, 1.7, 1.3, 1.5, 1.4, 1.8, 1.6, 1.1, 1.6, 1.5, 1.4, 1.1, 1.6, 1.5. If the data are assumed to be normally distributed, is the work of the new machinist more variable than the first?

21. Two methods, A and B, for teaching children to read are compared by considering the comprehension, measured on a 1 to 20 scale, of nine students using method A and twelve students using method B. Test, with $\alpha = 0.05$, that there is no difference

between the methods, if the following results, which are assumed to be normally distributed, are obtained.

Method A: 15, 10, 13, 9, 6, 17, 14, 13, 10.

Method B: 9, 7, 10, 7, 15, 20, 13, 19, 7, 11, 14, 12.

22. The weekly sales for ten gasoline stations (in thousands of dollars) before and after an advertising campaign are given below.

Station	Before	After
1	59	62
2	64	60
3	65	63
4	50	65
5	58	68
6	55	64
7	61	71
8	60	73
9	63	60
10	67	65

Use a 5% level of significance and a one-sided test to see whether the advertising campaign has increased sales.

23. Samples from two types of electric light bulbs have their life length observed, yielding the following results:

	Type I	Type II
Number in sample	25	48
Sample mean	1537 hours	1428 hours
Sample variance	400	350

If the data are assumed to be normally distributed, from populations with equal variances, test that the two types are not significantly different. Use $\alpha = 0.05$.

24. From two machines supposed to be the same, samples of parts are taken and measured with the following results:

	Machine A	Machine B
Sample size	30	30
Sample mean	4.3 inches	4.6 inches
Sample standard deviation	0.50	0.60

The data are assumed to be normal and the population variances are assumed to be equal. Test the supposition that the two machines are the same, with a 5% level of significance.

25. An examination in a business course was given to two sections of 100 and 120 students, respectively. The mean grade of the first section was 75 with a standard deviation of 15, while in the second section the mean grade was 80 with a standard deviation of 10. Is there a significant difference at the 1% level of significance between the two sections, if the data are assumed normal with equal population variances?

26. The average earnings of samples from two separate industries (in dollars per hour) are as follows:

	Industry A	Industry B
Sample size	17	14
Mean ($\bar{X}$)	2.52	3.04
Variance (s^2)	10.50	1.83

Use a 5% level of significance to test that the two industries have equal earnings. Assume that the variances are equal.

27. It is claimed that in a bushel of apples, 10% are defective. 150 apples are examined and 30 are found to be defective. What would you conclude about the claim, using a 5% two-sided test?

28. Machine parts turned out by an automatic process are known to be one percent defective on the average. A quality-control expert examines 200 similar parts and finds 4 defectives. Test whether the data is consistent with the manufacturer's claim. Use $\alpha = 0.05$.

29. A manufacturer of kitchen appliances such as refrigerators, stoves, dishwashers, etc., suspects that people like colored kitchen appliances and wants to know whether the sales of colored appliances is greater than the conventional white color. To test this, 1000 families are asked if they prefer a colored kitchen appliance, to which 600 answer "Yes." Test the manufacturer's hunch, using a 5-percent one sided test.

30. Suppose that 30% of a state's wage-earners earn more than $2.50/hour. It is suspected that the proportion is greater than 30% for a particular city. To test this,

200 workers in the city are polled, and 80 of them earn more than $2.50 per hour. Test the suspicion, using a 5% one-sided test.

31. A transistor manufacturer claims that 10% of his products are defective. A sample of fifteen transistors is examined and three are found to be defective. Would you reject his claim?

32. A poll of 400 people from Chicago showed that 250 favored Brand Y coffee, while 170 out of 350 in St. Louis favored the same brand. At the 0.01 level of significance, test the hypothesis that there is no difference in preferences in the two cities.

33. To compare the smoking habits of two areas, samples of sizes 1000 and 1500 people, respectively, are asked if they smoke. 300 in the first and 500 in the second answer "Yes." Test the hypothesis that there is no difference in the proportion of smokers for the two areas.

34. If we have N pairs of observations (for instance, the sales receipts before and after an advertising campaign in N cities), then the two samples are not independent, and the two sample techniques for comparing the means are not applicable. If we take the difference for each pair, (before—after), we will end up with N differences D_1, $D_2, \dots, D_N$. If these differences are assumed to come from a normal population with mean $\mu_1 - \mu_2$, and if $\bar{D}$ and s_D are the mean and standard deviation of the D's, then show that the test of $H_0: \mu_1 = \mu_2$ is based on $t = \bar{D}/(s_D/\sqrt{N})$, which has a t-distribution for N small and an approximate normal distribution for N large.

35. The before-and-after daily sales (in thousands of dollars) of 10 outlets of a chain store were measured before and after an advertising campaign, to see whether the campaign had an effect. The figures are given below:

Outlet	1	2	3	4	5	6	7	8	9	10
Before (B)	30	37	23	44	51	29	32	37	48	40
After (A)	42	35	41	50	48	59	55	53	49	62

Use a 5% level of significance to see whether the advertising campaign had an effect.

REGRESSION
AND CORRELATION 12

12.1. INTRODUCTION

So far we have been concerned with statistical procedures used in studying problems involving a *single variable*. In previous chapters, we discussed descriptive measures and probability distributions of this variable; we studied the problem of estimating the unknown population mean and variance; and we applied tests for making inferences about the population mean and variance of this variable. We now turn to methods for studying the joint behavior of *two variables*. In many problems of business and economics, a businessman is interested to know whether there is a relationship between two or more variables, and how this relationship can be used to predict the average value of one variable, called the dependent variable, based on the other related variable, called the independent variable. Important applications of relationships between two or more variables in business and economics involve their use in forecasting sales; to study the quantity of outputs as the quantity of input changes (such as labor, raw material, and other ingredients); to find how the volume of sales depends on advertising expenditures; or to determine the relationship between the various levels of income and consumption in a certain community. Problems of the above type that we encounter when we try to predict the behavior of one variable, based on the known behavior of another variable, are called *regression problems*. When only two variables are involved, we speak of simple regression (or linear regression) between two variables. In cases where more than two variables are involved, we speak of *multiple regression*.

In this chapter we will study linear regression between two variables, and we will learn how the regression equation (or the prediction equation) can be used to predict the dependent variable on the basis of the related independent variable. The regression equation may also be termed the *equation of average*

217

relationship. In Sections 4 and 5, we discuss the reliability of our predictions or forecasts, by using methods learnt in earlier chapters. In Section 6 we give another type of relationship between two variables, called correlation. Correlation simply measures the degree of association or closeness between two variables.

12.2. LINEAR REGRESSION

Suppose we are interested in finding a possible relationship between the values of two different variables; for example: Do high advertising expenditures (X) produce high volume of sales (Y)? Do high aptitude-test scores (X) tend to go with high job performance (Y)? Is the volume of sales (Y) related to different levels of income in a community (X)? To study a possible relationship between the pairs of values of X and Y, the first step is to collect data and then plot the values of X along the horizontal axis and the values of Y along the vertical axis. The resulting diagram may reveal a relationship between the two variables.

> **Scatter Diagram**
> The plot of a set of N pairs $(X_1, Y_1), (X_2, Y_2),\ldots,(X_N, Y_N)$ of values of X and Y is called a scatter diagram.

Example 1 The marketing research department of ABC Company wants to study the relationship between volume of sales and advertising expenditures. It has gathered the data shown in Table 12.1. Draw the scatter diagram.

TABLE 12.1

Advertising expenditures (Thousands of dollars) X	Volume of sales (Thousands of dollars) Y
5	40
7	50
10	60
12	65
15	70
20	80
25	92
30	100

Solution: The scatter diagram for the data (eight pairs of points) given in Table 12.1 is shown in Fig. 12.1.

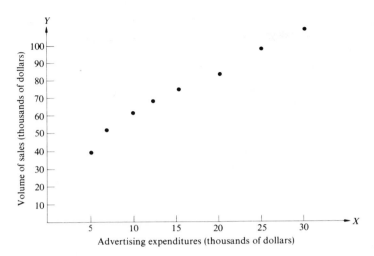

12.1   Scatter diagram of the data of Table 12.1.

You may note from Fig. 12.1 that we could make a reasonably good estimate of the relationship between the variable *advertising expenditure* (X) and the variable *volume of sales* (Y), by drawing a straight line. In general, various kinds of relationships are possible; the simplest and most widely used relationship is the linear relationship in two variables. If the scatter diagram reveals that a relationship between the two variables could be represented by a straight line, the next step is to find the equation of this line. This linear relationship (straight line) between the two variables has an equation of the form

$$\hat{Y} = a + bX,$$

where a is the Y-intercept and b is the slope of the line. Our problem here is to find a "best line" or a "best fit" to a given set of N pairs of points (X_1, Y_1), (X_2, Y_2),...,(X_N, Y_N). In other words, we want to determine the values of a and b in such a manner that the N pairs of values lie as close to the line as possible. Since there will be a difference between the actual Y-values and the corresponding values determined from the line, we shall denote the predicted value by $\hat{Y}$ for a given value of X. We may also note that, if the N pairs of points are only sample values, then the line fitted to these N pairs of points will simply be an *estimate* of the population regression line

$$Y = A + BX.$$

In other words, $\hat{Y}$ is an estimate of Y, a is an estimate of the intercept A and represents the value of the dependent variable when the independent variable X is zero, and b is an estimate of the slope B and represents the change in the dependent variable Y per unit change in the independent variable X. Note carefully that the regression line of Y on X

$$\hat{Y} = a + bX$$

is a line of the average predicted value of Y for a given value of X.

Prediction Equation or Regression Equation

Let the estimated or the predicted value of Y obtained from the line of best fit be $\hat{Y}$; then the equation

$$\hat{Y} = a + bX$$

is called the *prediction equation* (or the *regression equation*) of Y on X, or the *line of average relationship*, and is an *estimate* of the population regression equation

$$Y = A + BX.$$

The difference between the actual Y-values and the corresponding values $\hat{Y}$ predicted from the line is called an error or a residual or a deviation.

Error or Deviation

Let Y be the actual Y-value and $\hat{Y}$ be the predicted or the estimated value of Y; then the difference, denoted by e,

$$e = Y - \hat{Y},$$

is called the error (or the deviation) between the predicted and actual Y-value.

The errors, which may be positive, negative, or zero, are shown in Fig. 12.2.

In order to find the equation of the "best line" to the given set of points or pair of values, we apply the *method of least squares*. The method of least squares minimizes the sum of squares of deviations (or errors) between the actual Y-values and the predicted Y-values. Symbolically, we wish to minimize

$$\sum (Y - \hat{Y})^2 = \sum e^2.$$

Note that, for any least-squares line, it can be shown that the sum of positive deviations or errors cancels with the sum of negative deviations or errors. This means that the mean of deviations or errors is equal to zero. In view of this property, the least-squares line is the "best line," since by minimizing the sum of squares of errors we are, in fact, minimizing the variance of the errors.

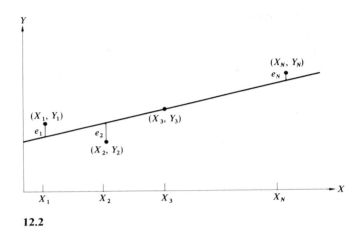

12.2

Method of Least Squares

The method of least squares is the method of estimating the "best line" to a given set of N pairs of points, in such a way that the sum of squares of the errors between the actual Y-values and the predicted Y-values is minimized.

The method of least squares may be applied in fitting a linear regression to more than two variables or any other curve (parabola, cubic, exponential), to a given set of N points.

Now we may find the values of a and b in such a way that $\Sigma(Y - \hat{Y})^2$ is minimized. The quantity $\Sigma(Y - \hat{Y})^2$ may be minimized using calculus, but the technique is beyond the scope of this book. The values of a and b are determined from the following two equations, called *normal equations*.

Normal Equations

Given N pairs of values of X and Y, the constants a and b of the equation

$$\hat{Y} = a + bX$$

are solutions of two linear equations, called *normal equations*,

$$aN + b \sum X = \sum Y,$$

$$a \sum X + b \sum X^2 = \sum XY.$$

Normal equations may be solved using a method of solving a system of linear equations. We may use the following formulas for computation of constants a and b of the prediction equation or the regression equation $\hat{Y} = a + bX$.

Formulas for Computation of a and b

The constants a and b of the equation $\hat{Y} = a + bX$ are given by

$$b = \frac{\sum XY - N\overline{X}\,\overline{Y}}{\sum X^2 - N(\overline{X})^2},$$

$$a = \overline{Y} - b\overline{X},$$

where $\overline{X}$ is the mean of the X-values and $\overline{Y}$ is the mean of the Y-values.

Alternatively, the above formula for the computation of b may be written as:

$$b = \frac{\sum (X - \overline{X})(Y - \overline{Y})}{\sum (X - \overline{X})^2}.$$

Example 2 (Refer to Example 1.)

a) Find the prediction equation (or the regression equation) of Y on X for the data of volume of sales and advertising expenditures given in Table 12.1.

b) Using the prediction equation obtained in (a), estimate the volume of sales for advertising expenditues of $16,000.

c) Graph the regression line obtained in (a) on the scatter diagram of data given in Table 12.1.

Solution: (a) We may use the formulas given earlier for the computation of the constants a and b of the regression line

$$\hat{Y} = a + bX.$$

Various calculations involved are shown in Table 12.2.

TABLE 12.2

Advertising expenditures (Thousands of dollars) X	Volume of sales (Thousands of dollars) Y	XY	X^2	Y^2	$\hat{Y}$
5	40	200	25	1,600	45.6535
7	50	350	49	2,500	50.2195
10	60	600	100	3,600	57.0685
12	65	780	144	4,225	61.6345
15	70	1,050	225	4,900	68.4835
20	80	1,600	400	6,400	79.8985
25	92	2,300	625	8,464	91.3135
30	100	3,000	900	10,000	102.7285
Total 124	557	9,880	2,468	41,689	

$$\bar{X} = \frac{124}{8} = 15.5 \qquad \bar{Y} = \frac{557}{8} = 69.625$$

Using the formula for the computation of the constant b of the regression line $\hat{Y} = a + bX$, we obtain

$$b = \frac{\sum XY - N\bar{X}\bar{Y}}{\sum X^2 - N\bar{X}^2} = \frac{9{,}880 - 8(15.5)(69.625)}{2{,}468 - 8(15.5)^2}$$

$$= \frac{9{,}880 - 8{,}633.5}{2{,}468 - 1{,}922} = 2.283.$$

Now, using the formula for the computation of the constant a of the regression line $\hat{Y} = a + bX$, we get

$$a = \bar{Y} - b\bar{X} = 69.625 - (2.283)(15.5)$$

$$= 34.2385.$$

Hence, the prediction equation or the regression equation of Y on X is

$$\hat{Y} = 34.2385 + 2.283X.$$

That is, the value of $a = 34.2385$ thousand dollars represents the sales volume when there are *no* advertising expenditures, and the value of $b = 2.283$ thousand dollars represents the increase in volume of sales for every thousand-dollar increase in advertising expenditures.

b) Using the above regression equation for $X = 16$, we find

$$\hat{Y} = 34.2385 + (2.283)(16)$$

$$= 70.7665.$$

c) To graph the regression line $\hat{Y} = 34.2385 + 2.283X$, we need only two points fairly far apart. Using the above regression equation, we find

$$\hat{Y} = 34.2385 + 2.283(0) = 34.2385 \qquad \text{when } X = 0;$$

$$\hat{Y} = 34.2385 + 2.283(30) = 102.7285 \qquad \text{when } X = 30.$$

The graph of the regression line

$$\hat{Y} = 34.2385 + 2.283X$$

is obtained by connecting two points $(0, 34.2385)$ and $(30, 102.7285)$ by a straight line. The scatter diagram of data given in Table 12.1. and the graph of the regression line $\hat{Y} = 34.2385 + 2.283X$ are shown in Fig. 12.3.

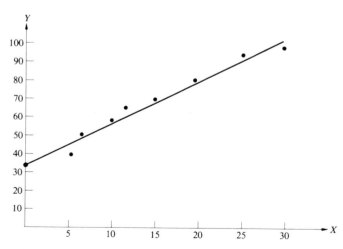

12.3   Prediction or regression line fitted to the data of Table 12.1.

So far we have considered the regression equation of Y on X. The regression equation of Y on X predicts the Y-values, given the X-values. In other words, X is the independent variable and Y is the dependent variable. There are many occasions when it is of interest to predict the X-values, given the Y-values. That is, now Y is the independent variable and X is the dependent variable. For example, in Example 1, the marketing research department of ABC Company might be interested in predicting the amount of advertising expenditures for a given volume of sales. That is, the company may like to know how much money is needed for advertising expenditures to achieve a certain level of volume of sales. The regression equation of X on Y may be written as

$$\hat{X} = c + dY.$$

To find the values of the constants c and d we apply the method of least squares to minimize $\sum(X - \hat{X})^2$. We may note that the roles of X and Y have interchanged. The values of the constants c and d are given by the following formulas.

Formulas for Computations of c and d

The constants c and d of the equation $\hat{X} = c + dY$ are given by

$$d = \frac{\sum XY - N\overline{X}\,\overline{Y}}{\sum Y^2 - N(\overline{Y})^2},$$

$$c = \overline{X} - d\overline{Y},$$

where $\overline{X}$ is the mean of the X-values and $\overline{Y}$ is the mean of the Y-values.

Example 3 (Refer to Example 2.) Find the regression line of X on Y for the data of Example 2. Use this equation to predict the amount of advertising expenditures needed for a volume of sales of 150 thousand dollars.

Solution: Using the formula for the computation of the constant d of the regression equation $\hat{X} = c + dY$, and calculations from Table 12.2, we obtain

$$d = \frac{\sum XY - N\overline{X}\,\overline{Y}}{\sum Y^2 - N(\overline{Y})^2} = \frac{9{,}880 - 8{,}633.5}{41{,}689 - 8(69.625)^2}$$

$$= \frac{1{,}246.5}{2{,}907.875} = 0.4287.$$

Now using the formula for the computation of the constant c of the regression line $\hat{X} = c + dY$, we get

$$c = \bar{X} - d\bar{Y}$$

$$= 15.5 - (0.4287)(69.625)$$

$$= -14.3482.$$

Hence, the regression equation of X on Y is

$$\hat{X} = -14.3482 + 0.4287\,Y.$$

Using the above regression equation for $Y = 150$, we find

$$\hat{X} = -14.3482 + (0.4287)(150)$$

$$= 49.9568.$$

12.3. ESTIMATE OF THE VARIANCE OF THE REGRESSION EQUATION

It has been pointed out earlier that the estimated regression equation

$$\hat{Y} = a + bX$$

is an estimate of the population regression equation

$$Y = A + BX.$$

The main use of the prediction equation is to predict a value of Y, given a value of X. The constants a and b are estimated from a sample of N pairs of observations, and they are subject to sampling variability. If we are to use the prediction equation to predict a future value of Y for a given value of X, we would like to know the degree of precision attached to $\hat{Y}$, the estimated value of Y. In other words, we want to know how well the prediction equation fits to a given set of N pairs of observations. The reader must have already noted that in finding the values of the constants a and b of the prediction equation the basic criterion was to minimize the sum of squares of deviations or errors between the actual Y-values and the predicted Y-values. If the sum of squares of errors between the actual Y-values and the predicted Y-values is zero, our prediction equation (or regression equation) fits perfectly to N pairs of sample observations. But unfortunately, as we have pointed out many times earlier in this book, most business decisions are made in the face of uncertainty, and so we have to learn how to make decisions in the face of uncertainty. If we are to use $\hat{Y}$ as an estimate of Y, we would like to know the degree of precision attached to $\hat{Y}$, the estimated value of Y. To answer this question of degree of precision attached to our predictions using the regression equation, we define below three quantities known as *sum of squares, sum of squares for regression*, and *sum of squares for errors*.

Sum of Squares

The sum of squares of the deviations of the actual Y-values from their mean $\overline{Y}$, denoted by SS, is defined as

$$SS = \sum (Y - \overline{Y})^2.$$

This sum of squares, when divided by the number of degrees of freedom $(N - 1)$, gives an estimate of the total variance of the actual Y-values about their mean $\overline{Y}$.

Sum of Squares for Regression

The sum of squares of the deviations of the predicted Y-values from their mean $\overline{Y}$, called the sum of squares for regression, denoted by SSR, is defined as

$$SSR = \sum (\hat{Y} - \overline{Y})^2.$$

The sum of squares for regression gives an estimate of variance of the predicted Y-values about the mean $\overline{Y}$ of the actual Y-values.

Sum of Squares for Errors

The sum of squares of the deviations of the actual Y-values from their corresponding predicted values on the regression line, denoted by SSE, is defined as

$$SSE = \sum (Y - \hat{Y})^2.$$

Since the constants a and b of the regression equation have been estimated from the data, the number of degrees of freedom associated with S.S.E. is $(N - 2)$. Thus, the sum of squares for errors, when divided by the number of degrees of freedom $(N - 2)$, gives an estimate of the variance of the actual Y-values from their predicted Y-values. It may be shown, using elementary algebra, that the total variance of Y-values is equal to the sum of the variance of the predicted Y-values, $\hat{Y}$, from their mean $\overline{Y}$ and the variance of the actual

Y-values from their corresponding predicted values $\hat{Y}$. The following relationship between these three sums of squares may easily be checked, using elementary algebra.

Relationship Between SS, SSR, and SSE.

If SS, SSR, and SSE are as defined above, then the following relationship holds:

$$SS = SSR + SSE$$

The partitioning of the sum of squares may be summarized in Table 12.3.

TABLE 12.3

Source of variation	Sum of squares	Degrees of freedom
Due to regression (SSR)	$\sum(\hat{Y} - \bar{Y})^2$	1
Deviations from regression (SSE)	$\sum(Y - \hat{Y})^2$	$N - 2$
Total (SS)	$\sum(Y - \bar{Y})^2$	$N - 1$

We note, from Table 12.3, that the degrees of freedom, as well as the sum of squares, are partitioned.

Example 4 Compute SS, SSR, and SSE for the data of advertising expenditures and volume of sales given in Example 1.

Solution: Using calculations given in Table 12.2, we find

$$SS = \sum(Y - \bar{Y})^2 = \sum Y^2 - N(\bar{Y})^2$$

$$= 41,689 - 8(69.625)^2 = 2,907.875;$$

$$SSR = \sum(\hat{Y} - \bar{Y})^2$$

$$= (-23.9715)^2 + (-19.4055)^2 + (-12.5565)^2$$
$$+ (-7.9905)^2 + (-1.1415)^2 + (10.2735)^2$$
$$+ (21.6885)^2 + (33.1035)^2 = 2,845.8006;$$

and

$$SSE = \sum (Y - \hat{Y})^2$$
$$= (-5.6535)^2 + (-0.2195)^2 + (2.9315)^2 + (3.3655)^2$$
$$+ (1.5165)^2 + (0.1015)^2 + (0.6865)^2 + (-2.7285)^2$$
$$= 62.1566.$$

To estimate σ^2, the variance of the random error for a regression line fitted to a sample of N pairs of X, Y values, we already know that the quantity SSE provides a measure of variation of actual Y-values from the predicted Y-values, $\hat{Y}$.

Estimate of the Variance of the Regression Line

The estimate of the variance of the regression line of Y on X, denoted by $s_{Y \cdot X}^2$, is given by the quantity

$$s_{Y \cdot X}^2 = \frac{SSE}{N-2} = \frac{\sum (Y - \hat{Y})^2}{N-2}.$$

The positive square root of the estimate of variance of the regression line is called the *standard error of the estimate*, or the *sample standard deviation of regression*.

Standard Error of Estimate

The standard error of estimate for a regression line of Y on X is given by the formula

$$s_{Y \cdot X} = \sqrt{\frac{\sum (Y - \hat{Y})^2}{N-2}}.$$

We note that the nearer the points lie to a regression line, the smaller will be the standard error of estimate.

Example 5 Find the standard error of estimate for the data of advertising expenditures and volume of sales given in Example 1.

Solution: Using the sum of squares of errors calculated in Example 4, we find

$$s_{Y \cdot X} = \sqrt{\frac{\sum (Y - \hat{Y})^2}{N-2}} = \sqrt{\frac{62.1566}{6}} = \sqrt{10.3595} = 3.2186.$$

12.4. CONFIDENCE INTERVALS CONCERNING *A* AND *B*

In Section 12.3 we have obtained an expression, called the *standard error of estimate*, to judge the accuracy of the regression equation fitted to a given N pairs of sample observations. The reader must have noticed by now that the accuracy of our predictions depends on two constants a and b which are estimated from a sample of N pairs of observations. It has been pointed out earlier that a is an estimate of A and b is an estimate of B, and both are subject to sampling variability. In fact, we would find several different estimates of A and B based on different sample sizes. Now the question is, if the businessman has estimates of A and B based on a particular sample of size N, what sort of confidence can he have in them? To answer this question, we can set up confidence intervals on A and B with a certain confidence coefficient such as 95% or 99%, etc. The procedure of setting confidence intervals on A and B is quite similar to those discussed in Chapter 10. However, the reader must note that we have to make an assumption that the errors are *independent* and are *normally distributed*. If we make this assumption, then the sampling distributions of different estimates of A and B will be normally distributed, and the methods developed for confidence intervals are applicable.

Now, to proceed further, we need expressions for the variances of the estimates a and b, which are given below:

Formulas for Variances of *a* and *b*

Estimates for variances of a and b are given by the following formulas:

$$s_a^2 = \frac{s_{Y \cdot X}^2 \left(\sum X^2 \right)}{N \sum (X - \overline{X})^2}$$

and

$$s_b^2 = \frac{s_{Y \cdot X}^2}{\sum (X - \overline{X})^2} \,.$$

We may construct a confidence interval for the parameter (intercept) A using the procedures of Chapter 10 for setting confidence intervals. A 95% confidence interval for the intercept A is given by

$$P \left[-t_{0.025} < \frac{a - A}{s_a} < t_{0.025} \right] = 0.95.$$

After some algebraic manipulations, the above expression may be written as

$$P[a - t_{0.025} s_a < A < a + t_{0.025} s_a] = 0.95.$$

Thus, for a given sample of size N, a confidence interval for the intercept A is given as follows:

Confidence Interval for the Intercept A

A 95% confidence interval for the parameter A lies between

$$a - t_{0.025}s_a \quad \text{and} \quad a + t_{0.025}s_a.$$

If we require, for instance, a 99% or 90% confidence interval, we should replace $t_{0.025}$ by $t_{0.005}$ or $t_{0.05}$, respectively. Remember that the values of t are found in Appendix Table 6 with $N-2$ degrees of freedom.

Example 6 (Refer to Example 2.) Find the 95% confidence interval for A for the data of advertising expenditures and volume of sales given in Table 12.1.

Solution: The value of A is estimated in Example 2 as $a = 34.2385$. Thus, the 95% confidence interval for A lies between

$$34.2385 - (2.447)s_a \quad \text{and} \quad 34.2385 + (2.447)s_a.$$

Using the data from Table 12.2, the standard deviation of a may be calculated as

$$s_a = \frac{s_{Y \cdot X} \sqrt{\sum X^2}}{\sqrt{N \sum (X - \bar{X})^2}} = \frac{(3.2186)\sqrt{2{,}468}}{\sqrt{8 \left[2{,}468 - 8(15.5)^2 \right]}} = 2.419.$$

Since there are $N - 2 = 8 - 2 = 6$ d.f., $t_{0.025} = 2.447$. Now, the 95% confidence interval for A lies between:

$$34.2385 - (2.447)(2.419) \quad \text{and} \quad 34.2385 + (2.447)(2.419)$$

or between

$$28.3196 \quad \text{and} \quad 40.1574.$$

That is, there is 95% probability that the value of A of our prediction equation lies between 28.3196 and 40.1574 thousand dollars.

We may construct a confidence interval for the parameter B (slope), using the procedures of Chapter 10 for setting confidence intervals. A 95% confidence interval for the parameter B is given by

$$P\left[-t_{0.025} < \frac{b - B}{s_b} < t_{0.025} \right] = 0.95.$$

After some algebraic manipulations, the above expression may be written as

$$P[b - t_{0.025}s_b < B < b + t_{0.025}s_b] = 0.95.$$

Thus, for a given sample of size N, a confidence interval for the slope B is given as follows:

Confidence Interval for the Slope B

A 95% confidence interval for the slope B lies between

$$b - t_{0.025}s_b \quad \text{and} \quad b + t_{0.025}s_b.$$

If we require, for instance, a 99% or 90% confidence interval, replace $t_{0.025}$ by $t_{0.005}$ or $t_{0.05}$, respectively.

Example 7 (Refer to Example 2.) Find the 95% confidence interval for the slope B of the prediction equation of Y on X for the data of advertising expenditures and volume of sales.

Solution: The value of B is estimated in Example 2 as $b = 2.283$ thousand dollars. Thus, the 95% confidence interval for B lies between

$$2.283 - (2.447)s_b \quad \text{and} \quad 2.283 + (2.447)s_b.$$

Using the data from Table 12.2, the standard deviation of b may be calculated as

$$s_b = \frac{s_{Y \cdot X}}{\sqrt{\sum (X - \bar{X})^2}} = \frac{3.2186}{\sqrt{2,468 - 8(15.5)^2}} = 0.137.$$

Now, the 95% confidence interval for B lies between

$$2.283 - (2.447)(0.137) \quad \text{and} \quad 2.283 + (2.447)(0.137),$$

or between 1.948 and 2.622. That is, there is a 95% probability that the value of B of our prediction equation lies between 1.948 and 2.622 thousand dollars. In other words, there is a 95% probability that the increase in volume of sales lies between 1.948 and 2.622 thousand dollars for every one-thousand-dollar increase in advertising expenditures given in Example 1.

12.5. CONFIDENCE INTERVAL FOR MEAN VALUE OF Y AND PREDICTION INTERVAL FOR AN INDIVIDUAL VALUE Y FOR A GIVEN VALUE OF X

Once the estimate of the regression line is obtained, we may then use it to predict the mean value of Y for a given value of X, or an individual value Y for a given value of X. Now the question is: What sort of confidence can the businessman have in the accuracy of these predictions? For instance, the marketing research department of ABC Company, in Example 1, may like to know a 95-percent confidence interval for the average volume of sales for a given amount of advertising expenditure; or the marketing research department may like a prediction interval for the volume of sales for a given amount of advertising expenditure. It has been pointed out in Section 4 that the estimate a of the intercept A and the estimate b of the slope B are both subject to sampling variability. This means that the prediction of the mean value of Y, for a given value of X, is also subject to sampling variability. Now to find a confidence interval for the mean value of Y for a given value of X, we need an expression for the variance of a point on the regression line $\hat{Y} = a + bX$. The regression line

$$\hat{Y} = a + bX$$

may be written as

$$\hat{Y} = \overline{Y} + b(X - \overline{X})$$

since $a = \overline{Y} - b\overline{X}$. Thus the variance of $\hat{Y}$ for a given value of X may be calculated as follows:

Formula for the Variance of $\hat{Y}$ for a Given Value of X

$$\operatorname{Var} \hat{Y} = \operatorname{Var} \overline{Y} + \operatorname{Var}\left[b(X - \overline{X}) \right]$$

$$= \operatorname{Var} \overline{Y} + (X - \overline{X})^2 \operatorname{Var} b$$

$$= \frac{s_{Y \cdot X}^2}{N} + \frac{(X - \overline{X})^2 s_{Y \cdot X}^2}{\sum (X - \overline{X})^2}.$$

We may now construct a confidence interval for the mean value of Y, for a given value of X, using the procedures of Chapter 10 for constructing confidence intervals. For a given sample of size N and a given value of X, a confidence interval for the mean value of Y is given as follows:

Confidence Interval for the Mean Value of Y

A 95-percent confidence interval for the mean value of Y lies between

$$\hat{Y} - t_{0.025}(\text{Standard error of } \hat{Y})$$

and

$$\hat{Y} + t_{0.025}(\text{Standard error of } \hat{Y}),$$

where

$$\text{Standard error of } \hat{Y} = \sqrt{\text{Var } \hat{Y}}$$

$$= s_{Y \cdot X} \sqrt{\frac{1}{N} + \frac{(X - \overline{X})^2}{\sum (X - \overline{X})^2}} \ .$$

If we require, for instance, a 99% or 90% confidence interval, we must replace $t_{0.025}$ by $t_{0.005}$ or $t_{0.05}$, respectively. As before the values of t are found in Appendix Table 6 with $N - 2$ degrees of freedom.

Example 8 (Refer to Example 2.) Find the 95 percent confidence interval for the population regression line for the data of advertising expenditures and volume of sales given in Table 12.1.

Solution: Using the estimate of the regression line obtained in Example 2, we may calculate the estimate of the *mean* of Y-values for a fixed value of $X = 5$, as follows:

$$\hat{Y} = 34.2385 + 2.283(5) = 45.6535.$$

The *variance* of $\hat{Y}$ for the same fixed $X = 5$ is given by:

$$\text{Standard error of } \hat{Y} = s_{Y \cdot X} \sqrt{\frac{1}{N} + \frac{(X - \overline{X})^2}{\sum (X - \overline{X})^2}}$$

$$= 3.2186 \sqrt{\frac{1}{8} + \frac{(5 - 15.5)^2}{546}}$$

$$= 1.8403.$$

Since there are $(N-2)=(8-2)=6$ degrees of freedom, $t_{0.025}=2.447$. Thus the 95-percent confidence interval for the mean value of Y lies between

$$\hat{Y}-2.447(\text{Standard error of } \hat{Y})$$

and

$$\hat{Y}+2.447(\text{Standard error of } \hat{Y}),$$

or between

$$45.6535-2.447(1.8403) \quad \text{and} \quad 45.6535+2.447(1.8403)$$

or between

$$41.1503 \quad \text{and} \quad 50.1567.$$

The 95-percent confidence interval for the mean of Y-values for the remaining X-values may similarly be computed. The computations for the 95-percent confidence intervals for the mean of Y-values are shown in Table 12.4 and the confidence intervals are plotted in Fig. 12.3.

TABLE 12.4.

X	$\hat{Y}$	$(X-\bar{X})$	$(X-\bar{X})^2$	Standard error of $\hat{Y}$	$\hat{Y}-$ [2.447 times (Standard error of $\hat{Y}$)]	$\hat{Y}+$ [2.447 times (Standard error of $\hat{Y}$)]
5	45.6535	-10.5	110.25	1.8403	41.1503	50.1567
7	50.2195	-8.5	72.25	1.6327	46.2243	54.2147
10	57.0685	-5.5	30.25	1.3671	53.7232	60.4138
12	61.6345	-3.5	12.25	1.2359	58.6103	64.6587
15	68.4835	-0.5	0.25	1.1400	65.6939	71.2731
20	79.8985	4.5	20.25	1.2958	76.7277	83.0693
25	91.3135	9.5	90.25	1.7341	87.0702	95.5568
30	102.7285	14.5	210.25	2.2987	97.1036	108.3534

A 95-percent confidence band for the population regression line may be constructed by joining the plotted points of these confidence intervals, as shown in Fig. 12.4; that is, we are 95-percent confident that the population regression line will fall within the confidence band, as shown in Fig. 12.4.

In many problems of business and economics, it is of interest to predict (or forecast) an individual value of Y for a given value of X. For instance, the marketing research department of ABC Company, in Example 1, may wish to predict the future volume of sales for a given amount of 40 thousand dollars for advertising expenditures. Now, it is important to know the *prediction interval* for an individual forecast for Y. A prediction interval for a future value of Y or an

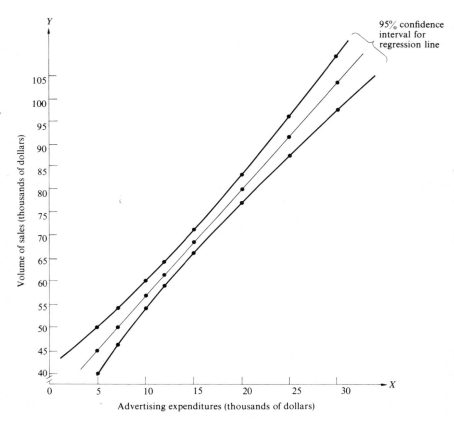

12.4   95-percent confidence interval for population regression line.

individual forecast for Y may be obtained in the same manner as the confidence interval for the mean value of Y, the only difference being that the standard error of an individual value for Y is larger. The standard error of an individual forecast for Y is obtained by combining the standard error of a point on the regression line with the standard error of the estimate, which is a measure of the sampling error in the regression line itself. The standard error of an individual forecast for Y may be calculated as follows:

Standard Error for an Individual Forecast for Y

$$\text{Standard error for an individual forecast for } Y = \sqrt{\left(\begin{array}{c}\text{Standard error}\\ \text{of estimate}\end{array}\right)^2 + \left(\begin{array}{c}\text{Standard error}\\ \text{of a point on the}\\ \text{regression line}\end{array}\right)^2}$$

$$= s_{Y \cdot X} \sqrt{1 + \frac{1}{N} + \frac{(X - \overline{X})^2}{\sum (X - \overline{X})^2}} \; .$$

We may now construct a prediction interval for a future value of Y for a given value of X, using the procedures of Chapter 10 for constructing confidence intervals. For a given sample of size N and a given value of X, a prediction interval of an individual forecast for Y is given as follows:

Prediction Interval of an Individual Forecast for Y

A 95-percent prediction interval of an individual forecast for Y lies between:

$$\hat{Y} - t_{0.025}(\text{Standard error for forecast } Y)$$

and

$$\hat{Y} + t_{0.025}(\text{Standard error for forecast } Y).$$

If we require, for instance, a 99% or 90% prediction interval, we should replace $t_{0.025}$ by $t_{0.005}$ or $t_{0.05}$, respectively.

Example 9 (Refer to Example 2.) Find the 95-percent prediction intervals for the individual Y-values for given X-values for the data of advertising expenditures and volume of sales given in Table 12.1.

Solution: We may note that the predicted mean value of Y and an individual predicted value of Y are the same for a fixed value of X. Thus for a given value of $X = 5$, $\hat{Y} = 45.6535$. The standard error of forecast Y for the same fixed $X = 5$

is given by

$$\text{Standard error of forecast} = s_{Y \cdot X}\sqrt{1 + \frac{1}{N} + \frac{(X - \bar{X})^2}{\sum (X - \bar{X})^2}}$$

$$= 3.2186\sqrt{1 + \frac{1}{8} + \frac{(5 - 15.5)^2}{546}}$$

$$= 3.7076.$$

Since there are $N - 2 = 8 - 2 = 6$ degrees of freedom, $t_{0.025} = 2.447$. Now the prediction interval for $X = 5$ lies between

$$\hat{Y} - 2.447(\text{Standard error of forecast})$$

and

$$\hat{Y} + 2.447(\text{Standard error of forecast})$$

or between

$$45.6535 - 2.447(3.7076) \quad \text{and} \quad 45.6535 + 2.447(3.7076)$$

or between

$$36.5810 \quad \text{and} \quad 54.7260.$$

The 95-percent prediction intervals for the individual Y-values for the remaining X-values may be calculated the same way. The computations for 95-percent prediction intervals for the individual Y-values are shown in Table 12.5, and the prediction intervals are plotted in Fig. 12.5.

TABLE 12.5

X	$\hat{Y}$	$(X - \bar{X})$	$(X - \bar{X})^2$	Standard error of forecast	$\hat{Y} - 2.447$ (Standard error of forecast)	$\hat{Y} + 2.447$ (Standard error of forecast)
5	45.6535	− 10.5	110.25	3.7076	36.5810	54.7260
7	50.2195	− 8.5	72.25	3.6090	41.3883	59.0507
10	57.0685	− 5.5	30.25	3.4969	48.5116	65.6254
12	61.6345	− 3.5	12.25	3.4477	53.1980	70.0710
15	68.4835	− 0.5	0.25	3.4145	60.1282	76.8388
20	79.8985	4.5	20.25	3.4697	71.4081	88.3889
25	91.3135	9.5	90.25	3.6560	82.3673	100.2597
30	102.7285	14.5	210.25	3.9552	93.0501	112.4069

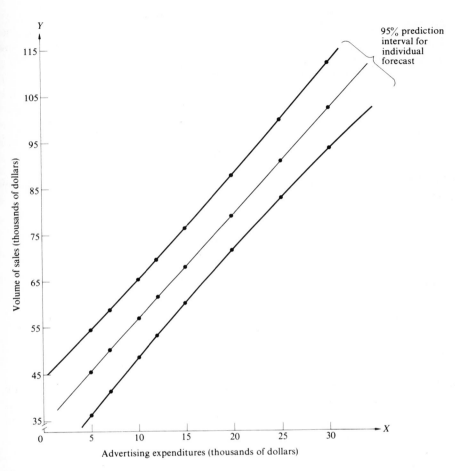

12.5   95-percent prediction interval for individual forecast.

12.6. CORRELATION

In the preceding sections we have discussed how regression analysis can be used
to determine the relationship between two variables in the form of an equation.
The regression equation can then be used as a *prediction equation* to predict the
dependent variable when the independent variable is known. In some problems,
however, the businessman might be interested in knowing only the *degree of
relationship* between two variables—for example, the relationship between sales
volume and advertising expenditures, the relationship between a family's income
and the consumption of a certain product, or the relationship between aptitude
score and productivity of workers in a factory, etc. Correlation analysis mea-

sures the *joint variation* between two variables when both are random. Correlation may be either positive or negative. If an increase in the value of one
variable is associated with an increase in the value of the other variable, the two
variables are said to be *positively correlated*. If an increase in the value of one
variable is associated with a decrease in the value of the other variable, the
variables are said to be *negatively correlated*. If there is no change in the value of
one variable as the other changes, we say that there is no correlation between
them; that is, they are *uncorrelated*. Note that "no correlation" does not mean
that two variables are statistically independent. When we say the two variables
have no correlation, we mean the two variables have no *linear correlation*; but
the variables may have a nonlinear relationship. A measure of linear correlation
between the two variables is called the *coefficient of correlation*, and is defined
below.

Coefficient of Correlation

Given N pairs of values of X and Y, the sample correlation coefficient,
denoted by r, is given by the formula

$$r = \frac{\sum(X-\bar{X})(Y-\bar{Y})}{\sqrt{\sum(X-\bar{X})^2}\sqrt{\sum(Y-\bar{Y})^2}} = \frac{\sum XY - N\bar{X}\bar{Y}}{\sqrt{\sum X^2 - N(\bar{X})^2}\sqrt{\sum Y^2 - N(\bar{Y})^2}}.$$

Example 10 (Refer to Example 2.) Compute the coefficient of correlation r for
the data of advertising expenditures and volume of sales given in Example 2.

Solution: Using the formula for the coefficient of correlation and the calculations from Table 12.2, we get:

$$r = \frac{\sum XY - N\bar{X}\bar{Y}}{\sqrt{\sum X^2 - N(\bar{X})^2}\sqrt{\sum Y^2 - N(\bar{Y})^2}}$$

$$= \frac{9,880 - 8(15.5)(69.625)}{\sqrt{2,468 - 8(15.5)^2}\sqrt{41,689 - 8(69.625)^2}}$$

$$= \frac{9,880 - 8,633.5}{\sqrt{546}\sqrt{2,907.875}}$$

$$= \frac{1,246.5}{(23.3666)(53.9247)} = 0.9893.$$

After having computed the value of the sample correlation coefficient r, we would like to know the meaning of this number in terms of the degree of correlation between two variables. In order to interpret the meaning of the coefficient of correlation, we shall make use of three quantities, given in Section 12.3, called sum of squares, sum of squares for regression, and sum of squares for errors. The following relationship between the coefficient of correlation r, the sum of squares for regression, and the total sum of squares may easily be checked by using formulas for the computation of SSR and SS given in Section 12.3.

Relationship Between r, SSR and SS

$$r^2 = \frac{\text{SSR}}{\text{SS}} = \frac{\sum(\hat{Y} - \bar{Y})^2}{\sum(Y - \bar{Y})^2}.$$

That is, r^2 is the proportion of total variance in Y which is due to linear association between X and Y. The quantity r^2 is called the *coefficient of determination*.

Coefficient of Determination

The square of the coefficient of correlation is called the *coefficient of determination*; it measures the proportion of total variance in Y which is due to linear association between X and Y.

Example 11 (Refer to Example 10.) The coefficient of correlation r for the data of advertising expenditures and volume of sales, given in Example 2, is calculated as $r = 0.9893$. Thus, the coefficient of *determination* may be calculated as

$$r^2 = (0.9893)^2 = 0.9787.$$

That is, 97.87% of the variance in the volume of sales is determined by the linear association between volume of sales and advertising expenditures, for the data given in Table 12.1.

The coefficient of determination r^2 is also helpful in setting limits on the coefficient of correlation r. If all the points of the scatter diagram lie on a line, then the prediction equation (or the regression equation) fits the data perfectly and the sum of squares of errors is zero; that is, S.S.E. $= 0$. We already know, from Section 12.3, that

$$\text{SS} = \text{SSR} + \text{SSE}$$

Now, if SSE = 0, this means that

$$SS = SSR$$

Therefore, using the relationship $r^2 = SSR/SS$, we find $r^2 = 1$.

On the other hand, if no linear relationship exists between the X and Y values, then SSR = 0. Therefore, using the relationship $r^2 = SSR/SS$, we find $r^2 = 0$.

In other words, r^2 must lie between 0 and 1. Now, taking the square root of $r^2 = 1$, we find that r lies between -1 and $+1$. We may state this property as follows:

Property of the Coefficient of Correlation

The coefficient of correlation r lies between -1 and $+1$.

Examples of $r = +1$, r approximately equal to -1, and r approximately equal to zero are shown in Figs. 12.6, 12.7, and 12.8, respectively.

12.7. TESTING THE SIGNIFICANCE OF THE CORRELATION COEFFICIENT

Like other statistics, the sample coefficient of correlation r is an *estimate* of the population coefficient of correlation ρ and is subject to *sampling variability*. In fact, we could find several different values of r based on different sample sizes. To test the reliability or significance of r, we need the sampling distribution of r. Assuming that the samples are drawn from a bivariate normal population, it can

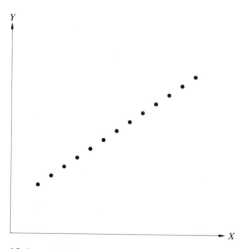

12.6 $r = +1$.

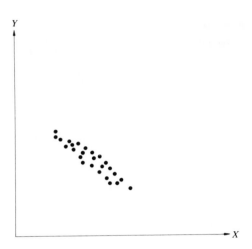

12.7   *r* approximately equal to -1.

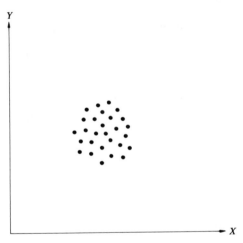

12.8   *r* approximately equal to 0.

be shown that the distribution of *r* is normal as the size of the sample increases. For $\rho = 0$, the distribution of *r* is symmetric and *r* is normally distributed even for smaller *N*. However, when $\rho \neq 0$, the distribution of *r* will be normally distributed only when *N* is very large. To test the hypothesis $\rho = 0$, that is, to determine whether *r* is significantly different from zero, we must note that $\rho = 0$ if and only if $B = 0$. Thus, to test the null hypothesis $H_0 : \rho = 0$ against the alternative $H_A : \rho \neq 0$ is equivalent to testing the null hypothesis $H_0 : B = 0$ against the alternative $H_A: B \neq 0$. The procedure for testing the hypothesis concerning B is quite similar to the procedures discussed in Chapter 11. We may test the hypothesis $\rho = 0$ directly by using the following:

Test of Hypothesis Concerning $\rho = 0$

If all samples of size N are drawn from a bivariate normal population with population coefficient of correlation $\rho = 0$, and if their sample coefficients of correlation are denoted by r, then

$$t = \frac{r}{\sqrt{\dfrac{1 - r^2}{N - 2}}}$$

has the t-distribution with $(N - 2)$ degrees of freedom. The decision to reject the null hypothesis $\rho = 0$ is made in terms of t at a given level of significance.

Example 12 A personnel manager of a company wants to know if a training course has any effect on a worker's ability to perform his job. Fifteen employees are randomly selected and the correlation coefficient between the performance before and after the training is found to be 0.49. Using a 0.05 level of significance, test the hypothesis that $\rho = 0$.

Solution

$$\text{Null hypothesis:} \qquad H_0: \rho = 0;$$
$$\text{Alternate hypothesis:} \qquad H_A: \rho > 0.$$

The test statistic t may be calculated as:

$$t = \frac{r}{\sqrt{\dfrac{1 - r^2}{N - 2}}} = \frac{0.49}{\sqrt{\dfrac{1 - (0.49)^2}{13}}} = 2.0267.$$

Since the alternative hypothesis is one-sided, we shall use a one-sided t-test. From Appendix Table 6, we find $t_{0.05} = 1.771$ for 13 degrees of freedom. Since the computed value of $t = 2.0267$ is greater than the table value $t_{0.05} = 1.771$, we reject the null hypothesis that $\rho = 0$.

SUMMARY

There are many problems in business and industry where the businessman is interested to know whether there is a relationship between two or more variables and how this relationship can be helpful in predicting the behavior of one variable, given that we know the other related variable. Problems concerning the prediction of a dependent variable (knowing the independent variable) are known as *regression problems*. To study a possible relationship between the pairs of X and Y values, we plot the values of X along the horizontal axis and the values of Y along the vertical axis. The resulting dot diagram is known as a *scatter diagram*. The method of least squares may be applied in fitting a straight line, exponential, or any other curve, to a given set of N points. The regression equation of Y on X predicts the Y-values, given the X-values. There are many occasions when it is of interest to predict the X-values given the Y-values. The regression equation of X on Y is different from the regression equation of Y on X. The variance of the random error for a regression line fitted to a sample of N pairs of X, Y values provides a measure of variance of the actual Y-values from the predicted Y-values. In order to test hypotheses and calculate confidence intervals concerning A and B, we make the assumption that the errors are independent and normally distributed. The linear regression equation with one independent variable may be extended to several independent variables; this is called *multiple linear regression*.

Regression analysis predicts the value of the dependent variable when the independent variable is known. *Correlation* analysis measures the joint variation of two variables when both are random. Correlation may be positive, negative, or zero. Note that "no correlation" does not always mean that the two variables are statistically independent. In the case of the bivariate normal, "no correlation" implies independence. When we say that the two variables have no correlation, we mean the two variables have no *linear* correlation; the variables *may* have a nonlinear relationship. A measure of linear correlation between the two variables is called the *coefficient of correlation*. The coefficient of correlation r lies between -1 and $+1$. The square of the coefficient of correlation is called the *coefficient of determination*; it measures the proportion of the total variance in Y that is due to the linear association between X and Y.

Words to Remember

Scatter diagram	Regression coefficient
Linear regression	Standard error of estimate
Regression equation	Sum of squares
Prediction equation	Sum of squares for errors
Error or deviation	Sum of squares for regression
"Best line"	Correlation
Least squares	Coefficient of correlation
Normal equations	Coefficient of determination

Symbols to Remember

A	Y-intercept of the population regression line
B	Slope of the population regression line
a	Estimate of A
b	Estimate of B
$\hat{Y}$	Predicted or estimated value of Y
e	Error or deviation between the predicted and actual Y-value
SS	Sum of squares
SSR	Sum of squares for regression
SSE	Sum of squares for errors
$s_{Y \cdot X}^2$	Estimate of the variance of the regression line of Y on X
$s_{Y \cdot X}$	Standard error of estimate
s_a^2	Variance of a
s_b^2	Variance of b
r	Sample correlation coefficient
ρ	Population correlation coefficient
r^2	Coefficient of determination

EXERCISES

1. The following table gives the test scores of 10 salesmen on an aptitude test and their weekly sales:

Salesman	Aptitude test score X	Weekly sales (Thousand dollars) Y
1	50	3.5
2	70	6.0
3	60	4.0
4	80	5.5
5	70	5.0
6	90	7.0
7	60	3.0
8	50	3.0
9	70	4.5
10	60	5.0

a) Find the equation of the regression line of sales on aptitude test scores.

b) Estimate the weekly sales volume of a salesman making a test score of 75.

c) Plot the scatter diagram, and graph the line.

2. A firm wants to study the relationship between the age of the machines in its manufacturing plant and the repair costs for the machines. The following table gives the ages in years and the repair costs per year for a random sample of ten machines in the manufacturing plant.

Age (years) X	Repair costs (dollars) Y
2	50
4	125
5	150
7	230
8	265
4	105
9	290
10	400
5	100
7	215

a) Find the equation of the regression line of repair costs on age of the machine.

b) Estimate the repair costs of a 12-year-old machine.

c) Plot the scatter diagram, and graph the line.

3. In a certain city, age and income were found to be related as follows:

Age (X)	31	47	25	60	65	37	34	40
Income \$(Y)	5280	6300	5700	7425	7470	6200	6350	7000

a) Plot a scatter diagram.

b) Find the linear regression equation of Y on X.

c) Estimate the income for age 50.

d) Does the regression line pass through $(\overline{X}, \overline{Y})$?

4. A furniture manufacturer wants to study the relationship between national disposable annual personal income and the national amount spent yearly on furniture. The data collected over the last 5 years is given below:

Disposable personal income (Million dollars) X	Amount spent on furniture (Million dollars) Y
7.0	2.0
10.0	3.5
12.0	4.2
15.0	5.0
18.0	6.3

a) Find the equation of the regression line of furniture sales on disposable personal income.

b) Estimate the furniture-sales volume when the disposable personal income is 20 million dollars.

c) Plot the scatter diagram, and graph the line.

5. The following table gives the weights of a chemical compound Y dissolved in 100 grams of water, at temperature $X°$ centigrade.

X	0	10	20	30	40	50	60
Y	55	59	65	70	75	81	86

a) Find the equation of the regression line of Y on X.

b) Use this equation to estimate Y when $X = 55$.

c) Draw the scatter diagram, and graph the line.

6. In a study of automobile repair costs, the following figures relating the ages (in years) of fourteen cars and their yearly repair costs are given.

| | Repair costs |
| Age (years) | (Dollars) |
X	Y
2	82
1	53
5	148
3	97
7	205
2	79
3	137
2	25
4	130
1	25
3	120
5	125
2	58
4	100

a) Plot a scatter diagram.

b) Find the linear regression of Y on X.

c) Estimate the repair costs for a 6-year-old car.

7. Find $s^2_{Y \cdot X}$ for the data of Exercise 2.

8. Find the standard error of estimate for the data of Exercise 1.

9. Calculate SS, SSR, and SSE, for the data of Exercise 3.

10. Refer to Exercise 9. Then,
   a) Verify that $SS = SSR + SSE$.
   b) Calculate $s^2_{Y \cdot X}$.

11. In Exercise 2, find a 95-percent confidence interval for A for the data of repair costs on age of the machine.

12. In Exercise 1, find a 99-percent confidence interval for A for the data of 10 salesmen's aptitude test scores and their weekly sales.

13. In Exercise 3, find a 95-percent confidence interval for the slope B for the data of age and income in a certain city.

14. In Exercise 5, find 95-percent confidence intervals for A and B.

15. In Exercise 4, find a 95-percent confidence interval for A for the data of furniture sales and disposable personal income.

16. Calculate the coefficient of correlation, r, for the data of test scores of salesmen and their weekly sales, given in Exercise 1. What is the significance of the value of r?

17. Calculate the coefficient of correlation, r, for the data of machine age and repair costs given in Exercise 2. Interpret this computed value of r.

18. Calculate the coefficient of correlation, r, for the data of age and income in a certain city, given in Exercise 3. Explain the meaning of this value of r.

19. Calculate the coefficient of correlation, r, for the data of disposable personal income and furniture sales given in Exercise 4. What information can the furniture manufacturer get from this data?

TIME SERIES
ANALYSIS 13

13.1. INTRODUCTION

A series of observations arranged chronologically is called a *time series*; the study of such series is called time series analysis. Time series analysis is applied to many problems of business and economics. Business executives, economists, and government officials are often faced with problems that require forecasts such as future sales, future revenue and expenditures, and the total business activity for the next decade. Time series analysis is a statistical method which helps the businessman understand the past behavior of economic variables, based on collections of observations taken at different time intervals. Having recognized the behavior or movements of a time series, the businessman tries to forecast the future of economic variables, on the assumption that the time series of such an economic variable will continue to behave in the same fashion as it has in the past. Thus, to a businessman, forecasting is the main purpose of the study of a time series.

The movements of a time series may be classified into the following four components:

1. The secular trend
2. Seasonal variations
3. Cyclical variations
4. Irregular variations.

The above four movements (or components) of a time series may be explained as follows.

The Secular Trend

The secular trend T is the long-term movement of a time series. It is also called the secular movement (or the secular variation).

That is, the growth (or decline) of a time series over a long period of time is called the secular trend; this applies to population growth, price levels of a certain commodity, growth in demand for the number of cars, etc. The following is an example of secular trend.

Example 1 Table 13.1 gives the United States' gross national product (GNP) from 1961 to 1970. The graph of this time series is shown in Fig. 13.1.

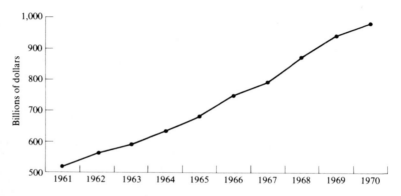

13.1 Gross National Product. (Source: *Business Statistics*, U.S. Department of Commerce.)

TABLE 13.1 United States' GNP, 1961–1970

Year	GNP (Billions of dollars)
1961	520.1
1962	560.3
1963	590.5
1964	632.4
1965	684.9
1966	749.9
1967	793.5
1968	865.7
1969	931.4
1970	974.1

> **Seasonal Variation**
>
> The seasonal variation S of a time series describes the movements (or variations) which recur at particular seasons of the year.

Seasonal variations follow a particular pattern over a certain period of time; examples are the increase in sales before Easter or Christmas, or the increase in sales of snow boots and winter clothes in winter, or the increase in unemployment of construction workers in winter, etc. The following is an example of seasonal variation in a time series.

Example 2 Table 13.2 gives the monthly retail sales in the United States from 1967 to 1970. The graph of this time series is shown in Fig. 13.2.

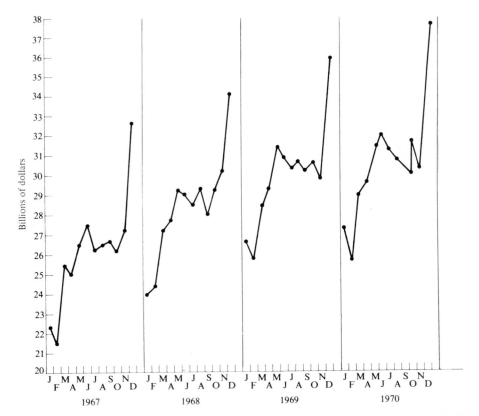

13.2 Monthly retail sales in the United States. (Source: *Business Statistics*, U.S. Department of Commerce.)

TABLE 13.2 Monthly retail sales in billions of dollars, 1967–1970

Month	1967	1968	1969	1970
		Year		
January	22.58	24.09	26.20	27.05
February	21.66	24.21	24.81	25.43
March	25.70	27.04	27.92	28.71
April	25.10	27.60	28.78	29.65
May	26.58	29.28	30.81	31.32
June	27.64	28.88	29.62	31.41
July	26.03	28.54	29.02	31.14
August	26.22	29.41	29.42	30.40
September	26.26	27.01	28.58	29.73
October	26.18	29.41	30.64	31.84
November	27.18	30.11	29.83	30.21
December	32.62	34.08	35.96	37.62

Cyclical Variation

The cyclical variation C of a time series describes long-term variations that move up and down about a trend line or a curve, after every few years.

The cyclical variations vary in length and time; and in the business community these variations are known as business cycles (or more commonly known as periods of prosperity, recession, depression, recovery, and back to prosperity). The following is an example of cyclical variation in a time series.

Example 3 Table 13.3 gives the number of (private residential) new housing units from 1955 to 1970. The graph of this time series is shown in Fig. 13.3.

Irregular Variation

The irregular variation I of a time series consists of variations which are random and unpredictable.

The irregular variations are caused by unforeseen causes such as wars, strikes, revolutions, floods, fires, etc.

We know now that the observed value Y of an economic time series may be decomposed into four components, the secular trend T, the seasonal variation S, the cyclical variation C, and the irregular variation I. The decomposition of the observed value Y may be done either through an additive model or through a multiplicative model. These are explained below.

TABLE 13.3 Number of private
residential new housing units,
1955–1970*

Year	New housing units (Thousands)
1955	18.24
1956	16.14
1957	14.73
1958	15.44
1959	19.23
1960	16.41
1961	16.18
1962	18.63
1963	20.38
1964	20.35
1965	20.35
1966	17.96
1967	17.88
1968	22.42
1969	25.94
1970	24.15

*Source: *Business Statistics*,
U.S. Department of Commerce.

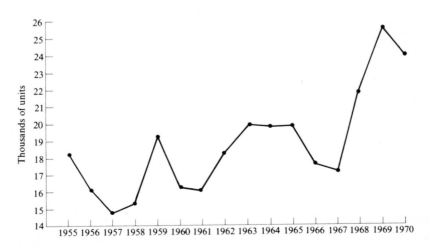

13.3

Additive Model of a Time Series

The additive model assumes that the observed value Y of a time series is equal to the *sum* of four components, the secular trend T, the seasonal variations S, the cyclical variations C, and the irregular variations I. That is,

$$Y = T + S + C + I.$$

The additive model assumes that the four components have no interaction and are *independent* of each other. That is, in the additive model, it is assumed that the four components are outcomes of independent causes and they do not affect each other. The four components in the additive model are measured in the same units as those of the original data.

Multiplicative Model of a Time Series

The multiplicative model assumes that the observed value Y of a time series is equal to the *product* of four components, the secular trend T, the seasonal variation S, the cyclical variation C, and the irregular variation I. That is,

$$Y = T \times S \times C \times I.$$

The multiplicative model assumes that the four components are *dependent on each other*. That is, in the multiplicative model, it is assumed that, although the four components may be due to different causes, they still affect each other. Normally, the component trend T, in the multiplicative model, is measured in the units of the original data, and other components are measured as *percentages*.

There are many other models for studying time series, but the main purpose of all these models is forecasting. In this chapter, we shall discuss the estimation of four components, T, S, C, and I, of a time series, assuming the multiplicative model, since this model is most often used in analyzing economic time series.

13.2. ESTIMATION OF SECULAR TREND

It has been explained in the previous section that the secular trend T is the long-term movement of a time series. This long-term movement may be a straight-line trend, or an exponential trend, or any other trend. To find out whether the trend is a straight line or not, the data of a given time series is plotted graphically in a scatter diagram, as in Chapter 12. If the scatter diagram of a given time series supports the hypothesis that the trend is a straight line, then we may proceed further to estimate this long-term movement of a time series as a straight-line trend. In this section, we shall explain the estimation of a straight line since most time series in business and economics follow a straight-line trend. The straight-line trend may be estimated by any of the following three methods.

1. The Method of Least Squares

The straight-line trend is best estimated by the method of least squares, which was discussed in Chapter 12. The student may recall that we can find the best prediction line by the *method of least squares*. Since the main purpose of studying a given time series in business and economics is forecasting, we would like to have the best estimate of the secular trend on which to base our forecasts. Following the method of least squares discussed in Chapter 12, the equation of a straight-line trend may be written as

$$\hat{Y} = a + bX,$$

where $\hat{Y}$ denotes the estimate of the observation Y of a time series for a given time period X. The periods of time such as days, weeks, and years are equally spaced. To estimate a and b in our trend equation, we may use either normal equations or the formulas for computation of a and b given in Chapter 12. For computational purposes, it will be easier if we transform the time variable X to simpler numbers. For example, one way would be to set $X = 0$ for the first year, $X = 1$ for the second, and $X = 2$ for the third, and so on for the remaining years. The procedure of finding the equation of the straight-line trend in a given time series is similar to that of finding the equation of the prediction line discussed in Chapter 12. We shall explain this procedure again with the help of the following example.

Example 4 The annual sales of ABC Company from 1961–1970, inclusive, are given in Table 13.4. Fit a straight-line trend to these data by the method of least squares.

TABLE 13.4 ABC Company annual sales, 1961–1970

Year	Sales (Millions of dollars)
1961	20
1962	22
1963	25
1964	29
1965	32
1966	34
1967	37
1968	41
1969	43
1970	46

Solution: The various calculations in the computation of the constants *a* and *b* of the straight-line trend $\hat{Y} = a + bX$ are given in Table 13.5.

TABLE 13.5 Computations of the constants *a* and *b* of the straight-line trend

Year	Sales (Millions of dollars)				Trend value $\hat{Y}$
	X	Y	X^2	XY	
1961	0	20	0	0	19.6182
1962	1	22	1	22	22.5697
1963	2	25	4	50	25.5212
1964	3	29	9	87	28.4727
1965	4	32	16	128	31.4242
1966	5	34	25	170	34.3757
1967	6	37	36	222	37.3272
1968	7	41	49	287	40.2787
1969	8	43	64	344	43.2302
1970	9	46	81	414	46.1817
Totals	45	329	285	1724	

$$\overline{X} = \frac{45}{10} = 4.5; \quad \overline{Y} = \frac{329}{10} = 32.9$$

Using the formulas for the computation of the constants *a* and *b* from Chapter 15, we obtain

$$b = \frac{\sum XY - N\overline{X}\,\overline{Y}}{\sum X^2 - N(\overline{X})^2} = \frac{1,724 - 10(4.5)(32.9)}{285 - 10(4.5)^2}$$

$$= \frac{1,724 - 1,480.5}{285 - 202.5} = 2.9515$$

and

$$a = \overline{Y} - b\overline{X}$$

$$= 32.9 - 2.9515(4.5)$$

$$= 19.6182.$$

Hence, the equation of the straight-line trend is

$$\hat{Y} = 19.6182 + 2.9515X,$$

(where the origin is July 1, 1961; *X* is in 1-year units; and *Y* is annual sales in millions of dollars). That is, the annual sales of the ABC Company increase, on the average, by 2.9515 million dollars each year. Now, this trend equation may be used to predict annual sales for future years. For example, to predict the annual sales for 1975, we substitute $X = 14$ (that is, 14 time units from the origin) in our prediction equation, and obtain:

$$\hat{Y} = 19.6182 + 2.9515(14)$$

$$= 60.9392 \quad \text{million dollars}$$

The trend values for the years 1961 through 1970 are calculated in the same way, using the above trend equation, and are listed in the last column of Table 13.5.

It may sometimes be of interest to a businessman to know the *monthly* trend increments. The trend equation for annual sales may be converted to a trend equation for monthly sales by simply dividing a by 12 and b by 144. The value of a so obtained is the value of the *trend of monthly sales* for the origin year $X = 0$, and the value of b so obtained is the *monthly trend increment* for monthly sales.

Example 5 Estimate the monthly trend equation for the data on annual sales of the ABC Company (given in Table 13.4).

Solution: From Example 4, we know that the straight-line trend equation for the data of annual sales, given in Table 13.4, is

$$\hat{Y} = 19.6182 + 2.9515X.$$

To convert this trend equation of annual sales so that we may deal with *monthly* sales, we divide 19.6182 by 12 and 2.9515 by 144. Thus, the monthly trend equation for monthly sales is

$$\hat{Y} = \frac{19.6182}{12} + \frac{2.9515}{144}X$$

$$= 1.6348 + 0.02X$$

(where the origin is January 15, 1961; X is in 1-month units; and Y is monthly sales in millions of dollars).

2. Method of Semi-Averages

The straight-line trend may also be estimated by the method of semi-averages. To estimate the straight-line trend by this method, the time series is divided into two equal parts and the average or the mean of *each part* is then calculated. (If a time series has an *odd* number of periods, the middle period is ignored.) The averages or means of the two equal parts are plotted against the midpoints of the two respective periods. The straight line so drawn joining these two points is called the trend line. The trend line obtained by the method of semi-averages is easy to calculate but is not as accurate as that estimated by the method of least squares.

Example 6 Estimate the straight-line trend by the method of semi-averages for the data of annual sales of the ABC Company given in Table 13.4.

Solution: The straight-line trend is estimated by the method of semi-averages as shown in Table 13.6.

TABLE 13.6

Year	Sales (Millions of dollars)	Total	Semi-average
1961	20		
1962	22		
1963	25	128	$\dfrac{128}{5} = 25.6$
1964	29		
1965	32		
1966	34		
1967	37		
1968	41	201	$\dfrac{201}{5} = 40.2$
1969	43		
1970	46		

The estimate of the constant a of our straight-line trend is 25.6. Therefore, the increase in a 5-year period is

$$40.2 - 25.6 = 14.6.$$

Hence, the annual sales increase is

$$b = \frac{14.6}{5} = 2.92.$$

Thus,

$$\hat{Y} = 25.6 + 2.92X$$

is the estimate of the straight-line trend by the method of semi-averages. The least-squares trend obtained in Example 5 and the semi-average trend obtained in Example 6 are both graphed in Fig. 13.4.

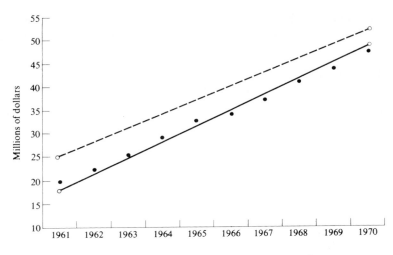

13.4   Least-squares trend, _____ ; semi-average trend, _ _ _ _ _ _ _ .

3. Method of Moving Averages

The method of moving averages takes into account the *averages* of the two, three, or more periods, instead of actual values corresponding to particular periods. By using the method of moving averages, we may be able to eliminate seasonal, cyclical, and irregular movements. That is, the method of moving averages helps us to eliminate the effects of short-term changes; thus smoothing out the fluctuations brings out the *trend* in a given time series. For example, to compute the three-year moving average, we find the average for the first three years and write this average against the middle year; then we find the average for the second, third, and fourth years, and write this average against the *third* year; and so on. The three-year averages are then plotted on graph paper, along with the given time series. A curve is then drawn through these points smoothing out the seasonal, cyclical, and other variations in a given time series. The student must note that this curve *may or may not* be a straight line. In fact, though it may not be a straight line, it does help us in showing the long-term trend in a given time series. The method of moving averages does not give us the *equation* of our trend line. It only helps us in smoothing out the time series by eliminating seasonal, cyclical, and other variations.

Example 7   The number of units of a product sold by a certain manufacturer is given in Table 13.7. Compute the three-year moving average.

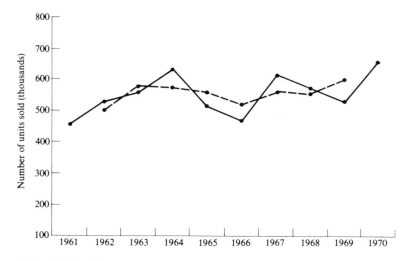

13.5   Original time series, _____ ; three-year moving average, _ _ _ _ _ _ .

TABLE 13.7

Year	Number of units sold (Thousands)
1961	475
1962	520
1963	550
1964	610
1965	500
1966	458
1967	625
1968	575
1969	510
1970	650

Solution: Calculation of the 3-year moving average is shown in Table 13.8. The graphs of the original time series and the 3-year moving average are shown in Fig. 13.5.

13.3. ESTIMATION OF SEASONAL VARIATION

It has been explained in Section 13.1 that the seasonal variation S of a time series refers to movements (or variations) which *recur at particular seasons* of the year. Estimation of the seasonal factor S is necessary in order for a businessman

TABLE 13.8

Year	Number of units sold (Thousands)	3-year moving total	3-year moving average
1961	475		
1962	520	1545	515.00
1963	550	1680	560.00
1964	610	1660	553.33
1965	500	1568	522.67
1966	458	1583	527.67
1967	625	1658	552.67
1968	575	1710	570.00
1969	510	1735	578.33
1970	650		

to determine monthly or quarterly sales forecasts. By knowing the seasonal factor S, the businessman can achieve better planning of his production, inventory, and service schedules. Although the businessman might be interested in hour-to-hour variations of a *daily* time series, or day-to-day variations of a *weekly* time series, it is more reasonable to discuss the *monthly* or the *quarterly* variations of a *yearly time series*. The seasonal variations are estimated in terms of the quantity called *seasonal index*, which is defined as follows.

Seasonal Index

The seasonal index S is the percentage of the average monthly or quarterly sales for the whole year, and measures the seasonal variations in a time series.

The seasonal index may be computed by several different methods. Three methods of computing the seasonal index, the *simple average* method, the *ratio-to-moving-average* method, and the *ratio-to-trend* method, are given below. (Even though the simple average method is seldom used in practice, we explain this method because it is simple and elementary.

1. The Simple-Average Method

You may recall that, in the previous section, we said that any observation Y of a time series is the *product* of four components, the secular trend T, the seasonal variation S, the cyclical variation C, and the irregular variation I. That is,

$$Y = T \times C \times S \times I.$$

Now, to calculate the *monthly* or the *quarterly* seasonal index S, we can eliminate the *irregular variations* I by finding the monthly (or the quarterly) average for each month (or each quarter) of the year. If the number of years is large, then, by finding the monthly or the quarterly average of the year, we have also eliminated the *cyclical* variation C. In other words, we are left with only *two* components, T and S. That is,

$$Y = T \times S.$$

But T may be estimated by finding the average of monthly or quarterly averages. Now, the seasonal variation may be calculated by dividing the monthly or the quarterly average by the *average* of monthly or quarterly averages, and multiplying it by 100. That is, the seasonal index S is computed by the simple-average method as follows:

Simple-Average Method

By the simple-average method, the monthly or the quarterly seasonal index S is given by the formula

$$S = \frac{\text{Monthly or quarterly average} \times 100}{\text{Average of monthly or quarterly averages}}.$$

Example 8 The monthly retail sales of a department store (in millions of dollars) for the years 1966 to 1970 are given in Table 13.9. Compute a seasonal index, using the simple-average method.

TABLE 13.9   A department store's monthly retail sales, 1966–1970

Month	1966	1967	1968	1969	1970
			Year		
January	2.0	2.5	2.3	3.0	3.5
February	2.3	2.7	2.9	3.3	3.7
March	2.8	3.0	3.3	4.1	4.3
April	3.5	3.9	4.2	4.0	4.5
May	4.0	4.3	4.5	5.0	4.1
June	4.6	5.0	4.6	5.5	5.3
July	3.9	4.2	4.0	5.1	4.7
August	3.2	3.8	3.7	4.7	4.2
September	3.0	3.5	3.2	4.3	5.0
October	3.7	4.0	4.3	5.0	5.6
November	4.2	4.8	5.0	5.8	6.9
December	5.0	6.0	6.8	8.0	8.5
Total	42.2	47.7	48.8	57.8	60.3

Solution: Calculations for computing a seasonal index, using the simple-average method, are shown in Table 13.10.

TABLE 13.10 Computation of a seasonal index using the simple-average method

Total of 5 years	Average	Percentage of monthly average to average of monthly averages (Seasonal index)
13.3	2.66	62.15
14.9	2.98	69.63
17.5	3.50	81.78
20.1	4.02	93.93
21.9	4.38	102.34
25.0	5.00	116.82
21.9	4.38	102.34
19.6	3.92	91.59
19.0	3.80	88.79
22.6	4.52	105.61
26.7	5.34	124.77
34.3	6.86	160.28
Totals	51.36	1200.03
Average	4.28	100.0025

If the total of monthly percentages is not equal to 1200, then the monthly percentages are adjusted, to make the total 1200. Since the total of the percentages in the present case is 1200.03, no adjustment is necessary.

2. The Ratio-to-Moving-Average Method

The moving-average method has been used in the previous section to estimate the secular trend in a given time series. To estimate the monthly seasonal index (or the quarterly seasonal index), a 12-month moving average (or a 4-quarter moving average) is calculated. The assumption is that the moving average so calculated estimates not only the *secular* trend but also the *cyclical* variation. Now, if we divide the actual data (or the actual observation) by the moving average, we will be left with only the seasonal variation S and the irregular variation I. Thus, the seasonal index may be calculated as follows.

The Ratio-to-Moving-Average Method

Let Y be the actual observation of a given time series. Then the monthly or the quarterly seasonal index S is given by:

$$\frac{\text{Actual observation } Y}{\text{Moving average}} = \frac{T \times S \times C \times I}{T \times C} = S \times I$$

and

$$S = \frac{S \times I}{I} = \text{Average of } S \times I \text{ for each month or for each quarter.}$$

Example 9 Compute a quarterly seasonal index for the data of monthly retail sales given in Table 13.6, using the ratio-to-moving-average method.

TABLE 13.11 Computation of a quarterly seasonal index using the ratio-to-moving-average

(1) Year and quarter	(2) Actual sales (Millions of dollars)	(3) 4-Quarter moving total	(4) 2-Quarter moving total of Col. 3	(5) 4-Quarter centered moving averages [(Col. 4) ÷ 8]
1966 1	7.1	–	–	–
2	12.1	–	–	–
3	10.1	42.2	85.5	10.6875
4	12.9	43.3	87.7	10.9625
		44.4		
1967 1	8.2		90.2	11.2750
2	13.2	45.8	93.5	11.6875
3	11.5	47.7	95.7	11.9625
4	14.8	48.0	96.1	12.0125
		48.1		
1968 1	8.5		96.6	12.0750
2	13.3	48.5	97.3	12.1625
3	10.9	48.8	100.5	12.5625
4	16.1	51.7	104.6	13.0750
		52.9		
1969 1	10.4		108.0	13.5000
2	14.5	55.1	112.9	14.1125
3	14.1	57.8	116.7	14.5875
4	18.8	58.9	117.2	14.6500
		58.3		
1970 1	11.5		116.4	14.5500
2	13.9	58.1	116.4	14.8000
3	13.9	60.3	–	
4	21.0	–	–	

Solution: Calculations for computing a quarterly seasonal index, using the ratio-to-moving-average method, are shown in Table 13.11.

Now we may find the percentage-to-moving average by dividing each of the quarterly actual sales by the corresponding 4-quarter centered moving average, and multiplying by 100. That is, divide each value in Column 2 by the corresponding value in Column 5 in Table 13.11, and multiply the quotient by 100. The results are given in Table 13.12.

TABLE 13.12

Year	Quarter				
	1	2	3	4	
1966	–	–	94.50	117.67	
1967	72.73	112.94	96.13	123.20	
1968	70.39	109.35	86.77	123.14	
1969	77.04	102.75	96.66	128.33	
1970	79.04	93.92	–	–	
Total	299.20	418.96	374.06	492.34	Total
Average	74.8	104.74	93.515	123.085	396.14
Seasonal index	75.53	105.76	94.43	124.28	400

Since the sum of the averages of percentage-to-moving-average is equal to 396.14, these averages are to be adjusted to get the required quarterly seasonal index. We multiply each average by 400/396.14 to make the sum of the averages equal to 400. The seasonal indices so obtained are given in the last row of Table 13.12.

3. The Ratio-to-Trend Method

The estimation of seasonal variation by the ratio-to-trend method requires the monthly (or the quarterly) trend values. The monthly or the quarterly trend values may be estimated by fitting a trend equation (by the method of least squares) to the data of a given time series. Then, each actual value of the data is divided by its corresponding trend value, and the ratios are expressed as percentages. The average percentage of each month (or each quarter) is the estimate of seasonal variation. The student should note that the estimate of seasonal variation by the ratio-to-trend method includes the cyclic variations as well.

The Ratio-to-Trend Method

Let Y be the actual observation and T be the trend value of a given time series; then

$$\frac{\text{Actual observation } Y}{\text{Trend } T} = \frac{T \times S \times C \times I}{T} = S \times C \times I$$

and

$$S \times C = \frac{S \times C \times I}{I} = \begin{array}{l}\text{Average of } S \times C \times I \\ \text{for each month or} \\ \text{for each quarter.}\end{array}$$

Example 10 (Refer to Example 8.) The monthly retail sales of a department store (in millions of dollars), for the years 1966 to 1970, are given in Table 13.9. Compute a monthly seasonal index, using the ratio-to-trend method.

Solution: To calculate the seasonal variation by this method, we first find the equation of the straight-line trend by the method of least squares. The calculations for the computation of the constants a and b of the straight-line trend equation

$$\hat{Y} = a + bX$$

are given in Table 13.13.

TABLE 13.13 Worksheet for the computation of the constants of a straight-line trend equation

Year	X	Y	X^2	XY
1966	0	42.2	0	0
1967	1	47.7	1	47.7
1968	2	48.8	4	97.6
1969	3	57.8	9	173.4
1970	4	60.3	16	241.2
Total	10	256.8	30	559.9

$$\bar{X} = \frac{10}{5} = 2; \qquad \bar{Y} = \frac{256.8}{5} = 51.36$$

Using the formulas for the computation of the constants a and b from Chapter 12, we obtain

$$b = \frac{\sum XY - N(\overline{X})(\overline{Y})}{\sum X^2 - N(\overline{X})^2}$$

$$= \frac{559.9 - 5(2)(51.36)}{30 - 5(2)^2}$$

$$= \frac{559.9 - 513.6}{10} = 4.63$$

and

$$a = \overline{Y} - b\overline{X}$$

$$= 51.36 - 4.63(2)$$

$$= 42.1.$$

Hence, the annual trend equation for the retail sales of a department store, given in Table 13.9, is

$$\hat{Y} = a + bX$$

$$= 42.1 + 4.63X$$

(where the origin is July 1, 1966; X is in 1-year units; and Y is annual sales in millions of dollars). That is, the annual sales of the department store increase, on the average, by 4.63 million dollars each year.

Now the annual trend equation may be converted to a *monthly* trend equation by dividing the constant a by 12 and the constant b by 144. Thus, the monthly trend equation is

$$\hat{Y} = \frac{42.1}{12} + \frac{4.63}{144}X$$

$$= 3.5083 + 0.0322X.$$

Before we calculate the monthly trend values, we shift the origin of the monthly trend equation to January 15. This may be done as follows:

$$\hat{Y} \text{ (January 1966)} = 3.5083 + (0.0322)(-5\tfrac{1}{2})$$

$$= 3.5083 - 0.1771 = 3.3312.$$

Thus the modified monthly trend equation is

$$\hat{Y} = 3.3312 + 0.0322 X$$

(where the origin is January 15, 1966; X is in 1-month units; and Y is average monthly sales in millions of dollars). Now this modified monthly trend equation may be used to calculate the monthly trend values. The monthly trend values so calculated are given in Table 13.14.

TABLE 13.14 Trend values, T

Month	Year				
	1966	1967	1968	1969	1970
January	3.3312	3.7176	4.1040	4.4904	4.8768
February	3.3634	3.7498	4.1362	4.5226	4.9090
March	3.3956	3.7820	4.1684	4.5548	4.9412
April	3.4278	3.8142	4.2006	4.5870	4.9734
May	3.4600	3.8464	4.2328	4.6192	5.0056
June	3.4922	3.8786	4.2650	4.6514	5.0378
July	3.5244	3.9108	4.2972	4.6836	5.0700
August	3.5566	3.9430	4.3294	4.7158	5.1022
September	3.5888	3.9752	4.3616	4.7480	5.1344
October	3.6210	4.0074	4.3938	4.7802	5.1666
November	3.6532	4.0396	4.4260	4.8124	5.1988
December	3.6854	4.0718	4.4582	4.8446	5.2310

We now divide the actual monthly sales given in Table 13.9 by the corresponding trend values given in Table 13.14. The ratios expressed as percentages are given in Table 13.15.

We may now find the monthly seasonal index by finding the average percentage of each month. (The average across the five columns in Table 13.15 will fill Column 1 of Table 13.16.) Since the *sum* of the average monthly percentages for 12 months is equal to 1194.4020, these averages are to be adjusted to get the required monthly seasonal index. We multiply each average by 1200/1194.4020 to make the sum of the average percentages equal to 1200. The seasonal indices so obtained are given in the second column of Table 13.16.

TABLE 13.15 Percentages: Actual monthly sales divided by corresponding trend values

Month	Year				
	1966	1967	1968	1969	1970
January	60.04	67.25	56.04	66.81	71.77
February	68.38	72.00	70.11	72.97	75.37
March	82.46	79.32	79.17	90.01	87.02
April	102.11	102.25	99.99	87.20	90.48
May	115.61	111.79	106.31	108.24	81.91
June	131.72	128.91	107.85	118.24	105.20
July	110.66	107.39	93.08	108.89	92.70
August	89.97	96.37	85.46	99.66	82.32
September	83.59	88.05	73.37	90.56	97.38
October	102.18	99.82	97.87	104.60	108.39
November	114.97	118.82	112.97	120.52	132.72
December	135.67	147.35	152.53	165.13	162.49

TABLE 13.16 Monthly seasonal index

Month	Average percentage	Seasonal index
January	64.3820	64.6837
February	71.7660	72.1024
March	83.5960	83.9878
April	96.4060	96.8578
May	104.7720	105.2631
June	118.3840	118.9389
July	102.5440	103.0246
August	90.7560	91.1814
September	86.5900	86.9958
October	102.5720	103.0527
November	120.0000	120.5624
December	152.6340	153.3494
Total	1194.4020	1200.0000

We may now use the seasonal index to remove the seasonal variations from the given data. This may be done by dividing the actual monthly (or quarterly) data by the corresponding monthly (or quarterly) index. The data so obtained are called *deseasonalized* (or *seasonally adjusted*) data.

Deseasonalized Data or Seasonally Adjusted Data

The *deseasonalized* or *seasonally adjusted* data are given by the formula

$$\text{Deseasonalized data} = \frac{\left(\begin{array}{c} \text{Actual monthly (or} \\ \text{quarterly) data} \end{array} \right)}{\left(\begin{array}{c} \text{Monthly (or quarterly)} \\ \text{seasonal index} \end{array} \right)}$$

Example 11 (Refer to Example 8.) Deseasonalize the monthly sales of the department store for the year 1970 (given in Table 13.9).

Solution: The monthly seasonal indices of the sales for the data given in Table 13.9 are shown in Table 13.16. We divide the monthly sales for 1970 by the corresponding seasonal index. The seasonally adjusted sales are given in Table 13.17.

TABLE 13.17   Deseasonalized sales for 1970

Month	Sales Y	Seasonal index S	Deseasonalized sales $\frac{Y}{S} = \frac{TSCI}{S} = TCI$
January	3.5	0.646837	5.4109
February	3.7	0.721024	5.1316
March	4.3	0.839878	5.1198
April	4.5	0.968578	4.6460
May	4.1	1.052631	3.8950
June	5.3	1.189389	4.4561
July	4.7	1.030246	4.5620
August	4.2	0.911814	4.6062
September	5.0	0.869958	5.7474
October	5.6	1.030527	5.4341
November	6.9	1.205624	5.7232
December	8.5	1.533494	5.5429

13.4. ESTIMATION OF CYCLICAL AND IRREGULAR VARIATIONS

The cyclical variations of a time series have already been defined in the first section of this chapter as the variations moving up and down about a trend line or a curve after every few years. So far, in this section, we have discussed the estimation of the secular trend T and the seasonal variations S. The irregular variations I may be removed by averaging over a number of years. The estimation of cyclical variations is quite difficult since the business cycles vary in length and time. The study of business cycles has attracted the attention of economists for a great number of years. (We shall make no attempt here to present the theory of business cycles. The student interested in this area should take higher courses in economics and statistics.)

The cyclical and irregular variations may be estimated by several different methods. The student may follow either of the following two methods to estimate the cyclical and irregular variations.

Method I

First, estimate the trend values T and then divide the actual data by the trend values:

$$\frac{Y}{T} = \frac{T \times S \times C \times I}{T} = S \times C \times I.$$

Now eliminate the seasonal variations by dividing the above quotient by the seasonal variation:

$$\frac{S \times C \times I}{S} = C \times I.$$

Method II

Remove the seasonal variations by dividing the actual data by the seasonal index:

$$\frac{Y}{S} = \frac{T \times S \times C \times I}{S} = T \times C \times I.$$

Now eliminate the trend T by dividing $T \times C \times I$ by T:

$$\frac{T \times C \times I}{T} = C \times I.$$

13.5. BUSINESS FORECASTING

It was stated, in the earlier sections of this chapter, that the main purpose of the study of a time series is forecasting. Having broken down a time series into four components, the secular trend T, the seasonal variation S, the cyclical variation

C, and the irregular variation I, the businessman can estimate each of these components with the methods explained in preceding sections. The estimates of these components will help him to forecast the future of economic variables.

However, in forecasting the future activity of a firm or an industry, we are making a major assumption: that the time series of economic variables will *continue to behave* in the same fashion as it has in the past. That is where the basic difficulty lies in forecasting the future of business activity, since the economic variables may *not* behave in the same fashion as in the past. The best the businessman can do is to use the available information cautiously and to employ these techniques only in combination with his own sound judgment.

We already know how to estimate the trend values and the seasonal index for a given time series. The forecasts for any given month (or quarter) may be obtained by estimating the trend value for the given month (or the given quarter) and *multiplying* it by the corresponding monthly (or quarterly) *seasonal index*.

Forecast for the Given Month or the Given Quarter

The forecast for the given month or the given quarter is given by the formula:

$$\begin{pmatrix} \text{Monthly} \\ \text{(or quarterly)} \\ \text{forecast} \end{pmatrix} = \begin{pmatrix} \text{Monthly} \\ \text{(or quarterly)} \\ \text{trend value} \end{pmatrix} \times \begin{pmatrix} \text{Monthly} \\ \text{(or quarterly)} \\ \text{seasonal index} \end{pmatrix}.$$

Example 12 (Refer to Example 8.) Find the sales forecasts for the year 1972 for the data of sales given in Table 13.9.

Solution: Recall that the modified monthly trend equation for the data given in Table 13.9 is

$$\hat{Y} = 3.3312 + 0.0322X.$$

Since the origin year is 1966, the number of months from January 1966 to December 1971 is 72. Thus, the trend value for January 1972 is:

$$\hat{Y}\,(\text{January } 1972) = 3.3312 + 0.0322(72)$$

$$= 5.6496 \text{ million dollars.}$$

To obtain the sales forecast for the month of January 1972, we multiply the trend value 5.6496 for January 1972 by the January seasonal index, 0.646837. Thus

$$\text{Sales forecast (January } 1972) = 5.6496 \times 0.646837$$

$$= \$3.6544 \text{ million.}$$

The sales forecasts for the remaining months of 1972 may be similarly calculated. The sales forecasts are given in Table 13.18.

TABLE 13.18 Sales forecasts for 1972

Month	Trend value (Millions of dollars)	Seasonal index	Forecast (Millions of dollars)
January	5.6496	0.646837	3.6544
February	5.6818	0.721024	4.0967
March	5.7140	0.839878	4.7991
April	5.7462	0.968578	5.5656
May	5.7784	1.052631	6.0825
June	5.8106	1.189389	6.9111
July	5.8428	1.030246	6.0195
August	5.8750	0.911814	5.3569
September	5.9072	0.869958	5.1390
October	5.9394	1.030527	6.1207
November	5.9716	1.205624	7.1995
December	6.0038	1.533494	9.2068

SUMMARY

The main purpose of the study of a time series to a businessman is forecasting. The movements of a time series may be classified into four components, the secular trend, the seasonal variations, the cyclical variations, and the irregular variations. Estimation of each of these components is necessary before making any reasonable forecast. The straight-line trend may be estimated by the method of least squares, the method of semi-averages, and the method of moving averages. The seasonal variations may be estimated in terms of the quantity called the *seasonal index*. The seasonal index can be computed by the simple-average method, by the ratio-to-moving average method, or by the ratio-to-trend method. The seasonal index so computed is then used to *deseasonalize* the given data. The cyclical and irregular variations may be estimated by the two methods given in Section 13.4. After having estimated each of these components, the businessman can use his own sound judgment to arrive at a reasonable forecast.

Words to Remember

Time series	Multiplicative model
Forecasting	Method of moving averages
Secular trend	Method of semi-averages
Seasonal variations	Ratio-to-moving average method
Cyclical variations	Ratio-to-trend method
Irregular variations	Deseasonalized data
Additive model	Seasonal index

Symbols to Remember

T Secular trend

S Seasonal variation, or seasonal index

C Cyclical variation

I Irregular variation

Y Observed value of a time series

EXERCISES

1. The yearly retail sales of two different department stores, in millions of dollars, for the years 1961–1970, are given in Table 13.19.

TABLE 13.19

Year	Department store A	Department store B
1961	40	50
1962	43	47
1963	47	45
1964	49	49
1965	50	46
1966	54	43
1967	56	48
1968	59	51
1969	61	53
1970	65	55

   a) Plot the sales of each department store against time.

   b) Fit an annual straight-line trend to the sales of department store A by the method of least squares.

   c) Fit an annual straight-line trend to the sales of department store B by the method of least squares.

   d) Compare the two trends obtained in (b) and (c).

2. a) Convert the annual straight-line trend obtained in Exercise 1(b) for department store A to a monthly trend.

   b) Convert the annual straight-line trend obtained in Exercise 1(c) for department store B to a monthly trend.

3. a) Using the annual straight-line trend obtained in Exercise 1(b) for department store A, make a forecast for 1973.

   b) Using the annual straight-line trend obtained in Exercise 1(c) for department store B, make a forecast for 1973.

4. a) Using the annual straight-line trend obtained in Exercise 1(b) for department store A, calculate the *trend* values for each year.

    b) Using the annual straight-line trend obtained in Exercise 1(c) for department store B, calculate the *trend* values for each year.

5. a) Estimate the straight-line trend by the method of semi-averages for the sales of department store A given in Exercise 1.

    b) Estimate the straight-line trend by the method of semi-averages for the sales of department store B given in Exercise 1.

6. Refer to the data of department store B in Exercise 1.

    a) Construct a 3-year moving average.

    b) Graph the 3-year moving average obtained in (a), along with the original data.

7. The quarterly sales of a retail store, in millions of dollars, for the years 1966–1970, are given in Table 13.20.

TABLE 13.20

	\multicolumn{4}{c}{Quarter}			
Year	1	2	3	4
1966	25	20	26	30
1967	26	22	28	35
1968	27	24	30	43
1969	24	23	31	45
1970	28	25	36	50

a) Draw a graph of the quarterly time series for each year.

b) Compute a quarterly seasonal index using the simple-average method.

8. Obtain a seasonal index by using the ratio-to-moving-average method for the data of quarterly sales given in Exercise 7.

9. a) Estimate the annual trend line for the data of sales given in Exercise 7.

    b) Convert the annual trend line obtained in (a) to a quarterly trend line.

10. a) Obtain a seasonal index, using the ratio-to-trend method for the data of sales given in Exercise 7.

    b) Find the seasonally adjusted sales in Exercise 7 for each quarter.

    c) Use the seasonal index obtained in (a) to forecast quarterly sales for 1973.

11. Refer to the data of gross national product given in Example 1.

    a) Plot this time series on graph paper.

    b) Find the annual straight-line trend by the method of least squares.

    c) Graph the trend equation obtained in (b) on the plot of the time series obtained in (a).

    d) Use the trend equation obtained in (b) to forecast the GNP for 1975.

12. Refer to the data of monthly retail sales given in Example 2.

   a) Obtain a seasonal index using the ratio-to-trend method.

   b) Find the seasonally adjusted sales for each year.

   c) Use the seasonal index obtained in (a) to forecast the monthly sales for 1973.

13. Refer to the data given in Example 7.

   a) Estimate the annual sales straight-line trend.

   b) Convert the annual sales straight-line trend obtained in (a) to a monthly sales trend.

14. a) Using the annual sales straight-line trend obtained in Exercise 13(a), calculate the trend values for each year.

   b) Using the annual sales straight-line trend obtained in Exercise 13(a), make a sales forecast for 1975.

15. The quarterly regular gasoline sales of a particular manufacturer, for the years 1968–1972 (in millions of gallons), are given in Table 13.21.

TABLE 13.21

Year	Quarter			
	1	2	3	4
1968	15	22	27	25
1969	23	31	39	33
1970	30	35	44	40
1971	39	45	53	48
1972	48	52	65	56

a) Graph the quarterly sales.

b) Compute the quarterly seasonal index, using the simple-average method.

c) Obtain a seasonal index, using the ratio-to-moving-average technique.

d) Compute the annual trend line.

e) Obtain a seasonal index using the ratio-to-trend method.

f) Find the seasonally adjusted sales for each quarter.

INDEX
NUMBERS 14

14.1. INTRODUCTION

In the previous chapter, we considered a technique called time series analysis for analyzing business and economic data collected over different time intervals. In this chapter, we shall discuss another technique, called *index numbers*, for the comparison of business and economic data collected over *different* time intervals. An index number is simply a ratio of two quantities, such as prices, values, or other economic variables. It is a device which helps the businessman to compare the changes in prices, production, and the total value of his products, with similar data collected in another period called the *fixed period* (or the *base period*). The student must have heard, at one time or another, the names of some of the well-known indices, such as the wholesale price index, the consumer price index, the index of industrial production, the index of department store sales, etc. These indices measure the relative change in the wholesale prices, the prices paid by the people as consumers, and the quantity produced by industrial workers. The three types of index numbers which may be of interest to the businessman may be classified as follows:

1. Price index numbers
2. Quantity index numbers
3. Value index numbers.

In this chapter, we shall discuss methods of constructing the price index numbers, the quantity index numbers, and the value index numbers. We shall also give a brief description of three well known indices, the Wholesale Price Index, the Consumer Price Index, and the Industrial Production Index. Finally, we shall explain how the Price Index or the Consumer Price Index can be used in *deflating* prices and income.

14.2. METHODS OF CONSTRUCTING PRICE-INDEX NUMBERS

In this section, we shall first give a very simple example of an index number involving only one item, called a *price relative* (which may also be called a *simple price index* or a *simple index number*). Then we shall introduce an index number involving a number of items, called the *simple aggregate of price relatives* (or the *simple aggregate price index*). Finally, we shall give two methods of constructing a weighted index by assigning weights to the relative importance of the various items involved.

Let the subscript 0 represent the base period and the subscripts 1, 2, 3, etc., represent the subsequent periods. Then the letters $p_0, p_1, \ldots, p_n$ represent the individual prices of an item in their respective periods. We shall denote an index number with a letter I, and the letters $I_0, I_1, I_2, \ldots$, will represent the index numbers for subsequent periods. We shall use $I_0 = 100$ as the index number for the base period. We may now define price relatives as follows:

Price Relative

Let p_0 and p_n be the price of an item in the base period and the given period respectively; then the price relative or the simple index number, I, is given by

$$\text{Price Relative (Index number } I) = \frac{p_n}{p_0} \cdot 100.$$

Example 1 Assume that the average consumer prices for one dozen eggs, over three years, are given as follows: 1968, $0.25; 1969, $0.35; 1970, $0.45. Taking 1968 as the base year, we can calculate the changes in prices as in Table 14.1.

TABLE 14.1

Year	Price	Price relatives (Index numbers)
1968	$0.25 ($p_0$)	$I_0 = \left(\dfrac{p_0}{p_0}\right)(100) = 100$
1969	$0.35 ($p_1$)	$I_1 = \left(\dfrac{p_1}{p_0}\right)(100) = \left(\dfrac{\$0.35}{\$0.25}\right)(100) = 140$
1970	$0.45 ($p_2$)	$I_2 = \left(\dfrac{p_2}{p_0}\right)(100) = \left(\dfrac{\$0.45}{\$0.25}\right)(100) = 180$

Here, $I_1 = 140$ means that there has been an increase of 40 percent in the price of eggs from 1968 to 1969; $I_2 = 180$ means that there has been an increase of 80 percent in the price of eggs from 1968 to 1970.

In Example 1, we have considered the *price relative* or the index number of a single item. In actual situations, we are interested to know how the price of a group of items in a given period compares with the price of the same group of items in another period. For example, we might like to compare the price, in two different periods, of a dozen eggs, a pound of butter, a half gallon of ice cream, a pound of cheese, and a pound of beef. This may be accomplished by finding a price index for a group of items called a *simple aggregate price index*, defined below:

Simple Aggregate Price Index

The simple aggregate price index is given by

$$\text{Simple aggregate price index} = \frac{\sum p_n}{\sum p_0} \cdot 100,$$

where $\sum p_n =$ Sum of prices of all items in the given period, and $\sum p_0 =$ sum of prices of all items in the base period.

Example 2 Table 14.2 gives the prices and the quantities of five items used by a certain family in the years 1968, 1969, and 1970. Compute a simple aggregate price index for these items, using 1968 as the base period.

TABLE 14.2

		Price		
Item	Quantity	1968	1969	1970
Eggs	Dozen	$0.25	$0.30	$0.40
Butter	Pound	0.65	0.70	0.75
Ice cream	Half-gallon	0.60	0.70	0.90
Cheese	Pound	0.40	0.45	0.50
Beef	Pound	0.60	0.85	0.95
Total		$2.50	$3.00	$3.50

Solution: Taking 1968 as the base year, we may compute the simple aggregate price index as in Table 14.3.

TABLE 14.3

Year	Simple aggregate price index
1968	$I_{68} = \dfrac{25+65+60+40+60}{25+65+60+40+60}(100) = \dfrac{250}{250}(100) = 100$
1969	$I_{69} = \dfrac{30+70+70+45+85}{25+65+60+40+60}(100) = \dfrac{300}{250}(100) = 120$
1970	$I_{70} = \dfrac{40+75+90+50+95}{25+65+60+40+60}(100) = \dfrac{350}{250}(100) = 140$

Here, $I_{69} = 120$ means that there has been an increase of 20 percent in the price of eggs, butter, ice cream, cheese, and beef, as a group, from 1968 to 1969. $I_{70} = 140$ means that there has been an increase of 40 percent in the price of eggs, butter, ice cream, cheese, and beef, as a group, from 1968 to 1970.

We may note, from Example 2, that, although the simple aggregate price index is easy to compute, it has the following two defects:

1. That the different items are expressed in different units, such as tons, pounds, gallons, bushels, etc., will affect the price index.

2. It gives no consideration to the economic importance of the items involved. That is, in computing a cost of living index by this method, an equal weight would be given to a loaf of bread and a pair of shoes.

We can overcome the first defect by averaging the simple price relatives over the number of items involved in the price index. In other words, we have given equal importance to each item and have thus tried to remove the variation due to different units. This index is called the *simple relative price index* and is given below:

Simple Relative Price Index

The simple relative price index is the arithmetic mean of the price relatives, and is given by

$$I = \frac{\Sigma(p_n/p_0) \cdot 100}{n},$$

where n is the number of items involved in the index.

Example 3 (Refer to Example 2.) Compute the simple relative price index for the data given in Table 14.1, using 1968 as the base year.

Solution: Taking 1968 as the base year, we may compute the simple relative price index as in Table 14.4.

TABLE 14.4

Item	Price Relative		
	1968 p_0/p_0	1969 p_1/p_0	1970 p_2/p_0
Eggs	100	$\left(\frac{30}{25}\right)(100)=120.00$	$\left(\frac{40}{25}\right)(100)=160.00$
Butter	100	$\left(\frac{70}{65}\right)(100)=107.69$	$\left(\frac{75}{65}\right)(100)=115.38$
Ice Cream	100	$\left(\frac{70}{60}\right)(100)=116.67$	$\left(\frac{90}{60}\right)(100)=150.00$
Cheese	100	$\left(\frac{45}{40}\right)(100)=112.50$	$\left(\frac{50}{40}\right)(100)=125.00$
Beef	100	$\left(\frac{85}{60}\right)(100)=141.67$	$\left(\frac{95}{60}\right)(100)=158.33$
Total	500	598.53	708.71
Average	$I_{68}=\frac{500}{5}=100$	$I_{69}=\frac{598.53}{5}=119.706$	$I_{70}=\frac{708.71}{5}=141.742$

It was mentioned earlier that the simple aggregate method gives no consideration to the economic importance of the items involved. Thus, to assign *equal* weights to each item involved in computing a price index is not a realistic assumption. In assigning weights to each of these items, we must take into consideration their relative importance. The weights may be determined by the price and the quantity of each item bought or sold in the base period. The weights so determined are then used in computing a price index called the *weighted price index*. We shall explain this procedure with the help of the following example.

Example 4 Suppose the annual consumption of the five items, eggs, butter, ice cream, cheese, and beef, in the base year 1968 are given in the third column of Table 14.5. Columns 4, 5, and 6 represent the prices of the five items in the years 1968, 1969, and 1970, respectively.

TABLE 14.5

(1) Item	(2) Unit	(3) Annual consumption 1968 q_0	Price (4) 1968 p_0	Price (5) 1969 p_1	Price (6) 1970 p_2
Eggs	Dozen	150	$0.25	$0.30	$0.40
Butter	Pound	100	0.65	0.70	0.75
Ice cream	Half-gallon	50	0.60	0.70	0.90
Cheese	Pound	100	0.40	0.45	0.50
Beef	Pound	200	0.60	0.85	0.95

In Table 14.2, the letter q_0 represents the annual consumption of each item in the base year 1968. Using the entries in column (3) as weights, we may compute the weighted index number as in Table 14.6.

TABLE 14.6

(1) Item	(2) Unit	(3) Annual consumption, 1968 q_0	(4) 1968 $q_0 p_0$	(5) 1969 $q_0 p_1$	(6) 1970 $q_0 p_2$
Eggs	Dozen	150	$37.50	$45.00	$60.00
Butter	Pound	100	65.00	70.00	75.00
Ice cream	Half-gallon	50	30.00	35.00	45.00
Cheese	Pound	100	40.00	45.00	50.00
Beef	Pound	200	120.00	170.00	190.00
Total			$292.50	$365.00	$420.00
The weighted aggregate price index			$\frac{292.50}{292.50}(100)=100$	$\frac{365.00}{292.50}(100)=124.79$	$\frac{420.00}{292.50}(100)=143.59$

Columns (4), (5) and (6) in Table 14.6 are obtained by multiplying the annual consumption of each item by its corresponding price in a given year. The sums in columns (4), (5), and (6) represent the total money spent on the five items, in a given year. The weighted aggregate price index is calculated by simply dividing the total money spent in each year times 100, by the total money spent in the *base* year. The weighted price index so obtained is given in the last row of Table 14.6.

The weighted price index calculated in Example 4, with the base year quantities as weights, is called the Laspeyres Price Index. The Laspeyres Price Index may be defined as follows:

Laspeyres Price Index

The Laspeyres Price Index is given by the formula

$$I = \frac{\Sigma p_n q_0}{\Sigma p_0 q_0} \cdot 100,$$

where

$\Sigma p_n q_0 =$ total money spent in the given year,

and

$\Sigma p_0 q_0 =$ total money spent in the base year.

Example 5 A manufacturer of large kitchen appliances wishes to construct a price index of three models, standard, medium, and deluxe, of refrigerators. The prices and the quantities sold for the years 1969 and 1970 are given in Table 14.7. Construct the Laspeyres Price Index, using 1969 as the base year.

TABLE 14.7

Model	Prices		Quantities sold	
	1969	1970	1969	1970
	p_0	p_n	q_0	q_n
Standard	$230	$250	1,700	2,000
Medium	275	300	1,200	1,500
Deluxe	320	350	1,000	1,200

Solution: The Laspeyres Price Index, using 1969 as the base year and the quantities sold in 1969 as the weights, may be calculated as in Table 14.8.

TABLE 14.8

Model	Prices		Quantities sold		Income	
	1969	1970	1969	1970	1969	1970
	p_0	p_n	q_0	q_n	$p_0 q_0$	$p_n q_0$
Standard	$230	$250	1,700	2,000	$391,000	$425,000
Medium	275	300	1,200	1,500	330,000	360,000
Deluxe	320	350	1,000	1,200	320,000	350,000
Total					$1,041,000	$1,135,000

Thus,

$$I_{70} = \frac{\sum p_n q_0}{\sum p_0 q_0} \cdot 100$$

$$= \frac{1,135,000}{1,041,000} \cdot 100 = 109.03.$$

The Laspeyres Price Index (or the *weighted aggregate* price index) for 1970 is 109.03. That is, there has been a 9.03 percent increase in the 1970 prices over the 1969 prices.

The weighted price index may also be calculated with the given year quantities as weights. The *averages* of the quantities sold over several years may also be used as weights, especially when the base year or the given year may not look like the normal year. The weighted price index calculated with the given year quantities as weights is called the Paasche Price Index. The Paasche Price Index may be defined as follows:

Paasche Price Index

The Paasche Price Index is given by the formula

$$I = \frac{\sum p_n q_n}{\sum p_0 q_n} \cdot 100,$$

where

$$\sum p_n q_n = \text{total money spent in the given year,}$$

and

$$\sum p_0 q_n = \text{total money spent in the base year.}$$

Example 6 (Refer to Example 5.) Compute the Paasche Price Index for the data given in Table 14.7, using 1969 as the base year.

Solution: The Paasche Price Index, using 1969 as the base year and the quantities sold in 1970 as the weights, may be calculated as in Table 14.9.

TABLE 14.9

Model	Prices		Quantities sold		Income	
	1969	1970	1969	1970	1969	1970
	p_0	p_n	q_0	q_n	$p_0 q_n$	$p_n q_n$
Standard	$230	$250	1,700	2,000	$460,000	$500,000
Medium	275	300	1,200	1,500	412,500	450,000
Deluxe	320	350	1,000	1,200	384,000	420,000
Total					$1,256,500	$1,370,000

Thus,

$$I_{70} = \frac{\sum p_n q_n}{\sum p_0 q_n} \cdot 100$$

$$= \frac{1,370,000}{1,256,500} \cdot 100 = 109.$$

The weighted price index computed by means of the Paasche formula is 109. That is, there has been a 9 percent increase in the 1970 prices over the 1969 prices.

14.3. METHODS OF CONSTRUCTING QUANTITY AND VALUE INDEXES

We have explained earlier that an index number is simply a ratio of two quantities such as prices, values, or other economic variables. In the previous section, we have given several methods for constructing price index numbers. In this section, we give methods for constructing a quantity or volume index and a value index that are analogous to those for price indexes. In calculating a *quantity index*, the prices are used as weights instead of the quantities. The formulas for computing a quantity index, using Laspeyres'method and Paasche's method, may easily be obtained by interchanging p and q in the Laspeyres Price Index and the Paasche Price Index. These formulas are given below.

Laspeyres Quantity Index

The Laspeyres Quantity Index is given by the Formula

$$I = \frac{\sum q_n p_0}{\sum q_0 p_0} \cdot 100,$$

where

$\sum q_n p_0 =$ total value of the items in the given year,

and

$\sum q_0 p_0 =$ total value of the items in the base year.

Example 7 (Refer to Example 5.) Compute the Laspeyres Quantity Index for the data given in Table 14.7, using 1969 as the base year.

Solution: The Laspeyres Quantity Index, using 1969 as the base year and the prices in 1969 as the weights, may be calculated as in Table 14.10.

TABLE 14.10

Model	Prices 1969 p_0	Prices 1970 p_n	Quantities sold 1969 q_0	Quantities sold 1970 q_n	Income 1969 $q_0 p_0$	Income 1970 $q_n p_0$
Standard	$230	$250	1,700	2,000	$391,000	$460,000
Medium	275	300	1,200	1,500	330,000	412,500
Deluxe	320	350	1,000	1,200	320,000	384,000
Total					$1,041,000	$1,256,500

Thus, the Laspeyres Quantity Index is given by

$$I_{70} = \frac{\sum q_n p_0}{\sum q_0 p_0} \cdot 100$$

$$= \frac{1,256,500}{1,041,000} \cdot 100 = 120.7.$$

The Paasche Quantity Index may similarly be obtained by interchanging p and q in the Paasche Price Index.

Paasche Quantity Index

The Paasche Quantity Index is given by the formula

$$I = \frac{\sum q_n p_n}{\sum q_0 p_n} \cdot 100,$$

where

$$\sum q_n p_n = \text{total value of the items in the given year,}$$

and

$$\sum q_0 p_n = \text{total value of the items in the base year.}$$

Example 8 (Refer to Example 5.) Compute the Paasche Quantity Index for the data given in Table 14.7, using 1969 as the base year.

Solution: The Paasche Quantity Index, using 1969 as the base year and the prices in 1970 as the weights, may be calculated as in Table 14.11.

TABLE 14.11

Model	Prices 1969 p_0	Prices 1970 p_n	Quantities sold 1969 q_0	Quantities sold 1970 q_n	Income 1969 $q_0 p_n$	Income 1970 $q_n p_n$
Standard	$230	$250	1,700	2,000	$425,000	$500,000
Medium	275	300	1,200	1,500	360,000	450,000
Deluxe	320	350	1,000	1,200	350,000	420,000
Total					$1,135,000	$1,370,000

Thus, the Paasche Quantity Index is given by

$$I_{70} = \frac{\sum q_n p_n}{\sum q_0 p_n} \cdot 100$$

$$= \frac{1,370,000}{1,135,000} \cdot 100 = 120.7.$$

That is, there has been a 20.7 percent increase in the 1970 sales over the 1969 sales.

To compare the total value of the goods or items in a given period with respect to another period, a *value index* may be constructed. The Value Index gives the percentage change in the value of goods or items compared with the base period. For example, a percentage ratio of the monthly sales of a department store to the average monthly sales in the base period is a value index. We give below a formula for computing a simple aggregate value index. Other value indexes using weights may be constructed in the same way as the price and quantity indexes.

Value Index

The Value Index is given by the formula

$$I = \frac{\Sigma p_n q_n}{\Sigma p_0 q_0} \cdot 100,$$

where

$\Sigma p_n q_n$ = total value of the items in the given year,

and

$\Sigma p_0 q_0$ = total value of the items in the base year.

Example 9 (Refer to Example 5.) Construct a value index for the data given in Table 14.7, using 1969 as the base year.

Solution: The simple aggregate value index, using 1969 as the base year, may be calculated as in Table 14.12.

TABLE 14.12

Model	Prices 1969 p_0	Prices 1970 p_n	Quantities sold 1969 q_0	Quantities sold 1970 q_n	Income 1969 $p_0 q_0$	Income 1970 $p_n q_n$
Standard	$230	$250	1,700	2,000	$391,000	$500,000
Medium	275	300	1,200	1,500	330,000	450,000
Deluxe	320	350	1,000	1,200	320,000	420,000
Total					$1,041,000	$1,370,000

Thus, a simple aggregate value index is given by

$$I = \frac{\sum p_n q_n}{\sum p_0 q_0} \cdot 100$$

$$= \frac{1,370,000}{1,041,000} \cdot 100 = 131.6.$$

That is, there has been an increase of 31.6 percent in the 1970 total value of sales over the total value of the 1969 sales.

14.4. THREE IMPORTANT INDEXES

We know now how index numbers are used to compare the price, quantity, or value of an item (or group of items) in a given period with respect to a base period. An index number is a valuable aid to the businessman in decision-making. Many important and widely used indexes are prepared by private businesses, financial institutions, and the government. The Federal government prepares thousands of such indexes, including the index of unemployment, the index of retail food prices, the index of hourly wages, the index of retail sales, the index of construction, etc. In this section, we shall discuss very briefly two well-known price indexes, the Wholesale Price Index and the Consumer Price Index, and one *quantity* index called the Index of Industrial Production.

1. The Wholesale Price Index

The Wholesale Price Index measures *wholesale* prices at primary markets. This index is constructed by the Bureau of Labor Statistics based on a sample of about 2,000 representative items. This price index is now calculated by means of the Laspeyres formula, with 1957–59 as the base period. The weights in the formula are determined from manufacturers' census data. These weights are constantly reviewed and modified. The Bureau publishes daily, weekly, and monthly series of this index, in a publication called *Survey of Current Business*. These series keep the businessman on top of general business conditions and help him in planning, management, and business forecasting.

2. The Consumer Price Index

The Consumer Price Index measures the change in the prices of goods and services purchased by people as *consumers*. It measures the average change in the cost of living of families of urban wage earners and salaried workers. (For this reason, this index was originally called the *cost-of-living index*.) The Consumer Price Index measures the change in the cost of living of an average family based on a fixed "market basket" consisting of about 400 consumption items. It now uses 1957–59 as the base period, although previously 1947–49 has been used as the base period. This index is based on a weighted average of the price relatives. The *quantities* of foods making up the market basket are *held fixed*; thus the index measures the change in the price of the (total) market basket. There are several other indexes published by the Bureau of Labor Statistics, such as those for food, housing, transportation, medical costs, apparel, footwear, etc. The Consumer Price Index is used for a variety of purposes by private and governmental agencies. It is now widely used by unions in adjusting wage contracts.

3. The Index of Industrial Production

The Index of Industrial Production measures the change in the physical volume of industrial output by the manufacturing, mining, gas, and electricity industries. The Index of Industrial Production is compiled and published monthly by the Federal Reserve Board. This index now uses 1957–59 as the base period, although previously the period 1947–49 had been used. This index is very comprehensive, and it is used most widely by the business community. It helps the businessman to measure the progress of *his* industry in relation to other competitive industries. Since the industries included in this index make up about one-third of the total national income, it is a good indicator of general business conditions.

14.5. DEFLATION OF PRICES AND INCOME

We have discussed so far several methods of constructing index numbers. In this section, we shall discuss how index numbers can be used to *adjust income and prices*. The business manager can use index numbers to make corrections or adjustments for changes in prices, wages of workers, and the analysis of real income of his business. This procedure for making corrections or adjustments in prices and income using a price index is called *deflation of prices and income*.

The basic function of a price index is that of *measuring* the changes in prices. In other words, it measures the purchasing power of money. We all know that the purchasing power of the dollar decreases as the prices of goods and

TABLE 14.13

Year	Consumer price index* (1957–59 = 100)	Year	Consumer price index (1957–59 = 100)
1957	98.0	1964	108.1
1958	100.7	1965	109.1
1959	101.5	1966	113.1
1960	103.1	1967	116.3
1961	104.2	1968	121.2
1962	105.4	1969	127.7
1963	106.7		

*Source: *Business Statistics*, U.S. Department of Commerce.

services increase. In other words, since the index number in the base period is 100, the purchasing power of a dollar in the given period is equal to 100 divided by the price index in the given period. This may be stated as follows:

Purchasing Power of a Dollar

The purchasing power of money (or of a dollar) is given by the formula

$$\text{Purchasing power of a dollar} = \frac{100}{\text{Price index}}. \quad ^*$$

Example 10 The Consumer Price Index for the years 1957–1969 is given in Table 14.13. Determine the purchasing power of the dollar for the subsequent years.

Solution: The purchasing power of the dollar (in 1957–59 dollars) for each year may be obtained by dividing \$1.00 by the Consumer Price Index for that year and multiplying it by 100. The purchasing power of the dollar in 1957–59 dollars is given in the last column of the following table. The entry 0.78 corresponding to the year 1969 in Table 14.14 means that the purchasing power of the 1969 dollar was worth only 78 cents in terms of the 1957–59 dollar.

TABLE 14.14

Year	Purchasing power of the dollar in 1957–59 dollars	Year	Purchasing power of the dollar in 1957–59 dollars
1957–59	1.00	1965	0.92
1960	0.97	1966	0.88
1961	0.96	1967	0.86
1962	0.95	1968	0.83
1963	0.94	1969	0.78
1964	0.93		

---

*The quotient obtained by this formula will be in the form of a percent; this would then need to be multiplied by the (value of) the money units. Thus the purchasing power of the dollar in 1960 (compared to the 1957–59 dollar) was only \$0.97 (cf. Table 14.13):

$$(100/103.1) \times 1.00 = 0.97$$

Purchasing power of the British pound (*in the U.S.*) was

$$(100/103.1) \times 2.86 = 2.77,$$

and similarly for other monetary units.

The real income of a company, or the real wages of workers, or the real price of an item can be determined using a Consumer Price Index. Since the Consumer Price Index in the base period is 100, the real wage (or the real income, or the real price) is simply equal to the wage (or the income, or the price) in a given period *multiplied by* the purchasing power of a dollar. In other words, the index in the base period, divided by the consumer price index in the given period, multiplied by the income (or wage, or price) in the given period determines the real income (or the real wage, or the real price). We shall explain this procedure of *deflation* of income and prices with the help of the following example.

Example 11 The mean hourly wages of the employees of a company and the Consumer Price Index for the years 1966–1970 are given in Table 14.15. Determine the hourly wages in 1966 dollars.

TABLE 14.15

Year	Hourly wage	Consumer price index
1966	$2.25	100
1967	2.30	103
1968	2.38	107
1969	2.40	109
1970	2.45	112

Solution: The hourly wages in 1966 dollars may be obtained by dividing the hourly wage for a particular year by the corresponding Consumer Price Index for that year, and multiplying it by 100. The hourly wages in 1966 dollars are given in the last column of Table 14.16.

TABLE 14.16

Year	Hourly wage	Consumer price index	Hourly wage in 1966 dollars
1966	$2.25	100	$2.25
1967	2.30	103	2.23
1968	2.38	107	2.22
1969	2.40	109	2.20
1970	2.45	112	2.19

SUMMARY

The main purpose of constructing index numbers for a businessman is to compare the prices, the production, and the total value of his goods in a given period with the same data for another period called the *base period*. The index numbers may be classified into three types:

1. Price index numbers
2. Quantity index numbers
3. Value index numbers.

In this chapter, we have discussed the methods of constructing index numbers of the three types mentioned above. A brief description of three well-known indices (the Wholesale Price Index, the Consumer Price Index, and the Index of Industrial Production) is also given. Finally, we have explained how the businessman can use index numbers to make corrections or adjustments for changes in prices, wages of workers, and the analysis of real income of his business, over different time intervals.

Words to Remember

Price relative	Laspeyres quantity index
Index number	Paasche quantity index
Simple aggregate price index	Simple aggregate value index
Simple relative price index	Wholesale price index
Weighted aggregate price index	Consumer price index
Laspeyres price index	Index of industrial production
Paasche price index	Deflation

Symbols to Remember

P	Price
Q	Quantity
I	Index number
Subscript 0	Base period
Subscripts 1, 2, 3, etc.	Subsequent periods

EXERCISES

1. What are index numbers? Describe the important problems involved in constructing index numbers.
2. Discuss the importance of index numbers in economic and business analysis.

3. Suppose you are hired by the manager of a large department store to construct a price index for the thousands of items it sells. Discuss how you would proceed in constructing such an index.

4. Explain the meaning and the usefulness of each of the following price indexes:

   a) A price relative.

   b) A simple aggregate price index.

   c) A simple relative price index.

   d) Laspeyres price index.

   e) Paasche price index.

5. Explain the meaning and usefulness of each of the following quantity and value indexes:

   a) Laspeyres quantity index.

   b) Paasche quantity index.

   c) Value index.

6. A manufacturing company purchases four different items of raw materials for its manufacturing process. Table 14.17 gives the prices and the quantities used for the four items for 1970, 1971, and 1972. Construct a weighted aggregate price index, using 1970 as the base year.

TABLE 14.17

Item	Unit	Annual consumption 1970	Prices 1970	1971	1972
A	Ton	75	$60.00	$62.00	$65.00
B	Pound	6,000	0.55	0.60	0.65
C	Gallon	1,500	1.25	1.50	1.75
D	Ton	50	40.00	43.00	45.00

TABLE 14.18

Year	Thousands of barrels per day	Year	Thousands of barrels per day
1963	553	1968	700
1964	610	1969	733
1965	625	1970	750
1966	600	1971	765
1967	627	1972	772

7. The production of crude oil and natural gas liquids of oil company X is given in Table 14.18. Convert the yearly production figures to index numbers, using 1963 as the base year.

8. The United States' Gross National Product (GNP) from 1961–1970 inclusive is given in Table 13.1. Convert these yearly sales figures to index numbers using 1961 as the base year.

9. The ABC Company would like to construct an index of the prices of its three products. The prices and quantities sold for the years 1965 and 1970 are shown in Table 14.19. Construct the Laspeyres price index, using 1965 as the base year.

TABLE 14.19

Product	Prices		Quantities sold	
	1965	1970	1965	1970
I	$20	$25	1,200	1,500
II	$15	$18	1,500	2,000
III	$10	$12	2,000	2,700

10. Compute the Paasche price index for the data given in Exercise 9, using 1965 as the base year.

11. Compute the Laspeyres quantity index for the data given in Exercise 9, using 1965 as the base year.

12. Compute the Paasche quantity index for the data given in Exercise 9, using 1965 as the base year.

13. Compute the value index for the data given in Exercise 9, using 1965 as the base year.

14. As a wage analyst with the ABC Company, you are asked to determine the real hourly wages in 1968 dollars. The mean hourly wages of the employees of the company and the Consumer Price Index for the years 1968–72 are given in Table 14.20. Determine the value of the employees' hourly wages in 1968 dollars.

TABLE 14.20

Year	Hourly wage	Consumer price index
1968	$3.25	100
1969	3.75	104
1970	4.15	107
1971	4.50	110
1972	4.75	112

15. Residential construction cost indexes, using 1957–59 as the base period, are given in Table 14.21.Using these construction cost indices, find the construction cost of a house in 1968 if, in 1960, the construction of the same house cost $15,000.

TABLE 14.21

Year	Construction Cost Index (1957-1959 = 100)*	Year	Construction Cost Inex (1957-1959 = 100)*
1957	98.3	1963	108.5
1958	99.2	1964	111.6
1959	102.5	1965	115.2
1960	104.2	1966	120.1
1961	104.5	1967	127.4
1962	106.3	1968	136.7

*Source: *Business Statistics*, U.S. Department of Commerce.

Using these construction cost indices, find the construction cost of a house in 1968 if, in 1960, the construction of the same house cost $15,000.

THE
CHI-SQUARE
TEST 15

Small values of χ^2 - good acceptance
large values of χ^2 - poor acceptance

15.1. INTRODUCTION

There are many business problems which require that we ascertain the frequency of events falling in specified categories. For example, in market research we count the number of people who prefer a particular brand of soap, toothpaste, coffee, etc.; in quality control we count the number of defectives produced by a machine in a certain period; in socioeconomic studies we count the number of children in families at various income levels. There are many situations of the above type where measurement data are changed to categorical data. In the preceding chapters, we dealt with measurement data. In this chapter, we shall discuss business problems which deal with categorical, qualitative, or enumerative data. The chi-square (χ^2) statistic is a useful tool in determining whether a frequency with which a given event has occurred is significantly different from what we expected. We can formulate the null hypothesis that there is *no* difference between the observed and expected frequencies, and the alternate hypothesis that there is a difference between the observed and expected frequencies. A test statistic which measures the discrepancy between K observed frequencies o_1, o_2,...,o_K and their corresponding expected frequencies e_1, e_2,...,e_K is called the chi-square (χ^2) statistic. We calculate χ^2 by means of the formula

$$\chi^2 = \sum_{i=1}^{K} \frac{(o_i - e_i)^2}{e_i}.$$

That is, we square the difference between the observed and the expected frequencies, and then divide the result by the expected frequencies; the chi-square (χ^2) statistic is the sum of the quotients for all the cells or categories.

The chi-square (χ^2) statistic

$$\chi^2 = \sum_{i=1}^{K} \frac{(o_i - e_i)^2}{e_i}$$

If $\chi^2 = 0$, there is perfect agreement between observed and expected frequencies. The greater the discrepancy between the observed and expected frequencies, the larger will be the value of χ^2. The sampling distribution of the χ^2 statistic given above is very closely approximated by the chi-square distribution discussed in Chapter 10. In order to test the significance of χ^2, the calculated value of χ^2 is compared with the table value for the given degrees of freedom at a certain level of significance. If the calculated value of χ^2 is greater than the table value, the difference between the observed and expected frequencies is considered significant. On the other hand, if the calculated value of χ^2 is less than the table value, the difference between the observed and expected frequencies is considered insignificant. The number of degrees of freedom is given by the total number of cells minus the number of cells whose expected frequency can be determined independently. The following conditions must be satisfied in applying the chi-square test.

1. Each observation or frequency must be independent of all other observations.

2. The sample size N must be reasonably large in order that the difference between actual and expected frequencies be normally distributed. In practice, we may say that N should be at least 50.

3. No expected cell frequency should be small. In practice, 5 is regarded as the very minimum and 10 is considered better. In cases where the cell frequencies fall below these limits, they are amalgamated into a single cell.

Example 1 A marketing research department of a firm has divided a certain sales region into six districts. It is believed that each district has the same sales potential. The actual number of units sold in a specified district are given in the following table. Test the hypothesis that the six districts have equal sales potential, using a level of significance of 0.05.

District	Units sold
1	12
2	18
3	15
4	25
5	22
6	28
Total	120

Solution: On the hypothesis that each district has the same sales potential, the expected number of units sold in each district is 20. To test the hypothesis that the six districts have equal sales potential, we calculate the χ^2 statistic. If the calculated value of χ^2 is equal to zero, there is perfect agreement between observed and expected frequencies. In other words, we are testing the null hypothesis that there is perfect agreement between the observed and expected frequencies; or, equivalently, we are testing that $\chi^2 = 0$. The χ^2 statistic is calculated in Table 15.1.

TABLE 15.1

District	Observed frequency, o	Expected frequency, e	$o - e$	$(o-e)^2/e$
1	12	20	-8	3.20
2	18	20	-2	0.20
3	15	20	-5	1.25
4	25	20	5	1.25
5	22	20	2	0.20
6	28	20	8	3.20
Total	120	120		$\chi^2 = 9.30$

Since the sum of the six observed frequencies in Table 15.1 must equal 120, the expected frequency for the sixth cell can be determined as soon as the expected frequencies for the first five cells are known. Thus the number of degrees of freedom is

$$\nu = K - 1 = 6 - 1 = 5.$$

Using a 0.05 level of significance, the value of $\chi^2_{0.95}=11.07$. Since the calculated $\chi^2=9.3$ is less than the table value $\chi^2_{0.95}=11.07$, the hypothesis that the six districts have equal sales potential cannot be rejected.

Example 2 A firm makes house appliances such as refrigerators, stoves, washers, dryers, etc. The marketing research department of this firm has surveyed 200 families with respect to their preference in colors for appliances. The results are given below:

Color preference	Number of families
White	35
Coppertone	82
Avocado	39
Harvest gold	44
Total	200

Test the hypothesis of equal preference for colors, using a level of significance of 0.01.

Solution: On the hypothesis of equal preference of a color, we expect the number of families preferring white, coppertone, avocado, and harvest gold to be 50 for each color. To test the hypothesis of equal preference in colors, we calculate the χ^2 statistic, as in Table 15.2.

TABLE 15.2

Color	Observed frequency, o	Expected frequency, e	$o-e$	$(o-e)^2/e$
White	35	50	-15	4.50
Coppertone	82	50	32	20.48
Avocado	39	50	-11	2.42
Harvest gold	44	50	-6	0.72
Total	200	200		$\chi^2=28.12$

Since the sum of the four observed frequencies in Table 15.2 must equal 200, the expected frequency for the fourth can be determined as soon as the expected frequencies for the first three cells are known. Thus the number of degrees of freedom is $\nu=K-1=3$. Using a 0.01 level of significance, we find the value of

χ^2 for 3 degrees of freedom from Appendix Table 7 as $\chi^2_{0.99} = 11.34$. Since the calculated $\chi^2 = 28.12$ is greater than the table value $\chi^2_{0.99} = 11.34$, the hypothesis of equal preference in colors is rejected.

Example 3 Daily demand for loaves of bread at a grocery store is given in the following table. Test the hypothesis that the number of loaves of bread sold does not depend on the day of the week.

Day of the week	Number of loaves sold
Monday	110
Tuesday	150
Wednesday	330
Thursday	280
Friday	430
Saturday	500
Total	1800

Solution: On the hypothesis of equal demand for loaves of bread on any day of the week, the expected number of loaves of bread sold on any day of the week is 300. To test the hypothesis of equal demand, we calculate the χ^2 statistic in Table 15.3.

TABLE 15.3

Day	Observed frequency, o	Expected frequency, e	$o - e$	$(o-e)^2/e$
Monday	110	300	-190	120.33
Tuesday	150	300	-150	75.00
Wednesday	330	300	30	3.00
Thursday	280	300	-20	1.33
Friday	430	300	130	56.33
Saturday	500	300	200	133.33
Total	1800	1800		$\chi^2 = 389.32$

The number of degrees of freedom is $v = K - 1 = 5$. Using a 0.01 level of significance, we find the value of χ^2 for 5 degrees of freedom from Appendix Table 7 as $\chi^2_{0.99} = 15.08$. Since the calculated value $\chi^2 = 389.32$ is greater than the table value, the hypothesis of equal demand of loaves of bread on any day of the week is rejected.

15.2. GOODNESS-OF-FIT TEST

The term "goodness of fit" is used for the comparison of observed sample distributions with expected probability distributions such as the binomial, Poisson, normal, etc. The chi-square statistic can be used to judge the divergence between the observed and expected frequencies. That is, the curve of the expected frequencies is superimposed on the curve of the observed frequencies, and the chi-square statistic determines whether the fit is good or not. We shall explain this procedure with the help of the following examples.

Example 4 (Refer to Example 8, Chapter 4.) Examine the closeness of fit of the normal distribution to the data of 136 observations of television viewers given in Table 4.1.

Solution: Using formulas for grouped data from Chapters 3 and 4, we calculate the mean and the standard deviation of the frequency distribution given in Table 2.1 as

$$\overline{X} = 3.17 \quad \text{and} \quad s = 0.943.$$

Assuming normality, the probability that an individual will fall in the first class is equal to the area to the left of 1.105 under the normal curve. The probability that an individual will fall in the last class is equal to the area to the right of 5.505. The probability of an individual falling in a class other than the first and last can be found by computing the Z-value for lower and upper class limits, and then finding the area between the two Z-values. Calculation of the probability of an individual falling within a class and its corresponding expected frequency are shown in Table 15.4.

Since the parameters μ and σ are estimated from the sample, the number of degrees of freedom is $\nu = 10 - 2 - 1 = 7$. Using a level of significance of 0.05, we see that the value of χ^2 for 7 degrees of freedom (from Appendix Table 7) is $\chi^2_{0.95} = 14.06$. The calculated $\chi^2 = 9.9881$ is less than the table value $\chi^2_{0.95} = 14.06$. Hence, our hypothesis that the data of 136 observations of television viewers follows a normal distribution cannot be rejected, and the normal fit is good.

Example 5 The number of customers arriving at a service desk per five-minute period is given in the following table. Fit a Poisson distribution and test the goodness of fit.

TABLE 15.4

Class limits	Observed frequency	$\dfrac{X - \bar{X}}{s}$	Probability	Expected frequency
0.705–1.105	1 ⎤ 7	−2.1898	0.0143	1.9448 ⎤ 5.2224
1.105–1.505	6 ⎦	−1.7656	0.0241	3.2776 ⎦
1.505–1.905	8	−1.3415	0.0517	7.0312
1.905–2.305	10	−0.9173	0.0887	12.0632
2.305–2.705	15	−0.4931	0.1333	18.1288
2.705–3.105	16	−0.0689	0.1600	21.7600
3.105–3.505	33	0.3552	0.1647	22.3992
3.505–3.905	22	0.7794	0.1455	19.7880
3.905–4.305	10	1.2036	0.1026	13.9536
4.305–4.705	7	1.6278	0.0635	8.6360
4.705–5.105	5 ⎤	2.0520	0.0314	4.2704 ⎤
5.105–5.505	2 ⎬ 8	2.4761	0.0136	1.8496 ⎬ 7.0176
5.505–5.905	1 ⎦	2.9003	0.0066	0.8976 ⎦
Total	136		1.0000	136

The χ^2 statistic is calculated in Table 15.5.

TABLE 15.5

Class	Observed frequency, o	Expected frequency, e	$o - e$	$(o - e)^2 / e$
1 and 2	7	5.2224	1.7776	0.6051
3	8	7.0312	0.9688	0.1335
4	10	12.0632	−2.0632	0.3529
5	15	18.1288	−3.1288	0.5400
6	16	21.7600	−5.7600	1.5247
7	33	22.3992	10.6008	5.0170
8	22	19.7880	2.2120	0.2473
9	10	13.9536	−3.9536	1.1202
10	7	8.6360	−1.6360	0.3099
11, 12 and 13	8	7.0176	0.9824	0.1375
Total	136	136		$\chi^2 = 9.9881$

Number of customers, X	Frequency, f
0	4
1	7
2	12
3	18
4	5
5	4
6 or more	0
Total	50

Solution: The estimate of the mean (μ) of the number of customers arriving at a service desk per five-minute period is:

$$\bar{X} = \frac{\sum_{i=0}^{6} f_i X_i}{\sum_{i=0}^{6} f_i}$$

$$= \frac{(0 \times 4) + (1 \times 7) + (2 \times 12) + (3 \times 18) + (4 \times 5) + (5 \times 4) + (6 \times 0)}{50}$$

$$= 2.50.$$

Hence the Poisson distribution fitted to the data is

$$f(X) = \frac{e^{-2.5}(2.5)^X}{X!}, \qquad X = 0, 1, 2, \ldots$$

The expected frequencies and the value of the χ^2 statistic are calculated in Table 15.6.

Since the parameter μ is estimated from the sample, the number of degrees of freedom is $\nu = 4 - 1 - 1 = 2$. Using a level of significance of 0.01, we find that the value of χ^2 for 2 degrees of freedom from Appendix Table 7 is $\chi^2_{0.99} = 9.21$. The calculated value $\chi^2 = 6.6418$ is less than the table value $\chi^2_{0.99} = 9.21$. Hence, our hypothesis that the arrival of customers follows a Poisson distribution cannot be rejected, and the Poisson fit is good.

TABLE 15.6

Number of customers, X	Observed frequency, o	Expected frequency, e		$o-e$	$(o-e)^2/e$
0	4 } 11	4.10 } 14.36		-3.36	0.7862
1	7	10.26			
2	12	12.83		-0.83	0.0537
3	18	10.69		7.31	4.9987
4	5 } 9	6.68 } 12.12		-3.12	0.8032
5	4	3.34			
6 or more	0	2.10			
Total	50	50			$\chi^2 = 6.6418$

15.3. CONTINGENCY TABLES: TEST OF INDEPENDENCE

In Section 15.1, we learned how the χ^2 statistic can be used to test the hypothesis of equal sales potential in each district; or to test the hypothesis of equal preference of colors in house appliances; or to test the hypothesis of equal demand for loaves of bread on all days of the week, etc. In other words, we have thus far classified an item or an individual into one of several classes according to one attribute. Such a classification is known as a one-way classification. There are many problems in which a businessman may wish to classify an item or an individual in several classes according to two attributes. Such a classification is known as a two-way classification. For example, an individual may be classified acording to sex and his preference of color for a house appliance, and we may wish to test the hypothesis concerning a relationship between sex and preference of color for a house appliance. An individual may be classified according to his income and the kind of house he buys, and we may wish to test the hypothesis concerning a relationship between an individual's income and the kind of house he buys. An item may be classified as defective or nondefective and a record made of the type of machine on which it is manufactured; we may then wish to test the hypothesis concerning a relationship between the number of defectives and the type of machine on which the items are manufactured, etc. The chi-square statistic defined in Section 15.1 can also be used to test the hypothesis of independence of two or more criteria of classification. To study problems of this type, we could take a random sample of size N and classify the items according to two criteria. The observed frequencies can then be presented in the form of a table, defined below, known as a contingency table.

Contingency Table

An $r \times c$ contingency table is an arrangement in which the data is classified into r classes $A_1, A_2, \ldots, A_r$ of attribute A, and c classes $B_1, B_2, \ldots, B_c$ of attribute B; one attribute is entered in rows and the other in columns.

Example 6 We make use of an $r \times c$ contingency table with r rows and c columns, where:

o_{ij} = observed frequency in the ith row and jth column;
$N_{i.}$ = total observed frequency in the ith row;
$N_{.j}$ = total observed frequency in the jth column;
N = the sample size.

$r \times c$ Contingency table

	Columns				
Rows	B_1	B_2	$\cdots$	B_c	Row totals
A_1	o_{11}	o_{12}	$\cdots$	o_{1c}	$N_{1.}$
A_2	o_{21}	o_{22}	$\cdots$	o_{2c}	$N_{2.}$
.	.	.	$\cdots$	.	.
.	.	.	$\cdots$	.	.
.	.	.	$\cdots$	.	.
A_r	o_{r1}	o_{r2}	$\cdots$	o_{rc}	$N_{r.}$
Column totals	$N_{.1}$	$N_{.2}$	$\cdots$	$N_{.c}$	N

Since expected frequencies in all cells except in the last row and the last column can be determined given the row total, column total, and the other cell frequencies, the number of degrees of freedom for a contingency table having r rows and c columns is

$$\nu = rc - (r-1) - (c-1) - 1 = (r-1)(c-1).$$

That is, $(r-1)(c-1)$ is the number of cells whose frequencies can be chosen arbitrarily.

Degrees of Freedom for an $r \times c$ Contingency Table

The number of degrees of freedom for an $r \times c$ contingency table is

$$\nu = (r-1)(c-1),$$

where r is the number of rows and c is the number of columns.

The χ^2 statistic can be used to test whether there is any relationship between the two attributes. That is, the null hypothesis is that the two attributes are independent:

$$\chi^2 = \sum_{i=1}^{r} \sum_{j=1}^{c} \frac{(o_{ij} - e_{ij})^2}{e_{ij}},$$

where o_{ij} = observed frequency in the ijth cell, and e_{ij} = expected frequency in the ijth cell.

To calculate e_{ij}, the expected frequency in the ijth cell, we must note that we are testing the null hypothesis that the two attributes are independent. Under this assumption, the probability that an individual (or item) belongs to the ijth cell is equal to the probability that an individual (or item) belongs in the ith row, multiplied by the probability that an individual (or item) belongs in the jth column. Now the expected frequency in the ijth cell may be calculated as:

$$e_{ij} = (\text{Sample size}) \begin{pmatrix} \text{Probability that an} \\ \text{individual or item} \\ \text{belongs to the } i\text{th row} \end{pmatrix} \begin{pmatrix} \text{Probability that an} \\ \text{individual or item} \\ \text{belongs to the } j\text{th column} \end{pmatrix}.$$

$$= (N) \left(\frac{N_{i.}}{N} \right) \left(\frac{N_{.j}}{N} \right) = \frac{N_{i.} \times N_{.j}}{N}$$

That is, the expected frequency of individuals or items in the ijth cell is equal to the total of the ith row multiplied by the total of the jth column divided by the total sample size.

As before, in order to test the hypothesis of independence, the calculated value of χ^2 is compared with the table value for given degrees of freedom at a certain level of significance. If the calculated value of χ^2 is greater than the table value, the hypothesis of independence is rejected.

Example 7 A real-estate company wants to know whether there is any relationship between a family's income and the type of house the family buys. The data collected during the last year are classified below in the form of a contingency table.

Type of house	Family income			Total
	Under $10,000	$10,000–$15,000	Over $15,000	
A	28	47	55	130
B	45	63	27	135
C	75	35	5	115
Total	148	145	87	380

Test the hypothesis that family income and house status are statistically independent.

Solution: On the hypothesis that family income and house status are statistically independent, the expected frequencies may be calculated as:

$$e_{11} = \frac{130 \times 148}{380} = 50.63, \quad e_{12} = \frac{130 \times 145}{380} = 49.61, \quad e_{13} = \frac{130 \times 87}{380} = 29.76,$$

$$e_{21} = \frac{135 \times 148}{380} = 52.58, \quad e_{22} = \frac{135 \times 145}{380} = 51.51, \quad e_{23} = \frac{135 \times 87}{380} = 30.91,$$

$$e_{31} = \frac{115 \times 148}{380} = 44.79, \quad e_{32} = \frac{115 \times 145}{380} = 43.88, \quad e_{33} = \frac{115 \times 87}{380} = 26.33.$$

The χ^2 statistic is calculated in Table 15.7.

TABLE 15.7

Classification	Observed frequency, o	Expected frequency, e	$o - e$	$(o-e)^2/e$
A and under $10,000	28	50.63	−22.63	10.1149
A and $10,000–15,000	47	49.61	−2.61	0.1373
A and over $15,000	55	29.76	25.24	21.4065
B and under $10,000	45	52.58	−7.58	1.0927
B and $10,000–15,000	63	51.51	11.49	2.5630
B and over $15,000	27	30.91	−3.91	0.4946
C and under $10,000	75	44.79	30.21	20.3761
C and $10,000–15,000	35	43.88	−8.88	1.7970
C and over $15,000	5	26.33	−21.33	17.2795
Total	380	380		$\chi^2 = 75.2616$

The number of degrees of freedom for a 3×3 contingency table is

$$\nu = (r-1)(c-1) = (3-1)(3-1) = 4.$$

Using a level of significance of 0.01, the value of χ^2 for 4 degrees of freedom from Appendix Table 7 is $\chi^2_{0.99} = 13.27$. The calculated value $\chi^2 = 75.2616$ is greater than the table value $\chi^2_{0.99} = 13.27$. Hence, the hypothesis of independence is rejected and the family income does have an effect on house status.

Example 8 A product is manufactured independently on three machines M_1, M_2, and M_3. A quality-control engineer wants to test whether the numbers of defective items manufactured on the three machines are the same. The numbers of defective and nondefective items, in random samples of 100 items drawn from each machine, are given below:

Type	Machine			Total
	M_1	M_2	M_3	
Defective (D)	3	5	8	16
Nondefective (ND)	97	95	92	284
Total	100	100	100	300

Test the hypothesis that the numbers of defective and nondefective items manufactured on three machines are the same.

Solution: On the hypothesis that the three machines manufacture the same number of defective and nondefective items, the expected frequencies may be calculated as

$$e_{11} = \frac{16 \times 100}{300} = 5.33, \qquad e_{12} = \frac{16 \times 100}{300} = 5.33, \qquad e_{13} = \frac{16 \times 100}{300} = 5.33,$$

$$e_{21} = \frac{284 \times 100}{300} = 94.67, \qquad e_{22} = \frac{284 \times 100}{300} = 94.67, \qquad e_{23} = \frac{284 \times 100}{300} = 94.67$$

The χ^2 statistic is calculated in Table 15.8.

TABLE 15.8

Classification	Observed frequency, o	Expected frequency, e	$o - e$	$(o-e)^2/e$
D and M_1	3	5.33	-2.33	1.0186
D and M_2	5	5.33	-0.33	0.0204
D and M_3	8	5.33	2.67	1.3375
ND and M_1	97	94.67	2.33	0.0573
ND and M_2	95	94.67	0.33	0.0012
ND and M_3	92	94.67	-2.67	0.0753
Totals	300	300		$\chi^2 = 2.5103$

The number of degrees of freedom for a 2×3 contingency table is

$$\nu = (r-1)(c-1) = (2-1)(3-1) = 2.$$

Using a level of significance of 0.05, the value of χ^2 for 2 degrees of freedom from Appendix Table 7 is $\chi^2_{0.95} = 5.99$. The calculated value $\chi^2 = 2.5103$ is less than the table value $\chi^2_{0.95} = 5.99$. Hence, the hypothesis that the three machines manufacture the same number of defective and nondefective items cannot be rejected.

Example 9 Two employees, A and B, of the market research department of a company draw samples by different sampling techniques to estimate the number of persons falling in three income groups, poor, middle-class, and well-to-do. The results are as follows:

Employee	Poor	Middle-class	Well-to-do	Total
A	66	40	14	120
B	35	28	7	70
Total	101	68	21	190

Test whether the sampling techniques are statistically independent from the income groups.

Solution: On the hypothesis that the sampling techniques of both the employees are the same, the expected frequencies may be calculated as:

$$e_{11} = \frac{120 \times 101}{190} = 63.8, \quad e_{12} = \frac{120 \times 68}{190} = 42.9, \quad e_{13} = \frac{120 \times 21}{190} = 13.3,$$

$$e_{21} = \frac{70 \times 101}{190} = 37.2, \quad e_{22} = \frac{70 \times 68}{190} = 25.1, \quad e_{23} = \frac{70 \times 21}{190} = 7.7.$$

The χ^2 statistic is calculated in Table 15.9.

TABLE 15.9.

Classification	Observed frequency, o	Expected frequency, e	$o - e$	$(o-e)^2/e$
A and poor	66	63.8	2.2	0.0759
A and middle-class	40	42.9	-2.9	0.1960
A and well-to-do	14	13.3	0.7	0.0368
B and poor	35	37.2	-2.2	0.1301
B and middle-class	28	25.1	2.9	0.3351
B and well-to-do	7	7.7	-0.7	0.0636
Totals	190	190		$\chi^2 = 0.8375$

The number of degrees of freedom for a 2×3 contingency table is

$$\nu = (r-1)(c-1) = (2-1)(3-1) = 2.$$

Using a 0.05 level of significance, the value of χ^2 for 2 degrees of freedom from Appendix Table 7 is $\chi^2_{0.95} = 5.99$. The calculated value $\chi^2 = 0.8375$ is less than the table value $\chi^2_{0.95} = 5.99$. Hence, our hypothesis that the sampling techniques of both the investigators are the same cannot be rejected. That is, the present data do not show that the techniques used by employees A and B are different.

SUMMARY

The chi-square statistic is a useful tool in determining whether a frequency with which a given event has occurred is significantly different from that which we expected. The sampling distribution of the chi-square statistic is very closely approximated by the chi-square distribution. In this chapter, we have discussed the use of the chi-square statistic in many business problems which deal with categorical, qualitative, or enumerative data. The chi-square statistic can be used to test the "goodness of fit" of observed sample distributions by means of expected probability distributions such as the binomial, Poisson, normal, etc. The chi-square statistic may also be used to test the hypothesis of independence of two or more criteria of classification.

Words to Remember

Chi-square statistic
Observed frequency
Expected frequency

Goodness-of-fit test
Contingency table
Test of independence

Symbols to Remember

			Degrees of freedom	
χ^2	Chi-square statistic	(A) Multinomial	$\nu = k-1$	
o	Observed frequency	(B) Normal (μ, σ given)	$\nu = k-1$	
e	Expected frequency	(μ, σ not)	$\nu = k-3$	
$N_i.$	Total observed frequency in the ith row	BINOMIAL (p given)	$\nu = k-1$	
$N_{.j}$	Total observed frequency in the jth column	not	$\nu = k-2$	
r	Number of rows			
c	Number of columns	CONTINGENCY TABLES	$\nu = (r-1) \times (c-1)$	
ν	Degrees of freedom			

EXERCISES

1. The weekly sales for five grocery stores are given in the following table. Test the hypothesis that the five stores have equal sales potential, using a level of significance of 0.05.

Store	Sales (Thousands of dollars)
A	121
B	432
C	94
D	252
E	77
Total	976

2. An automobile dealer of a local agency in city A claims that sales of station wagons, hardtops, and sedans occur with equal frequency in city A. He gives as evidence the following record of sales during the last six months.

Type	Number sold
Station wagons	30
Hardtops	51
Sedans	45
Total	126

Test his claim at the 5% level of significance.

3. A marketing research department of a colored television manufacturing company has chosen five cities. It is believed that each city has the same sales potential. The actual number of colored television sets sold by the company in each city, in a six-month period, is given in the following table. Test the hypothesis that the five cities have equal sales potential, using a level of significance of 0.05.

City	Number of sets sold
A	150
B	180
C	250
D	230
E	190
Total	1,000

4. The weekly receipts for ten weeks for the transit company of city A are given in Exercise 6 of Chapter 2. Test the hypothesis that the weekly receipts of this company are all equal, using a level of significance of 0.05.

5. Three salesmen of a company, Smith, Brown, and Jones, each has been assigned a sales district for house-to-house sales calls. It is believed that each district has the same sales potential. The number of items sold during the last 30 days are given in the following table.

Salesman	Number of items sold
Smith	125
Brown	180
Jones	85

Test whether the salesmen's abilities to sell differ significantly. Use $\alpha = 0.05$.

6. The ages of heads of households buying dishwashers in a particular city are classified as follows.

Age of person	Number of people
20–30	48
30–40	64
40–50	36
50–60	32
Total	180

Using a 0.05 level of significance, test whether age is a factor in buying a dishwasher.

7. Examine the closeness of fit of the normal distribution to the age distribution of 72 employees given in Exercise 1 of Chapter 2.

8. Examine the closeness of fit of the normal distribution to the measurements of diameters of 44 machine parts given in Exercise 2 of Chapter 2.

9. Examine the closeness of fit of the normal distribution to the frequency distribution of wages of 500 employees in a certain factory, given in Exercise 26 of Chapter 3.

10. Examine the closeness of fit of the normal distribution to the frequency distribution of life lengths of 200 light bulbs made by a certain manufacturer, given in Exercise 27 of Chapter 3.

11. Examine the closeness of fit of the normal distribution to the data of 300 pairs of shoes sold at a certain shoe store, as given in Exercise 9 of Chapter 2.

12. The numbers of telephone calls arriving during two-minute periods at a telephone switchboard of a company are given in the following table. Fit a Poisson distribution and test the goodness of fit.

Number of telephone calls, X	Frequency, f
0	3
1	5
2	15
3	20
4	7
5 or more	0
Total	50

13. A typist of a company commits the following numbers of mistakes per page in typing 100 pages. Fit a Poisson distribution and test the goodness of fit.

Mistakes per page, X	Number of pages, f
0	42
1	33
2	14
3	6
4	4
5	1
Total	100

14. At a certain airport, the airport authority believes that the number of airplane arrivals for each 30-minute period follows a Poisson distribution with $\mu = 3.0$. The numbers of actual airplane arrivals for fifty 30-minute periods, randomly selected, are given in the following table. Test the airport authority's belief that airplane arrivals are Poisson with mean $\mu = 3.0$.

Number of airplanes	Number of 30-minute periods
0	2
1	8
2	15
3	20
4	3
5	2
6 or more	0
Total	50

15. The numbers of machines broken down per week during a one-year period, in a manufacturing plant of a certain company, are given in the following table. Fit a Poisson distribution and test the goodness of fit.

Number of machines broken down	Number of weeks
0	22
1	15
2	8
3	5
4	2
5 or more	0
Total	52

16. The numbers of customers per 10-minute period at the wrapping counter of a large store are as follows.

Number of customers, X	Frequency, f
0	2
1	4
2	7
3	13
4	6
5	3
6	1

Fit a Poisson distribution to the data and test the goodness of fit.

17. A marketing research department of a toothpaste manufacturing company wishes to know the effect on sales of adding a mint flavor to its toothpaste. To test this idea, a sample of 100 families is each given a tube of mint-flavored toothpaste and a tube of plain toothpaste. Reactions are listed below.

Toothpaste	Liked	Did not like	Total
Plain	40	60	100
Mint-flavored	60	40	100

Examine the effect of mint flavor in the toothpaste on its sales. Use $\alpha = 0.01$.

18. A survey is conducted by an automobile manufacturer to study the relationship between the number of cars per household and family income. To test this relationship, the researchers interview a random sample of 1500 households; the results are given below.

Number of cars per household	Under $10,000	$10,000–15,000	Over $15,000	Total
0	100	25	3	128
1	575	235	137	947
2	32	163	230	425
Total	707	423	370	1500

Test the hypothesis that family income and the number of cars per household are statistically independent. Use $\alpha = 0.05$.

19. A firm produces two kinds of water heaters. A year's record shows that the following numbers of heaters were sold by two salesmen:

Type	Salesman A	B
Type I	240	80
Type II	180	100

Does this record support the claim that the salesman's ability to sell depends on the type he is selling? Use $\alpha = 0.05$.

20. The General Automobile Corporation interviews prospective candidates (generally those who have just finished their apprenticeship) for jobs as skilled automobile technicians. On the basis of each applicant's record, a panel of twenty people from personnel, general management, and the technical departments judge him to be either an excellent, good, fair, or poor prospect for the job. Those applicants judged as poor are not considered further. Those applicants judged excellent, good, or fair are hired and put on a three-month training course within the company. At the end of the training, each applicant is judged as superior, very good, good, or fair. Classification of 500 applicants hired by the company is given in the following table.

After-training evaluation	Before-training evaluation			Total
	Excellent	Good	Fair	
Superior	150	15	5	170
Very good	50	60	15	125
Good	40	45	30	115
Fair	10	30	50	90
Total	250	150	100	500

Examine the effect of the company's further training in increasing the applicant's rating. Use $\alpha = 0.01$.

21. It is believed by a company's manufacturing department that a training program will help reduce the number of defective articles produced. To test this, the numbers of defective articles produced by a team of trained and a team of untrained personnel are counted and recorded as shown below.

Type	Trained personnel	Untrained personnel	Total
Defectives	10	11	21
Nondefectives	90	64	154
Total	100	75	175

Test, at the 0.05 level of significance, that the training program has no effect.

THE ANALYSIS
OF VARIANCE 16

16.1. INTRODUCTION

In Chapter 11, we discussed the problem of testing the difference between the means of two populations, using an independent sample from each population. In particular, we considered testing H_0: $\mu_1 = \mu_2$ against H_A: $\mu_1 \neq \mu_2$. There are often situations where the businessman wishes to test the equality of the means of several populations. He might, for instance, want to compare the output of several machines, or compare the daily sales of several stores.

The technique used to compare the means of several populations is called the *analysis of variance*; basically, it compares the variation between population means with the variation within populations. Such analysis is based on the fact that, even if there were no difference between the population means, not all of the observations would be the same because of the random nature of the measurements. Hence, we think of the variation within a population, which, after all, has only one mean, as the *inherent variation* in the experiment, and try to estimate this variance. We also obtain a similar estimate for the variability between the sample means, allowing for the fact that the variance between means is smaller than that between the original observations. If the variabilities within populations and between means are of the same order of magnitude, we conclude that there is no difference between the population means. If the variation between means is significantly larger than the within-population variation, then we conclude that there is a difference between the population means.

The analysis of variance is performed by considering, let us say, K populations with unknown means $\mu_1, \mu_2, \ldots, \mu_K$, and from each taking a random sample. The null hypothesis is H_0: $\mu_1 = \mu_2 = \ldots = \mu_K$; and the alternative hypothesis is H_A: At least one of the equalities is violated.

> **Analysis of Variance**
>
> Analysis of variance is a technique for comparing the unknown means $\mu_1, \mu_2, \ldots, \mu_K$ of K populations, by taking an independent random sample from each population and comparing the variation *between* populations with the variation *within* populations.

Example 1 In order to compare the abilities of job applicants from four different universities, 4, 3, 4, and 5 applicants were chosen, respectively, from universities A, B, C, and D, and given a written test graded on the basis of 100. The results were as follows.

	University				
	A	B	C	D	
	67	71	88	83	
	52	94	66	95	
	83	90	69	91	
	74		73	99	
				92	
Total	276	255	296	460	1287
Mean	69	85	74	92	80.4375

In due course, we shall test whether the job applicants from the different universities do equally well on the written test.

In Example 1, where the populations are distinguished on the basis of a single criterion (e.g., different universities), we speak of a *one-factor analysis of variance*.

16.2. ONE-FACTOR ANALYSIS OF VARIANCE

We have explained that, in order to compare the effects of a single criterion (or a single factor, as it usually is called), we use the one-factor analysis of variance. Before proceeding, it will be necessary to define some notation for use in this section.

Notation

    K = the number of populations;

    μ_i = the (unknown) mean of population i;

    N_i = the number of observations in sample i;

    N = the total number of observations;

    X_{ij} = the jth observation in the ith sample;

    T_i = the sum of observations in sample i;

    $\overline{X}_i$ = the mean of observations in sample i;

    T = the total of all observations;

    $\overline{X}$ = the mean of all observations.

Pictorially, this may be shown as follows:

	\multicolumn{4}{c}{Sample from population}			
	1	2	$\cdots$	K
	X_{11}	X_{21}	$\cdots$	X_{K1}
	X_{12}	X_{22}	$\cdots$	X_{K2}
	.	.	$\cdots$	.
	.	.	$\cdots$	.
	.	.	$\cdots$	.
	X_{1N_1}	X_{2N_2}	$\cdots$	X_{KN_K}
Totals	T_1	T_2	$\cdots$	T_K T
Means	$\overline{X}_1$	$\overline{X}_2$	$\cdots$	$\overline{X}_K$ $\overline{X}$

Example 2 In Example 1, we have:

$$N_1 = 4, \quad N_2 = 3, \quad N_3 = 4, \quad N_4 = 5, \quad N = 16;$$

$$X_{11} = 67, \quad X_{12} = 52, \quad \text{etc.,} \quad X_{21} = 71, \quad \text{etc.,} \quad \ldots, \quad X_{45} = 92;$$

$$T_1 = 276, \quad T_2 = 255, \quad T_3 = 296, \quad T_4 = 460, \quad T = 1287;$$

$$\overline{X}_1 = 69, \quad \overline{X}_2 = 85, \quad \overline{X}_3 = 74, \quad \overline{X}_4 = 92, \quad \overline{X} = 80.4375.$$

We stated above that the analysis-of-variance tests the means of several populations by comparing the variation between populations with the variation within populations. Thus, our first step is to develop expressions for the "within" and the "between" variation.

The total variation, the variation between means, and the variation within samples, are measured by three quantities, called *sums of squares*, whose formulas are as follows.

Total Sum of Squares

The total variability of the observations is measured by an expression called the *total sum of squares*, denoted by SST, whose formula is as follows:

$$\text{SST} = \sum_{i=1}^{K} \sum_{j=1}^{N_i} \left(X_{ij} - \overline{X} \right)^2.$$

Between Sum of Squares

The variability between means is measured by an expression called the *between sum of squares*, denoted by SSB, whose formula is as follows:

$$\text{SSB} = \sum_{i=1}^{K} \sum_{j=1}^{N_i} \left(\overline{X}_i - \overline{X} \right)^2.$$

Within Sum of Squares

The variability within samples is measured by an expression called the *within sum of squares*, denoted by SSW, whose formula is as follows:

$$\text{SSW} = \sum_{i=1}^{K} \sum_{j=1}^{N_i} \left(X_{ij} - \overline{X}_i \right)^2.$$

When SSW is divided by $N - K$ (called the *within degrees of freedom*), the result is called the *within mean squares*, and denoted by MSW. Similarly, the *between mean square* denoted by MSB, is SSB divided by its degrees of freedom $K - 1$. MSW and MSB, respectively, measure the within-population variance and the between-population variance.

Short-cut or computational formulas for SSW and SSB are given below:

$$\text{SSW} = \sum_{i=1}^{K} \sum_{j=1}^{N_i} X_{ij}^2 - \sum_{i=1}^{K} \left(\frac{T_i^2}{N_i} \right),$$

$$\text{SSB} = \sum_{i=1}^{K} \left(\frac{T_i^2}{N_i} \right) - \frac{T^2}{N}.$$

Example 3 For the data of Example 1,

$$T_1^2 / N_1 = (276)^2 / 4 = 19,044,$$

$$T_2^2 / N_2 = (255)^2 / 3 = 21,675,$$

$$T_3^2 / N_3 = (296)^2 / 4 = 21,904,$$

$$T_4^2 / N_4 = (460)^2 / 5 = 42,320,$$

$$T^2 / N = (1287)^2 / 16 = 103,523.0625;$$

$$\text{SSB} = 19,044 + 21,675 + 21,904 + 42,320 - 103,523.0625$$

$$= 1,419.9375,$$

$$\text{SSW} = 67^2 + 52^2 + \cdots + 92^2 - 104,943 = 106,185 - 104,943$$

$$= 1,242.$$

If we wish to test H_0: $\mu_1 = \mu_2 = \ldots = \mu_K$, and if the null hypothesis is true, then we would expect $\bar{X}_1, \bar{X}_2, \ldots, \bar{X}_K$ to be similar to each other and not very different from $\bar{X}$. This, in turn, would imply that MSB would be "small." By "small" we mean not very different from MSW. If, on the other hand, H_0 is not true, and at least one value of μ is different from the others, then MSB will be large in comparison with MSW. A fundamental question is "What do we mean by small and large?"

It can be shown that, if the K populations are normal, with common variance, and if the observations are mutually independent, then the ratio MSB/MSW follows what is called an F-distribution with $K-1$ degrees of freedom in the numerator and $N-K$ degrees of freedom in the denominator.

The F-distribution has an asymmetrical shape. A typical distribution is shown in Fig. 16.1, although the actual shape depends on two quantities, the degrees of freedom in the numerator and the degrees of freedom in the denominator.

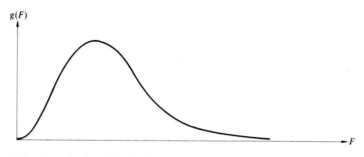

16.1 A typical F-distribution.

When dealing with the t and χ^2-distributions, we saw that, for instance, 5 percent of the area lay to the right of $t_{0.95}$ and $\chi^2_{0.95}$, respectively. Similarly, for the F-distribution, we define a value called $F_{0.95}$, where 5 percent of the area under the distribution lies to the right of $F_{0.95}$, as in Fig. 16.2. This distribution is tabulated in Appendix Table 9(a), with ν_1 and ν_2 denoting the numerator and denominator degrees of freedom, respectively.

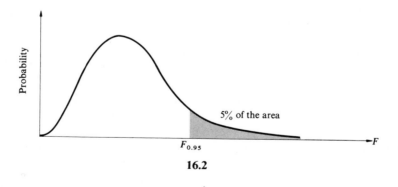

16.2

If one is interested in testing the null hypothesis with, say, a 5 percent level of significance, then it can be shown that the hypothesis should be rejected if the ratio MSB/MSW is greater than $F_{0.95}$, where $F_{0.95}$ is read from the table with the appropriate degrees of freedom. The results may be summarized in an analysis-of-variance table similar to Table 16.1.

TABLE 16.1. Analysis-of-variance table

Source of variance	Sum of squares	Degrees of freedom	Mean squares	F-ratio
Between means	$SSB = \sum_{i=1}^{K}\left(\dfrac{T_i^2}{N_i}\right) - \dfrac{T^2}{N}$	$K-1$	$MSB = SSB/(K-1)$	MSB/MSW
Within means	$SSW = \sum_{i=1}^{K}\sum_{j=1}^{N_i} X_{ij}^2 - \sum_{i=1}^{K}\left(\dfrac{T_i^2}{N_i}\right)$	$N-K$	$MSW = SSW/(N-K)$	
Total	$SST = \sum_{i=1}^{K}\sum_{j=1}^{N_i}\left(X_{ij} - \bar{X}\right)^2$	$N-1$		

Example 4 The analysis-of-variance table for the data of Example 1 is as follows:

Source of variation	Sum of squares	Degrees of freedom	Mean squares	F-ratio
Between universities	1419.9375	3	473.3125	4.57
Within universities	1242.0000	12	103.5	
Total	2661.9375	15		

If we wish to test that the job applicants from the four universities have the same abilities (as measured by the given written test) with a significance level of 0.05, then (from Appendix Table 9(a)) $F_{0.95}$, with 3 degrees of freedom in the numerator and 12 degrees of freedom in the denominator, is 3.49. Thus, we reject the null hypothesis and conclude that a job applicant's ability depends on the university from which he comes.

Example 5 The PQR Company wishes to compare the selling abilities of four salesmen, A, B, C, and D, for a particular three-week period. The total sales for three one-week periods for the four salesmen are given below in dollars.

	Salesmen				
	A	B	C	D	
Week 1	500	570	510	360	
Week 2	450	650	650	460	
Week 3	610	700	550	500	
Total	1560	1920	1710	1320	6510
Mean	520	640	570	440	542.5

Solution: It is easy to show that

$$\sum_{i=1}^{4} \sum_{j=1}^{3} X_{ij}^2 = 3,638,300,$$

$$\sum_{i=1}^{4} \left(\frac{T_i^2}{N_i} \right) = 3,595,500,$$

$$\frac{T^2}{N} = 3,531,675;$$

$$SSW = 3,638,300 - 3,595,500 = 42,800,$$

$$SSB = 3,595,500 - 3,531,675 = 63,825.$$

The analysis-of-variance table is as shown here.

Source of variation	Sum of squares	Degrees of freedom	Mean squares	*F*-ratio
Between salesmen	63,825	3	21,275	3.98
Within salesmen	42,800	8	5,350	
Total	106,625	11		

The value of $F_{0.95}$ with 3 and 8 degrees of freedom is 4.07. Since the F-ratio is smaller than 4.07, we cannot conclude that there are any differences between the selling abilities of the salesmen.

16.3. TWO-FACTOR ANALYSIS OF VARIANCE

The analysis of variance can also be used to study the effects of two or more factors on a variable of interest in many business problems. For example, we might be interested in learning about the effects of different advertising techniques, at different levels of expenditure, on the sales of a product; or we might wish to investigate the ability of our salesmen in different sales districts; or we might investigate the effects of different display methods on different shopping days of the week. The analysis-of-variance techniques with two factors, called the two-factor analysis of variance, can be applied to problems of the above type to measure the effects of each level of one factor (or variable) at each level of another factor (or variable). More specifically, by a judicious choice of the levels of factors, we can measure the effect of each factor separately; and further, we can also measure the effects of *combinations* of the two factors acting together.

In this text, we shall consider experiments consisting of only two factors, which for convenience we shall call rows and columns. We shall also assume that the rows and columns do not interact; that is, that the effect of rows will be the same for each column and that the effect of columns will be the same for each row. An example of the two-factor analysis of variance with three rows and four columns is given below.

Example 6 The PQR Company wishes to study the differences in the selling abilities of its four salesmen, A, B, C, and D, as well as the differences in its three sales districts, S_1, S_2, and S_3, all of the same size. The weekly sales in dollars for the four salesmen are given below:

District	Salesmen			
	A	B	C	D
S_1	550	450	700	500
S_2	300	350	550	400
S_3	350	550	400	600

The above example may easily be extended to r rows and c columns. Let X_{ij} denote an observation in the ith row and jth column. Because the notation must be expanded slightly from that in Section 16.2, we will use the following notation for the two-factor analysis of variance.

Notation

r = the number of rows,

c = the number of columns,

N = the total number of observations,

X_{ij} = the observation in the ith row and jth column,

$\overline{X}_{i.}$ = the mean of observations in row i,

$\overline{X}_{.j}$ = the mean of observations in column j,

$\overline{X}$ = the mean of all observations,

$T_{i.}$ = the total (or sum) of observations in row i,

$T_{.j}$ = the total (or sum) of observations in column j,

T = the total (or sum) of all observations.

Pictorially, the above notation may be shown as follows:

			Columns					
Rows	1	2	$\cdots$	j	$\cdots$	c	Total	Mean
1	X_{11}	X_{12}	$\cdots$	X_{1j}	$\cdots$	X_{1c}	$T_{1.}$	$\overline{X}_{1.}$
2	X_{21}	X_{22}	$\cdots$	X_{2j}	$\cdots$	X_{2c}	$T_{2.}$	$\overline{X}_{2.}$
.	.	.		.		.	.	.
.	.	.		.		.	.	.
i	X_{i1}	X_{i2}	$\cdots$	X_{ij}	$\cdots$	X_{ic}	$T_{i.}$	$\overline{X}_{i.}$
.	.	.		.		.	.	.
.	.	.		.		.	.	.
r	X_{r1}	X_{r2}	$\cdots$	X_{rj}	$\cdots$	X_{rc}	$T_{r.}$	$\overline{X}_{r.}$
Total	$T_{.1}$	$T_{.2}$	$\cdots$	$T_{.j}$	$\cdots$	$T_{.c}$	T	
Mean	$\overline{X}_{.1}$	$\overline{X}_{.2}$	$\cdots$	$\overline{X}_{.j}$	$\cdots$	$\overline{X}_{.c}$		$\overline{X}$

Example 7 Using the above notation, the data in Example 6 may be displayed as follows:

District	Salesmen				Total	Mean
	A	B	C	D		
S_1	550	450	700	500	$T_1. = 2200$	$\bar{X}_1. = 550$
S_2	300	350	550	400	$T_2. = 1600$	$\bar{X}_2. = 400$
S_3	350	550	400	600	$T_3. = 1900$	$\bar{X}_3. = 475$
Total	$T_{.1} = 1200$	$T_{.2} = 1350$	$T_{.3} = 1650$	$T_{.4} = 1500$	$T = 5700$	
Mean	$\bar{X}_{.1} = 400$	$\bar{X}_{.2} = 450$	$\bar{X}_{.3} = 550$	$\bar{X}_{.4} = 500$		$\bar{X} = 475$

In the two-factor analysis of variance, the total variability in the data is divided into three components; one component is attributable to one factor, a second component is due to the second factor, while the third component is due to chance (or *experimental error*). Each of these components, called *sums of squares*, when divided by the appropriate number of degrees of freedom, is an estimate of the variance of that component. The procedure for calculating the sum of squares for the two-factor analysis of variance is quite similar to the one used in the one-factor analysis of variance.

Total Sum of Squares

The variability of the observations is measured by an expression called the *total sum of squares*, denoted by SST, whose formula is given as follows:

$$SST = \sum_{i=1}^{r} \sum_{j=1}^{c} \left(X_{ij} - \bar{X} \right)^2$$

$$= \sum_{i=1}^{r} \sum_{j=1}^{c} X_{ij}^2 - \frac{T^2}{rc} .$$

The total sum of squares SST, when divided by its degrees of freedom $rc - 1$, gives an estimate of the total variance.

Sum of Squares for Rows

The variability between row means is measured by an expression called the *row sum of squares*, denoted by SSR, and is given by the following formula:

$$\text{SSR} = c \sum_{i=1}^{r} \left(\overline{X}_{i.} - \overline{X} \right)^2$$

$$= \frac{\sum_{i=1}^{r} T_{i.}^2}{c} - \frac{T^2}{rc} \,.$$

The sum of squares for rows, SSR, when divided by its degrees of freedom, $(r-1)$, is called *row mean square*, and is denoted by MSR.

Sum of Squares for Columns

The variability between column means is measured by an expression called the *column sum of squares*, denoted by SSC, and is given by the following formula:

$$\text{SSC} = r \sum_{j=1}^{c} \left(\overline{X}_{.j} - \overline{X} \right)^2$$

$$= \frac{\sum_{j=1}^{c} T_{.j}^2}{r} - \frac{T^2}{rc} \,.$$

The sum of squares for columns, SSC, when divided by its degrees of freedom, $(c-1)$, is called the *column mean square*, and is denoted by MSC.

Sum of Squares for Error

The sum of squares for error is measured by an expression called the *error sum of squares*, denoted by SSE, and is given by the following formula:

$$\text{SSE} = \text{SST} - \text{SSR} - \text{SSC}.$$

The sum of squares for error, SSE, when divided by its degrees of freedom, $(r-1)(c-1)$, is called the *error mean square*, and is denoted by MSE.

Example 8 Calculate SST, SSR, SSC, and SSE for the data of Example 7.

Solution

$$SST = \sum_{i=1}^{3} \sum_{j=1}^{4} X_{ij}^2 - \frac{(5700)^2}{(3)(4)}$$

$$= 2,865,000 - 2,707,500 = 157,500;$$

$$SSR = \frac{1}{4} \sum_{i=1}^{3} (T_{i.}^2) - \frac{(5700)^2}{(3)(4)}$$

$$= \frac{11,010,000}{4} - \frac{32,490,000}{12}$$

$$= 2,752,500 - 2,707,500 = 45,000;$$

$$SSC = \frac{1}{3} \sum_{j=1}^{4} (T_{.j}^2) - \frac{(5700)^2}{(3)(4)}$$

$$= \frac{8,235,000}{3} - 2,707,500$$

$$= 2,745,000 - 2,707,500 = 37,500;$$

$$SSE = SST - SSR - SSC$$

$$= 157,500 - 45,000 - 37,500 = 75,000.$$

If we wish to test the null hypothesis that the row effects are all equal, we compute the ratio MSR/MSE, called the *F*-ratio, and compare it with the table value of *F*. The degrees of freedom associated with this ratio are $(r-1)$ and $(r-1)(c-1)$. If the computed value of the *F*-ratio is greater than the table value, we reject the null hypothesis that the row effects are equal.

Similarly, if we wish to test the null hypothesis that the column effects are equal, we compute the ratio MSC/MSE, called the *F*-ratio, and compare it with the table value of *F*. The degrees of freedom associated with this ratio are $(c-1)$ and $(r-1)(c-1)$. If the computed value of the *F*-ratio is greater than the table value, we reject the null hypothesis that the column effects are equal.

The above discussion on the two-factor analysis of variance may be summarized in an analysis-of-variance table similar to the table for the one-factor analysis of variance.

Source of variation	Sum of squares	Degrees of freedom	Mean squares	F-ratio
Between-row means	SSR	$r-1$	$MSR = \dfrac{SSR}{r-1}$	$\dfrac{MSR}{MSE}$
Between-column means	SSC	$c-1$	$MSC = \dfrac{SSC}{c-1}$	$\dfrac{MSC}{MSE}$
Error	SSE	$(r-1)(c-1)$	$MSE = \dfrac{SSE}{(r-1)(c-1)}$	
Total	SST	$rc-1$		

Example 9 The analysis-of-variance table for the data of Example 6 is as follows:

Source of variation	Sum of squares	Degrees of freedom	Mean squares	F-ratio
Between districts	45,000	2	22,500	1.80
Between salesmen	37,500	3	12,500	1.00
Error	75,000	6	12,500	
Total	157,500	11		

Now to test the null hypothesis that there are no differences between the sales districts, we compare the F-ratio $MSR/MSE = 1.80$ with the table value of F. The degrees of freedom associated with this ratio are 2 and 6. From Appendix Table 9(a), we find that the value of $F_{0.95}$ with 2 and 6 degrees of freedom is 5.14. Since the F-ratio 1.80 is smaller than 5.14, we cannot conclude that there are any differences between the sales districts. Similarly, to test the null hypothesis that there are no differences between the selling abilities of the salesmen, we compare the F-ratio $MSC/MSE = 1.00$ with the table value of F. The degrees of freedom associated with this ratio are 3 and 6. From Appendix Table 9(a), we find that the value of $F_{0.95}$ with 3 and 6 degrees of freedom is

4.76. Since the F-ratio 1.00 is smaller than 4.76, we cannot conclude that there are any differences between the selling abilities of the salesmen.

SUMMARY

There are often situations where the businessman wishes to test the equality of several population means. He might, for instance, want to compare the output of several machines, or the daily sales of several stores. The technique used to compare the means of several populations is called the analysis of variance; basically, it compares the variation *between* population means with the variation *within* populations. In order to compare the effects of a single criterion or a single factor, we use the one-factor analysis of variance. The analysis-of-variance technique can also be used to study the effects of two or more factors on a variable of interest in many business problems. In the two-factor analysis of variance, we have assumed that the rows and columns do not interact, that is, that the effect of the rows is the same for each column, and that the effect of the columns is the same for each row. The methods discussed in this chapter give a very brief introduction to the analysis of variance.

Words to Remember

Analysis of variance	Within sum of squares
One-factor analysis of variance	Sum of squares for rows
Two-factor analysis of variance	Sum of squares for columns
Total sum of squares	Sum of squares for error
Between sum of squares	F-ratio

Symbols to Remember

K	Number of populations
μ_i	Mean of the ith population
N_i	Number of observations in sample i
N	Total number of observations
T_i	Sum of observations for sample i
$\overline{X}_i$	Mean of observations of sample i
T	Sum of all observations
$\overline{X}$	Mean of all observations
SST	Total sum of squares
SSB	Between sum of squares
SSW	Within sum of squares

MSW	Within mean square
MSB	Between mean square
r	Number of rows
c	Number of columns
SSR	Row sum of squares
SSC	Column sum of squares
SSE	Error sum of squares

EXERCISES

1. To study the variation of the retail prices of a certain commodity in four different cities, five shops are chosen at random. The prices observed (in cents) are given in the following table:

	Cities		
A	B	C	D
45	50	47	43
43	47	42	40
40	42	44	48
44	48	49	42
42	45	41	44

   Construct the analysis-of-variance table. Using a 0.05 level of significance, test whether there is a significant difference in the retail prices in four different cities.

2. The weekly earnings (in dollars) of five workers, drawn at random from three different industries, are given as follows:

	Industry	
A	B	C
120	150	270
150	130	200
200	220	190
170	300	160
250	170	220

Construct the analysis-of-variance table. Using a 0.05 level of significance, test whether or not there is a significant difference in the earnings of workers in the three industries.

3. Traditionally, applicants for positions at the RST Company came from five universities. In order to compare the abilities of the students, four from each university were chosen and given a written test, marked on the basis of 100. The results are shown below:

University				
A	B	C	D	E
87	73	58	65	85
73	70	63	68	71
70	89	60	58	77
84	78	59	72	81

Use an analysis of variance to compare the abilities of the students from the different universities, with a 1-percent level of significance.

4. A company wishes to compare the effects of three advertising brochures. To do this, nine sales districts are chosen and divided into three groups; each group is to receive one of the types of brochure. The response on which the brochures will be graded is the time, in weeks, until sales *increase significantly*, as judged by the sales manager. The responses are given below:

	Brochure		
	1	2	3
Number of	7.2	2.3	6.3
weeks for	5.1	3.7	5.8
effectiveness	6.7	4.1	7.0

Is there any difference in the effect of the three brochures? Use a 5-percent level of significance.

5. A particular type of machine part is supplied by three manufacturers. To compare the diameters of the parts, samples are taken from each manufacturer's product and measured, giving the following results:

	Manufacturer		
	1	2	3
Diameter	1.3	1.4	1.7
measure-	1.5	1.8	1.7
ments	1.2	1.7	1.9
	1.7	1.6	2.0
		1.8	1.8
			2.1

Perform the analysis of variance, and decide whether there is any difference in the diameters of the parts supplied by the three manufacturers. Use a 5-percent level of significance.

6. An experiment was conducted to investigate the differences in mean time to assemble three different mechanical devices, D_1, D_2, and D_3. Each device was assembled four times, and the times (in minutes) are recorded below:

	Device		
	D_1	D_2	D_3
Time to	34	27	38
assemble	27	22	42
	31	29	37
	30	20	35

Using the analysis of variance, decide whether there is any significant difference between the three devices.

7. Cable wire is manufactured by three firms. A test is performed to determine whether the mean breaking strength is the same. Five pieces of wire are picked up at random from each firm, and put under tension. The results are given in the following table:

	Firm		
	A	B	C
Breaking	10	12	5
strength	8	7	8
	12	15	9
	11	14	7
	9	10	10

Construct the analysis-of-variance table. Using a 0.05 level of significance, test whether there is a significant difference between the mean breaking strength of cables manufactured by the three firms.

8. The marketing research department of a toothpaste manufacturing company wishes to study the effect of mint flavor in the toothpaste on its sales in four different cities of equal size. The sales for one week (in thousands of dollars) are given below:

	City			
	A	B	C	D
Plain	10	12	15	13
Mint-flavored	8	14	12	9

Perform a two-factor analysis of variance, testing for city differences and whether or not the mint flavor makes a difference. Use $\alpha = 0.05$.

9. The following table gives the number of units of production per day turned out by four different employees, using four different types of machines:

Employee	Type of Machine			
	M_1	M_2	M_3	M_4
E_1	40	36	45	30
E_2	38	42	50	41
E_3	36	30	48	35
E_4	46	47	52	44

a) Test the hypothesis that the mean production is the same for the four machines.

b) Test the hypothesis that the four employees do not differ with respect to mean productivity.

10. The following table gives the mileage per gallon of two brands of gasoline on three different makes of automobiles:

Gasoline	Automobile		
	1	2	3
A	22	18	19
B	19	17	20

Perform a two-factor analysis of variance, using the level of significance $\alpha = 0.05$.

STATISTICAL
QUALITY
CONTROL 17

17.1. INTRODUCTION

In this chapter, we shall be concerned with industrial manufacturing processes with repetitive operations, where the quality of the product is of interest. We wish to control the quality of the product in order to reduce loss due to wastage, and to ensure a uniformly high quality in the product, which will satisfy the customers' requirements.

Consider the following example of a repetitive industrial process.

Example 1 A machine for packaging dry cement is supposed to put one hundred pounds of cement into each sack. Because of such factors as humidity, flow of cement to and from the machine, quality of cement, operator effects, and many others, there will be a variation in weight from sack to sack. If this variation is small, the process (of filling the sacks) will be called "in control." If the sacks are too light or too heavy, the process will be called "out of control." By some method, usually but not always based on economy, there will be limits set, such that if the sacks' weights are within the limits, the process will be in control. For instance, the sack-filling process might be "in control" if the sack weights are between 99 pounds and 101 pounds.

In the past, in order to check the sack weights, it would have been necessary to weigh each sack and adjust the weight of those which fell outside the "control limits" of 99 and 101 pounds. (One spoke of "*inspecting* the control into the product.") Now clearly, in Example 1, it would be impractical to inspect every sack, especially if the number of sacks was large. During the 1930's, a body of knowledge called *statistical quality control* (SQC) was developed by Dr. W. A. Shewhart. The purpose of SQC was to *control* the quality, instead of *inspecting* it into the product. There are two main aspects of SQC, both of which will be discussed here. They are:

341

a) *Process control*, using control charts, whereby the proportion of unsatisfactory product is kept from becoming excessive; and

b) *Acceptance sampling*, whereby no lots or batches should be accepted if they contain an excessive number of unsatisfactory items.

It is important, in any manufacturing process, to understand that *any two things* produced by ostensibly the same process will not be *exactly the same*. (The differences may be too small to detect visually, but nevertheless these differences do exist.)

We shall find it convenient to make a distinction between the types of measurements which we can obtain. Measurements of continuous data, such as tensile strength, pressure, temperature, horsepower, flash point, percentage of impurity, amounts of additive, etc., are called *variables*. Variable measurements on ostensibly the same parts are usually assumed to have a normal distribution. Measurements which classify an object simply as good or bad, acceptable or unacceptable, live or dead, up or down, etc., are called *attributes*. The similarity between attributes and binomial random variables is clear enough.

17.2. SOURCES OF VARIATION IN A MANUFACTURING PROCESS

We shall be concerned with two sources of variation in any industrial process:

a) Random (or chance) variation; and

b) Nonrandom or assignable variation.

In Example 1, some sources of random variation might be the humidity, the quality of the cement, and the internal machine friction. In general, random variation cannot be eliminated but *must be taken into account when product specifications are set*. Some examples of nonrandom variation in Example 1 are operator clumsiness, the sack openings being too small, and the machine being improperly set. These factors can be detected and eliminated to ensure a more uniform product.

We are concerned with sources of variation because nonrandom variation can affect the distribution of observations. In particular, it can change the mean and variance of a particular characteristic. For instance, in Example 1, if the sack openings are too small, then not enough cement will go into each sack, and the sack will be too light. The small sack openings will have the effect of shifting the distribution mean *downwards*. Again, in Example 1, a clumsy machine operator may make some sacks too light and some too heavy, thus increasing the *variance* of the distribution.

Example 2 Consider that we are interested in measuring the diameters of bolts made by a particular process. We have enough past data to know that the diameters follow a normal distribution, with mean 0.25 inches and standard deviation 0.01 inches. For simplicity, suppose that the specifications are to accept the bolts with diameters between 0.22 and 0.28 inches. This corresponds to the mean *plus or minus* three standard deviation, so that, from Chapter 8, we know that 99.74 percent of all bolts will be acceptable. Now suppose that something happens to the machine to cause the population mean to increase to 0.26. These two situations are shown in Fig. 17.1.

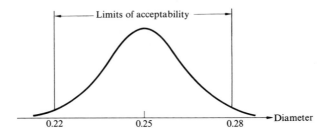

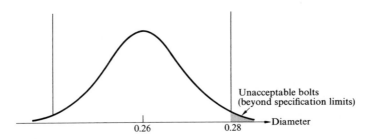

17.1   The relationship between a shift in the mean and the percentage defective.

We see that a shift in the mean from 0.25 to 0.26 has increased the percentage of bolts which are unacceptable. The student should be able to show that the percentage acceptable decreases from 99.74% to 97.73%.

Example 3 Reconsider Example 2, but suppose that something happens to the process which leaves the mean the same but *increases* the standard deviation to 0.015. This is shown in Fig. 17.2.

We see that an increase in the standard deviation from 0.01 to 0.015 causes an increase in the percentage of bolts which are unacceptable. In fact, the student should be able to show that the percentage acceptable decreases from 99.74% to 95.46%.

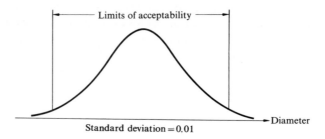

Standard deviation = 0.01

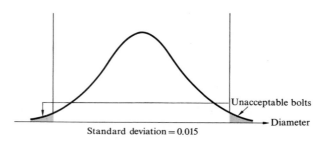

Standard deviation = 0.015

17.2 The relationship between a change in the standard deviation and the percentage defective.

Examples 2 and 3 illustrate that, if the specifications are based on the random variation of the process, and if a shift of the population mean *or* variance takes place, then this may cause a decrease in the number of acceptable parts.

The purpose of process quality control, then, is to detect and identify nonrandom sources of variation, so that they can be corrected.

17.3. CONTROL CHARTS FOR VARIABLES

The primary objective of process control is to keep the process within the control limits; and the statistical tool used for doing this is called a *control chart*. A control chart is a graphical method for showing the results of a sequence of samples. Figure 17.3 is a typical control chart.

When dealing with a variable (continuous data), one is generally interested in controlling the mean and standard deviation of the process; thus, there are two main process control charts for variables. The first is called an $\bar{X}$-chart, for looking at the sample *means*; the second is called an R-chart, for looking at the sample *ranges*. We may remember, from Chapter 4, that range can be used as a measure of dispersion. Although it is not a good estimator in general, it is adequate for small samples and has found acceptance in quality-control circles. Further, it is simple to calculate and therefore appropriate for use by untrained personnel. We now give some examples to illustrate the use of the $\overline{X}$- and R-charts.

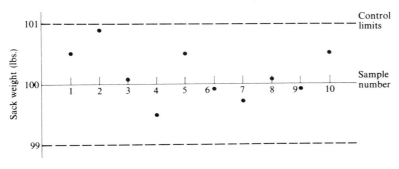

17.3   A typical control chart.

Example 4   A new peach-canning machine has been set up to fill cans with peaches so that the gross weight of each can is two pounds. In fact, each can will be of a slightly different weight due to factors such as lack of uniformity in the can thickness, internal machine variation, the type of peach being canned, the size of peaches, etc. In order to keep the quality (in this case, gross weight) under control, a sample of five cans is randomly chosen every hour, and weighed, and the mean and range of the sample is calculated. The first ten samples yielded the results shown in Table 17.1.

TABLE 17.1. Gross weights of cans of peaches

Sample	Weights (in pounds)					$\overline{X}$	Range (R)
1	1.93,	1.97,	2.04,	2.00,	1.96	1.98	0.11
2	2.02,	1.90,	1.92,	2.05,	2.01	1.98	0.15
3	2.10,	2.14,	2.07,	1.93,	2.09	2.07	0.21
4	1.87,	1.87,	2.07,	1.94,	2.01	1.95	0.20
5	2.07,	2.00,	2.04,	2.03,	1.95	2.02	0.12
6	2.04,	1.97,	2.11,	2.02,	1.95	2.02	0.16
7	2.13,	2.08,	1.92,	1.98,	2.14	2.05	0.22
8	1.90,	1.98,	1.96,	2.07,	2.04	1.99	0.17
9	1.98,	2.02,	2.10,	2.04,	1.96	2.02	0.14
10	1.96,	2.07,	2.04,	1.96,	1.97	2.00	0.11
Total						20.08	1.59

The center line of the control chart, denoted by $\overline{\overline{X}}$, is the mean of the sample means; thus

$$\overline{\overline{X}} = \frac{1.98 + 1.98 + \ldots + 2.00}{10} = 2.01.$$

The mean of the sample ranges, denoted by $\overline{R}$, is

$$\overline{R} = \frac{0.11 + 0.15 + \ldots + 0.11}{10} = 0.16.$$

We shall next need the value of A, from Appendix Table 10; for $N = 5$, this is 0.577. The upper control limit (UCL) and the lower control limit (LCL) are calculated as follows:

$$\text{UCL} = \overline{\overline{X}} + A\overline{R} = 2.01 + (0.577)(0.16) = 2.10;$$

$$\text{LCL} = \overline{\overline{X}} - A\overline{R} = 2.01 - (0.577)(0.16) = 1.92.$$

Values of $\overline{\overline{X}}$, the UCL, the LCL, and the sample means are shown in Fig. 17.4.

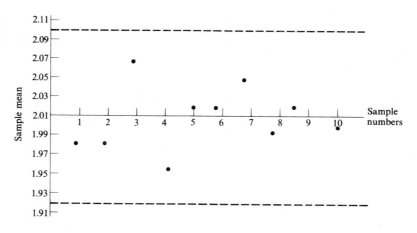

17.4 Control chart for means from Table 17.1.

The control chart for Example 4 was based on a past record of ten samples, for illustrative purposes. In actual practice, it is advisable to use a greater number of samples, in order that the process mean and standard deviation may have time to stabilize. A suggested figure is 25 or 30 samples.

Now the UCL and LCL represent the mean *plus and minus* three standard deviations (3-sigma limits), so that the probability that a sample mean falls between the UCL and the LCL is approximately 0.9974. The probability that a point falls *outside* these limits is 0.0026. Some indications that a process is out of control are as follows:

a) A single point is outside the 3-sigma limits.

   The probability of a point being outside the 3-sigma limits, if the process *is* in control, is 0.0026.

b) *Several* points are *near* the control limits, especially *successive* points.

   Some quality control experts draw 2-sigma and 3-sigma limits on a control chart. Several successive points beyond the 2-sigma limits indicate that the process should be carefully watched.

c) A run of several points is on *one side* of the center-line.

   The probability that a single point is (say) above the center line is $\frac{1}{2}$. The probability that six successive points would lie above the center line is only 0.016.

d) A trend in the points exists.

We may summarize the construction, use, and interpretation of $\overline{X}$-control charts as follows.

Construction, Use, and Interpretation of $\overline{X}$-Charts

1. From an adequate number of past samples, each of size N, calculate the mean of the sample means, $\overline{\overline{X}}$, and the mean of the sample ranges, $\overline{R}$.
2. From Appendix Table 10, read A for the appropriate value of N.
3. Calculate:

$$\text{UCL} = \overline{\overline{X}} + A\overline{R} \qquad \text{and} \qquad \text{LCL} = \overline{\overline{X}} - A\overline{R}.$$

4. Draw the control chart showing $\overline{\overline{X}}$, UCL, and LCL.
5. Take samples of size N at random times throughout the process run, and plot the sample means on the control chart.
6. The process will be considered out of control if (a) *one* point is beyond the 3-sigma limits, (b) *several* points are between the 2-sigma and 3-sigma limits, (c) several points are on *one side* of the center line, (d) there is a *trend* in the points. Otherwise, the process will be considered to be *in control*.

The R-chart shows variations in the sample ranges, and is therefore used to detect changes in the dispersion of the distribution. It looks much the same as the $\overline{X}$-chart except that the center line, UCL, and LCL are based on the *ranges* instead of $\overline{X}$. The center line is $\overline{R}$, the mean of the sample ranges. We next need the values of B_1 and B_2 from Appendix Table 11. The UCL and LCL are calculated below:

$$\text{UCL} = B_2 \overline{R} \quad \text{and} \quad \text{LCL} = B_1 \overline{R}.$$

Example 5 Reconsider Example 4, concerning the gross weights of cans of peaches. The ranges of the first ten samples of size five are shown in Table 17.1. Construct a control chart for the ranges, and plot the ten ranges from Table 17.1.

Solution: From Example 4, the value of $\overline{R}$ is:

$$\overline{R} = \frac{0.11 + 0.15 + \dots + 0.11}{10} = 0.16.$$

The values of B_1 and B_2 are found in Appendix Table 11 to be 0 and 2.115. Thus,

$$\text{UCL} = (2.115)(0.16) = 0.34,$$

$$\text{LCL} = (0)(0.16) = 0.00.$$

Values of $\overline{R}$, UCL, LCL, and the sample ranges are shown in Fig. 17.5.

Note that the UCL and the LCL are not the same distance from the center line. This is because the sample range cannot be less than zero. The UCL and LCL represent the mean range *plus and minus* three standard deviations.

The interpretation of the R-chart is similar to that for the $\overline{X}$-chart. The process can be considered *out of control* with respect to *dispersion* if (a) a *single point* is outside the 3-sigma limits, (b) *several* points are near the control limits, (c) several points lie on *one side* of the center line, or (d) there is a *trend* in the points.

17.4. CONTROL CHARTS FOR ATTRIBUTES

When each measurement is classified as success or failure, good or bad, marked or unmarked, defective or nondefective, etc., we say that we are dealing with *attributes*. Control charts dealing with attributes are called *fraction defective charts* (or *P*-charts).

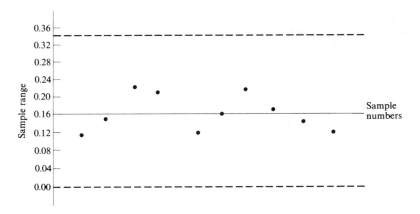

17.5 Control chart for ranges from Table 17.1.

Example 6 Consider setting up a *P*-chart for a process which makes aluminum window frames. The frames are made by a continuous foundry process, and each day one hundred frames are taken and inspected. Suppose that the results of the past thirty days' inspections are shown in Table 17.2.

TABLE 17.2. Results of 30 days inspections of aluminum window frames

Day	Number rejected	Day	Number rejected	Day	Number rejected
1	16	11	15	21	6
2	15	12	19	22	8
3	15	13	16	23	11
4	19	14	16	24	12
5	31	15	30	25	20
6	21	16	25	26	28
7	20	17	30	27	20
8	31	18	17	28	21
9	24	19	31	29	45
10	16	20	22	30	47

Set up a control chart for the fraction defective, and plot the thirty points of Table 17.2 on it.

Solution: Let us call the size of each sample *N*. Thus $N = 100$. We first wish to calculate *p*, the overall fraction defective. This is the total number of defectives divided by the total number of parts inspected, or

$$p = \frac{16 + 15 + \ldots + 45 + 47}{100 + 100 + \ldots + 100} = \frac{647}{3000} = 0.216.$$

The upper control limit (UCL) and lower control limit (LCL) are calculated below:

$$\mathrm{UCL} = p + 3\sqrt{\frac{p(1-p)}{N}} = 0.216 + 3\sqrt{\frac{(0.216)(0.784)}{100}} = 0.334$$

$$\mathrm{LCL} = p - 3\sqrt{\frac{p(1-p)}{N}} = 0.216 - 3\sqrt{\frac{(0.216)(0.784)}{100}} = 0.098.$$

Values of p, UCL, LCL, and the sample fraction defectives are shown in Fig. 17.6.

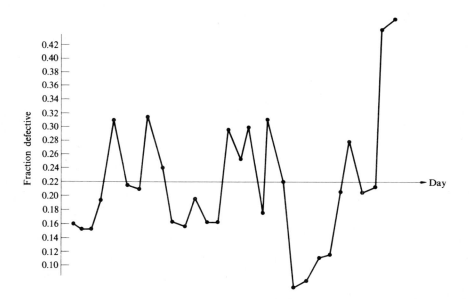

17.6 Control chart for fraction defectives from Table 17.2.

The UCL and LCL represent the mean fraction defective *plus and minus* three standard deviations, so that, if the process is in control, approximately 99.74 percent of all sample fraction defectives should be included within those limits. It will be noticed, in Fig. 17.6, that two points lie outside the control limits, indicating that the process was out of control on days 29 and 30. In trying to find the reason for the process going out of control, suppose we discover that on those two days, the shop foreman was breaking in some new employees. This might lead us to recommend that new employees be given a training period before being allowed to operate the window-frame assembly process.

Since the two high figures for days 29 and 30 are due to assignable causes, we should then eliminate (discard) them and *recalculate* the control limits.

Example 7 Using the data of Table 17.2, recalculate the control limits for the fraction defective if the data for days 29 and 30 are eliminated (since they are due to assignable causes).

Solution

$$p = \frac{16+15+\ldots+20+21}{100+\ldots+100} = \frac{555}{2800} = 0.198;$$

$$UCL = 0.198 + 3\sqrt{\frac{(0.198)(0.802)}{100}} = 0.318,$$

$$LCL = 0.198 - 3\sqrt{\frac{(0.198)(0.802)}{100}} = 0.078.$$

The student should make a graph, showing p, UCL, LCL, and the 28 points, and see that all points lie within the control limits.

The interpretation of the fraction defective chart is the same as for the variable chart. The same four indications that the process is out of control should be kept in mind. We summarize the results as follows:

Construction, Use and Interpretation of P-Charts

1. From an adequate number of past samples, each of size N, calculate the overall fraction defective p.

2. Calculate:

$$\text{UCL} = p + 3\sqrt{\frac{p(1-p)}{N}} \ ,$$

$$\text{LCL} = p - 3\sqrt{\frac{p(1-p)}{N}} \ .$$

3. Draw the control chart showing p, UCL, and LCL.

4. Plot the fraction defectives for samples of size N, taken randomly throughout the process run.

5. The process will be considered out of control if (a) *one* point is beyond the 3-sigma limit, (b) *several* points are between the 2-sigma and 3-sigma limits, (c) *several* points are on *one side* of the center line, (d) there is a *trend* in the points. Otherwise, the process will be considered *in control*.

17.5. ACCEPTANCE SAMPLING

We have discussed one of the two main aspects of quality control, namely, *process control*, using control charts. The second major aspect is called *acceptance sampling*, whereby a group of items, usually called a lot, is inspected and either accepted as meeting specifications or rejected because it does *not* meet specifications. Lots consist of a number of items, such as bearings, bushings, window frames, washers, or countless other such objects. One way of inspecting a lot is to perform 100-percent sampling; that is to say, each item in the lot is examined. (Such a sampling procedure does not guarantee that the lot will be perfect, since errors and mistakes will be made due to boredom and fatigue, which are unmeasurable and unpredictable.) *Acceptance sampling* involves choosing a random sample from the lot, inspecting it, and accepting or rejecting the lot based on the number of defectives found in the sample. Such a procedure is less expensive than 100-percent sampling and, in many cases, provides a better outgoing-product quality.

We should realize at the outset that the purpose of acceptance sampling is to determine a course of action, not to control quality. One simply *accepts* or *rejects* the sampled lot. Sampling plans are expressed in terms of the sample size n, and the acceptance number c. A sample of size n is taken, and if the number of defectives is less than or equal to c, we accept the lot. If the number of defectives is greater than c, we reject the lot.

Example 8 For the plan, $n = 100$, $c = 3$, we accept the lot if the number of defectives is 0, 1, 2, or 3. If the number of defectives is 4 or greater, we reject the lot.

Sampling plans provide a particular probability of accepting lots of given quality. If the lot has a fraction p defective, then there is a probability P of accepting the lot. Depending on the values of p and P that one is interested in, one chooses an appropriate plan. The relationship between p and P is expressed by an operating characteristic curve (or OC curve). Each sampling plan has its own OC curve. In Fig. 17.7, we show the OC curve for the plan $n = 100$, $c = 2$.

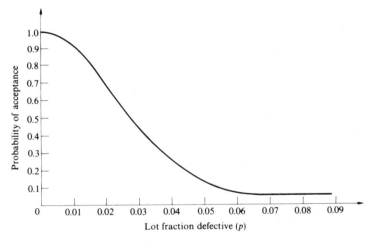

17.7   OC Curve for the plan $n = 100$, $c = 2$.

We see from Fig. 17.7 that if a lot contains 2 percent defectives, there is a probability of 0.67 of accepting the lot. If the lot is 5-percent defective, there is a probability of 0.12 of accepting it. If we interpret probability as a long-run relative frequency, then we are saying that, if we repeatedly test lots which are (say) 5-percent defective, we will accept 12% of them in using the plan $n = 100$, $c = 2$.

A third parameter (besides n and c) to describe a sampling plan is N, the lot size. If n is large compared to N, then clearly the lot size will have an effect on the discriminating ability of the plan. If N is large compared to n, then the lot sizes may all be regarded as infinite and need not be considered in the calculations. For simplicity, here we shall assume that N, the lot size, is large compared to n, the sample size. This assumption is generally valid if N is at least ten times as large as n.

We may next ask, "How does varying n and c affect the discriminating ability of a plan?" We answer this as follows:

a) If we hold the acceptance number c constant and increase n, the probability of accepting a "bad" lot decreases. Conversely, decreasing n for a constant acceptance number c will increase the probability of accepting a "bad" lot. One can show, for instance, that for $c = 2$, a lot that is 2% defective will have probabilities of acceptance equal to 0.37, 0.67, and 0.92, respectively, where $n = 150$, 100, and 50.

b) For a fixed sample size, if the acceptance number is reduced, the probability of accepting "bad" lots will decrease. For a fixed sample size, increasing the acceptance number will increase the probability of accepting "bad" lots. Figure 17.8 shows P versus p for $n = 100$, where $c = 0$, 1, and 2.

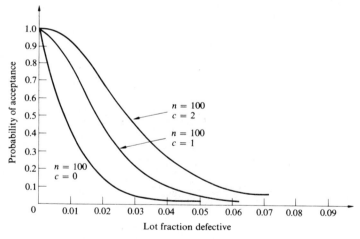

17.8   Comparing acceptance numbers for fixed n.

Note from Fig. 17.8 that the probabilities of acceptance for a lot that is 2-percent defective are 0.40, 0.68, and 0.86, for acceptance numbers 1, 2, and 3, respectively.

We sometimes wish to design a sampling plan so that the OC curve passes through two prechosen points (one point is not enough to select a unique plan). Thus, a user may require the probability of rejecting a good lot to be α, while the probability of accepting a poor lot can be β. We state this information succinctly below:

1. Define a "good" lot. This is known as the *acceptable quality level* (AQL).
2. State the probability of rejecting a good lot. This is denoted by α.
3. Define a "bad" lot. This is known as the *rejectable quality level* (RQL) or the *lot tolerance percent defective* (LTPD).
4. State the probability of accepting a poor lot. This is denoted by β.

In order to find n and c, we will consider the case where:

1. AQL $= 0.04$ (4%);
2. $\alpha = 0.10$;
3. LTPD $= 0.15$ (15%);
4. $\beta = 0.10$.

Turn to the chart of Appendix Table 12, where probability is marked on the vertical scale and np is marked on the horizontal scale. Proceed with the following steps:

1. Read $\beta = 0.10$ on the vertical scale.
2. Move horizontally until $c = 1$. Read the corresponding value of np. In our case, $np = 3.9$.
3. Read $1 - \alpha = 0.90$ on the vertical scale.
4. Move horizontally until $c = 1$. Read the corresponding value of np. In our case, $np = 0.55$.
5. Calculate (from step 2) $n = 3.9/\text{LTPD} = 3.9/0.15 = 26$.
6. Calculate (from steps 4 and 5) $p = 0.55/26 = 0.021$.
7. Compare 0.021 with the desired AQL, which in this case is 0.04. In this case, the computed value is less than the desired value.
8. Repeat this process for other values of c. The student should show that the plan which comes closest to our requirements specifies $n = 43$, $c = 3$.

Example 9 Reconsider Example 6, where window frames are being produced and inspected. The management feels that a day's run will be considered good if there are 3 percent or fewer defectives, while a day's run will be considered bad if there are more than 10 percent defectives. They are willing to accept a probability of 0.05 of rejecting a good day's run and a probability of 0.12 of accepting a bad day's run. Select a sampling plan which will meet their specifications.

Solution

$$\text{AQL} = 0.03; \qquad \alpha = 0.05; \qquad \text{LTPD} = 0.10; \qquad \beta = 0.12.$$

Using Appendix Table 12, we note:

1. Read $\beta = 0.12$ on the vertical scale.
2. For $c = 1$, $np = 3.6$.
3. Read $1 - \alpha = 0.95$ on the vertical scale.
4. For $c = 1$, $np = 0.36$.
5. $n = 3.6/\text{LTPD} = 3.6/0.10 = 36$.
6. $p = 0.36/n = 0.36/36 = 0.010$.

This process is repeated and summarized in Table 17.3.

TABLE 17.3 Selecting a sampling plan for example 9, given that:

$$\text{LTPD} = 0.10, \qquad \text{AQL} = 0.03, \qquad \alpha = 0.05, \qquad \beta = 0.12$$

(1)	(2)	(3)	(4)	(5)
Acceptance number c	np (for $\beta = 0.12$) from chart	np (for $1 - \alpha = 0.95$) from chart	$n = (2) \div$ LTPD	$p = (3) \div (4)$
1	3.6	0.36	36	0.010
2	5.1	0.83	51	0.016
3	6.3	1.40	63	0.022
4	7.7	1.90	77	0.025
5	8.9	2.65	89	0.030

Since, for $c = 5$, the value of p in column (5) comes closest to the desired value of AQL, we use the plan $n = 89$, $c = 5$.

17.6. TYPES OF ACCEPTANCE SAMPLING AND USE OF TABLES

The plans discussed in Section 17.5 are all *single-sample plans*. That is, a single sample is drawn from a lot and, based on the number of defectives in the sample, the lot is either accepted or rejected. Such plans are designated by two numbers, n and c. A *double sampling plan* is designated by five numbers n_1, n_2, c_1, c_2, and c_3, where c_1 is less than c_2, and c_2 is less than or equal to c_3.

The double sampling procedure is as follows: A sample of size n_1 is taken from a lot. If it contains c_1 or fewer defectives, the lot is accepted. If it contains c_2 or more defectives, the lot is rejected. If the number of defectives is greater than c_1 but less than c_2, a second sample of size n_2 is taken from the lot. If the combined number of defectives is less than or equal to c_3, the lot is accepted; otherwise, it is rejected.

A double sampling plan has certain advantages over single sampling. The total amount of inspection is usually less in double sampling because very good lots are accepted quickly and very bad lots are rejected quickly. Also, double sampling gives a doubtful lot a second chance. Like single-sample plans, double-sample plans have an associated OC curve. In order to select a double sampling plan, one needs the following information:

1. The definition of a good lot (AQL);
2. The probability of rejecting a good lot, α;
3. The definition of a bad lot (LTPD),
4. The probability of accepting a bad lot, β;
5. The relationship between n_1 and n_2.

Note that there are *five* requirements for a double sampling plan against four for a single sampling plan. Point 5, concerning n_1 and n_2, is usually expressed by saying that n_2 is some constant multiple of n_1. The actual derivation of such sampling plans is more complicated than for single sampling plans, and we shall not discuss their derivation here. The interested student is referred to *Quality Control and Industrial Statistics* by A. J. Duncan (Irwin, Inc., Homewood, Ill.: 1965).

An extension of double sampling leads to triple sampling, quadruple sampling, and finally to *sequential sampling*, where items are taken one at a time (randomly chosen, of course) from a lot. After each item is examined and counted as defective or nondefective, there are three possible decisions which can be made; they are:

1. Accept the lot;
2. Reject the lot;
3. Examine another item before deciding.

Sequential sampling is mentioned briefly by A. J. Duncan, whose book is referred to above, and more fully in the book *Sequential Analysis* by A. Wald (Wiley, Inc., New York: 1947).

There are several tables that one can use for sampling purposes. We mention just one table here, namely, the Military Standards entitled MIL-STD-105D. These tables are a joint effort of an American-British-Canadian group to derive common standards for the three countries. The MIL-STD-105D tables cover single-, double-, and multiple-sampling plans, which are specified by their AQL. For fraction defective plans, the AQL's run from 0.10% to 10%. There are three levels of inspection called Level I, Level II, and Level III. Level II is called normal inspection; Level III is used when more discrimination is required, and Level I is used when less discrimination is required. The choice of level depends on the product being inspected. Inexpensive items might be inspected at Level I while complex items might be inspected at Level III. The choice of what level to use is usually a management decision, made before sampling starts and not changed afterwards except for very special reasons. The use of the MIL-STD-105D is as follows:

1. Select the code letter in Appendix Table 13, corresponding to the size of the lot.
2. Select the sampling level (I, II, or III).
3. Using Appendix Table 13 (single-sampling) or Appendix Table 14 (double-sampling), read the sample size and acceptance number for the required AQL.

Example 10 We wish to select a normal sampling plan (Level II) for a lot of size 600. From Appendix Table 13, the sample size code letter is J. If our desired AQL is 1.5%, then we see that we should take a sample of size 80, accept the lot if there are three or fewer defectives, and reject it with more than four defectives.

SUMMARY

This chapter is concerned with *statistical quality control* (SQC). The two main aspects of SQC are (1) *process control*, using control charts, and (2) *acceptance sampling*. The purpose of *process control* is to ensure that the quality of items produced by a repetitive process satisfies specified levels of quality. This is done for (a) variables and (b) attributes, by means of control charts. The two main charts for variables are the $\bar{X}$-charts (for checking on the *mean* of a process) and the R-charts (for checking on the *dispersion* of a process). For attributes, we discussed the P-chart for checking on the *fraction defective*. For all the charts discussed, we essentially specify *mean value plus and minus three-sigma limits*, and mention four criteria whereby the process should be considered out of control. An out-of-control process is generally checked to see *why* it is out of control, and the process is readjusted if necessary.

Acceptance sampling involves sampling from a lot and accepting or rejecting the lot on the basis of the number of defectives in the sample. For every sampling plan, there is an associated OC curve, and one discriminates between plans on the basis of these curves. In order to select a plan, one may specify (1) the AQL, (2) the probability of rejecting good lots, (3) the LTPD, and (4) the probability of accepting bad lots. We show how to calculate the sample size and the acceptance number for single-sample plans. For double and sequential sampling plans, references are given. Finally, the use of an important set of tables, MIL-STD-105D, is discussed and an example is given.

Words to Remember

Statistical quality control	UCL
Random variation	LCL
Assignable variation	Acceptance sampling
Control charts	OC curve
Control limits	Fraction defective
Variable	Acceptance number
Attribute	AQL
$\overline{X}$-chart	LTPD
R-chart	Double sampling
P-chart	Sequential sampling

Symbols to Remember

$\overline{\overline{X}}$	The mean of sample means
$\overline{R}$	The mean of sample ranges
$\overline{\overline{X}} + A\overline{R}$	Upper control limit for chart for means
$\overline{\overline{X}} - A\overline{R}$	Lower control limit for chart for means
$B_2\overline{R}$	Upper control limit for chart for ranges
$B_1\overline{R}$	Lower control limit for chart for ranges
P	Overall fraction defective
n	Sample size (acceptance sampling)
c	Acceptance number
N	Lot size
α	Probability of rejecting a good lot
β	Probability of accepting a poor lot
AQL	Acceptance quality level
LTPD	Lot tolerance percent defective

EXERCISES

1. Consider a process with a normal distribution, with mean 100 and variance 25. The product is acceptable if it lies within the mean plus or minus 2.5 standard deviations. How much of the product will be acceptable if the process remains in control?

2. In Exercise 1, how much of the product will be acceptable if the process mean changes to 97 and the variance remains constant?

3. In Exercise 1, how much of the product will be acceptable if the process variance increases to 30 while the mean stays the same (that is, $\overline{X} = 100$)?

4. In order to keep control of the inside diameter of a manufactured machine part, a sample of 10 parts is taken every hour. The means and ranges of the last fifteen samples are as follows.

Mean $\bar{X}$	Range R	Mean $\bar{X}$	Range R	Mean $\bar{X}$	Range R
0.52	0.03	0.50	0.11	0.57	0.10
0.53	0.10	0.53	0.08	0.52	0.07
0.57	0.09	0.51	0.05	0.52	0.12
0.49	0.08	0.48	0.09	0.51	0.11
0.48	0.12	0.54	0.03	0.49	0.09

Set up a control chart for means and for ranges, plotting center line, UCL, and LCL. Then plot the points on the chart.

5. A bottle-making process is set to produce bottles weighing 32 ounces. Set up an $\bar{X}$-chart and an R-chart for a process having the means and ranges of 10 samples as follows, where each sample consists of the weights of eight bottles.

Sample	1	2	3	4	5	6	7	8	9	10
Mean	33	30	31	33	34	30	29	32	31	33
Range	3	5	3	4	2	3	5	4	4	5

6. An $\bar{X}$-chart and an R-chart are being designed. The quality-control expert takes samples of size 10, and computes the mean and range for each. The mean of the means is $\bar{\bar{X}} = 20$ and the mean of the ranges is $\bar{R} = 3$. Set up the $\bar{X}$-chart and the R-chart showing center line, UCL, and LCL.

7. What is the probability that a point lies between the two-sigma and the three-sigma limits?

8. Show that the probability of having six consecutive points above the center line is 0.016.

9. What is the probability that 6 out of 10 points lie above the center line? [*Hint:* Use the binomial formula with $p = \frac{1}{2}$.]

10. In order to test batches of a particular type of bearing, a sample of 150 is chosen each day and tested for size, using a gauge. Each piece is either accepted or rejected.

The figures for the past 20 batches are given below.

Batch	No. of items rejected	Batch	No. of items rejected
1	13	11	17
2	12	12	16
3	14	13	12
4	15	14	15
5	12	15	13
6	16	16	12
7	15	17	14
8	18	18	18
9	12	19	16
10	13	20	12

Set up a P-chart for the fraction defective.

11. Select an acceptance sampling plan for which AQL=0.03, $\alpha=0.05$, LTPD=0.20, and $\beta=0.15$.

12. Select an acceptance sampling plan for which AQL=0.06, $\alpha=0.10$, LTPD=0.15, and $\beta=0.10$.

13. Suppose, in Example 9, that β is reduced from 0.12 to 0.08. Select a sampling plan which satisfies the new requirements.

14. A manufacturer wants to inspect a 4000-piece lot using MIL-STD-105D, Level II. The specified AQL levels are:

   a) 0.65;

   b) 2.50;

   c) 10.00.

   What are the sample sizes and acceptance numbers under single- and double-sampling plans?

15. The following table gives the mean life length and the range (in hours) of 12 samples, of five electric bulbs each, drawn every hour from a manufacturing process.

   a) Draw the $\bar{X}$- and R-charts.

b) Is the process under control?

Sample number	Mean ($\bar{X}$)	Range (R)
1	632.18	125
2	685.50	208
3	573.37	136
4	595.15	163
5	612.00	184
6	580.32	179
7	590.16	200
8	520.20	215
9	705.50	236
10	610.63	198
11	550.00	154
12	620.15	148

16. A company manufactures electric-saw motors. To test the quality of motors, a quality-control engineer inspects the motors in daily samples of 50. The following table gives the number of defective motors inspected for ten working days.

Day	Number of defectives
1	4
2	3
3	2
4	5
5	4
6	3
7	2
8	1
9	4
10	3

Construct a P-chart for the fraction defective.

NONPARAMETRIC
STATISTICS 18

18.1. INTRODUCTION

Many of the statistical techniques that can be used in analyzing business data discussed so far have an underlying assumption that the population from which the sample is drawn has a normal distribution. In some cases this assumption is not too restrictive, since some techniques, most notably those based on the t-test, are robust; that is, they do not depend too critically on this assumption. Other techniques may be made applicable by an appropriate transformation of the data. When the samples are large, the assumption of normality is not too critical since in many cases the test statistic will be approximately normal due to the application of the central-limit theorem.

It is of practical interest, however, to discuss other techniques which do *not* depend on the underlying assumption of population normality, even for small samples. We shall call such techniques *nonparametric techniques*, although some authors refer to them as ranking methods (or ordering techniques). These techniques are widely used in marketing, organizational behavior, and similar areas such as personnel management, where ranked data are the best that one can expect. Of the large number of available nonparametric tests which could be discussed, we have chosen six for inclusion in this text. These six cover many of the principal problems that a business statistician is likely to meet. The names of the tests (usually named after their inventors) and the areas in which they can be applied are as follows:

Test	Area of application
Runs test	Test of randomness of data
Wilcoxon's test for paired data	Comparison of the means of two paired samples
Wilcoxon's rank-sum test	Comparison of the means of two independent samples
Kruskal–Wallis's test	Comparison of the means of several independent samples
Friedman's test	Comparison of the means of several related samples
Spearman's rank-correlation coefficient	Rank-correlation of two samples

18.2. TEST OF RANDOMNESS: THE RUNS TEST

The basic assumption in all statistical methods for estimation and hypothesis-testing is that the samples obtained are *random*. The researcher may wish to test the randomness of a given sample; that is, he may wish to test whether or not the sample observations are independent and identically distributed. For example, an economist may like to know the randomness of stock-market prices of a certain stock over a period of time; a manufacturer may like to know the quality of an article produced by machine and checked several times during the day. There are several tests available to judge the randomness of a given sample. We present in this section one such test, known as the *runs test*, based on the number of runs.

To illustrate the runs test, consider a manufacturing process where the quality-control supervisor observes the sequence of defective items, D, and nondefective items, N, produced by the machine. A sequence of defective items, D followed and preceded by nondefectives, N, is called a *run*. The number of defectives D and nondefectives N in a run is called the *length* of a run. For example, consider the following sequence of defective items D and nondefectives N:

$$\text{NNN} \quad \text{DD} \quad \text{NNNNN} \quad \text{DD} \quad \text{NNN.}$$

We find five runs of varying lengths. The first run consists of three nondefective items, the second run consists of two defective items, the third run consists of five nondefective items, the fourth run consists of two defective items, and the fifth run consists of three nondefective items. In general, we shall be concerned with observations which can be classified into one of two categories, say success (S) or failure (F), high (H) or low (L), true (T) or false (F), and so forth.

> **Run**
>
> A sequence of elements of one kind, followed and preceded by elements of another kind, is called a *run*. The number of elements of a particular kind in a run is called the *length* of a run.

Example 1 Consider the diameter in inches of nine bolts made by a machine:

$$1.3, \quad 1.6, \quad 1.5, \quad 1.2, \quad 1.4, \quad 1.1, \quad 1.7, \quad 1.8, \quad 1.9.$$

Let each measurement be designated as low, L, if smaller than the median, and as high, H, if larger than the median. Since the median of the above set of measurements is 1.5, the designation of L's and H's in the above set of measurements is:

$$\text{L} \quad \text{H} \quad \text{LLL} \quad \text{HHH}.$$

Let N_1 be the number of L's, N_2 be the number of H's and r be the total number of runs. From the above sequence we find

$$N_1 = 4, \quad N_2 = 4, \quad \text{and} \quad r = 4.$$

The sampling distribution of the total number of runs, r, can be obtained by mathematical methods, but is beyond the scope of this book. The distribution of r depends on N_1 and N_2, where N_1 is the number of L's smaller than the median or some other convenient reference point, and N_2 is the number of H's larger than the median or some other reference point. Appendix Table 15 gives critical values of r for a 5-percent two-sided test. The smaller integer is the left-tail critical value and the larger integer is the right-tail critical value. The runs test, for testing the hypothesis of randomness, consists in counting the number of runs. Too few or too many runs in a sequence would indicate nonrandomness. To test the null hypothesis of randomness at $\alpha = 0.05$, we count the number of runs in a sample and compare it with table values given in Appendix Table 15. If the number of runs, r, in a given sample falls between the two values given in Appendix Table 15, we accept the null hypothesis of randomness.

Example 2 Suppose a salesman of the Ajax Company contacts 15 customers on a certain day. He makes a note of sales (S) and no sales (N) in sequence as follows:

$$\text{NNNN} \quad \text{SS} \quad \text{N} \quad \text{SSSS} \quad \text{NN} \quad \text{SS}.$$

Does he have reason to suspect nonrandomness in this sequence?

Solution: The null hypothesis to be tested is that the N's and S's occur at random; and the alternative hypothesis is that the order of N's and S's deviates from randomness. Let N_1 be the number of sales and N_2 be the number of nonsales, and r be the total number of runs. From the above sequence, we find

$$N_1 = 8, \qquad N_2 = 7, \qquad \text{and} \qquad r = 6.$$

From Appendix Table 15, we find 4 and 13 as the critical values for $N_1 = 8$, $N_2 = 7$, at $\alpha = 0.05$. Since $r = 6$ falls in the acceptance region 4 and 13, we accept the null hypothesis that the sales have occurred in random order.

18.3. WILCOXON'S TEST FOR PAIRED DATA

The Wilcoxon test for paired samples is a test of the null hypothesis that two population medians are the same, against the alternative that they are unequal. We assume that there are N pairs of observations, and that each pair is independent of every other pair. It is not necessary that the two observations in a pair be independent. For populations 1 and 2, with medians Md(1) and Md(2), we shall let M be the number of pairs of observations and N the number of nonzero differences D_i. In order to test:

$$H_0: \mathrm{Md}(1) = \mathrm{Md}(2),$$

$$H_A: \mathrm{Md}(1) \neq \mathrm{Md}(2),$$

we rank the absolute value of the differences $|D_i|$ as follows. Replace the smallest value of $|D_i|$ by 1, the next smallest by 2, and so on up to the largest, which is replaced by N. (Tied differences are given by the *average* of the ranks that would have been given if no ties were present. Thus if two values tied for fourth place, they would each be given the rank of 4.5 ($\frac{1}{2}$ of $4+5$).) The "signs" of the original differences are then restored, and the sum of the positive ranks, called $T(+)$ is the test statistic. The critical region for $T(+)$ is given in Appendix Table 16. For example, if $N = 10$, and we want a two-sided test at a 0.05 level of significance, then reject H_0 if $T(+)$ is less than 8 or greater than 47. For the one-sided alternative $H_A: \mathrm{Md}(1) < \mathrm{Md}(2)$, reject H_0 at the 0.05 level of significance if $T(+) < 10$, and for $H_A: \mathrm{Md}(1) > \mathrm{Md}(2)$, reject H_0 if $T(+) > 45$.

Example 3 In a small advertising study to compare the amount of time spent watching television by middle-class and underprivileged children, twelve pairs of children, one middle-class and one underprivileged per pair, matched on the basis of age, were observed for a week, and the amount of time (in hours) spent watching television was recorded. The test of hypothesis was that the two sets of children watched the same amount, versus the alternative that they watched different amounts of time. Test these hypotheses at the 0.05 level of significance if the differences for each of the twelve pairs are as presented in Table 18.1.

TABLE 18.1

Pair	Difference	Absolute difference	Rank	Pair	Difference	Absolute difference	Rank
1	− 10	10	8	7	+ 4	4	2.5
2	− 8	8	6.5	8	+ 6	6	4.5
3	+ 14	14	10	9	− 8	8	6.5
4	− 6	6	4.5	10	− 16	16	11
5	− 2	2	1	11	− 4	4	2.5
6	− 12	12	9	12	+ 18	18	12

Solution: Adding those ranks whose differences have a plus sign, we have that $T(+) = 10 + 2.5 + 4.5 + 12 = 29$. Now if the null hypothesis is true and the two sets of children do indeed watch the same amounts of television, then the differences between the pairs will sometimes be positive and sometimes negative. Thus, the sum of positive ranks will be neither very large nor very small, but will in fact be moderate.

We can calculate what is meant by "moderate" by finding those values outside of which $T(+)$ has a small probability of falling; in fact, this has been done for us in Appendix Table 16.

Since for a 0.05 significance level, we reject H_0 if $T(+) \leq 13$ or $T(+) \geq 65$ (from Appendix Table 16), and since $T(+) = 29$, we accept H_0 and conclude that there is no evidence to say that the two sets of children watch different amounts of television.

Example 4 In order to compare the number of hours lost by absenteeism in two departments of a company (sales and production departments), figures (expressed as a percentage of total employee hours worked) were kept for fifteen weeks and are shown in columns (2) and (3) of Table 18.2.

TABLE 18.2

(1) Week	(2) Sales	(3) Production	(4) Difference	(5) Absolute difference	(6) Rank	(7) Ranks with signs restored
1	10.3	9.4	+0.9	0.9	3	+3
2	12.1	7.2	+4.9	4.9	12	+12
3	37.6	21.0	+16.6	16.6	15	+15
4	4.1	2.7	+1.4	1.4	4	+4
5	13.2	15.3	−2.1	2.1	6	−6
6	9.7	6.4	+3.3	3.3	9	+9
7	9.8	5.7	+4.1	4.1	10	+10
8	15.4	11.2	+4.3	4.3	11	+11
9	7.0	7.2	−0.2	0.2	1	−1
10	35.7	29.1	+6.6	6.6	14	+14
11	3.1	4.9	−1.8	1.8	5	−5
12	20.5	18.0	+2.5	2.5	7.5	+7.5
13	17.8	12.7	+5.1	5.1	13	+13
14	9.0	9.7	−0.7	0.7	2	−2
15	12.3	9.8	+2.5	2.5	7.5	+7.5

Test the hypothesis that the percentages absent in the two departments differ, using $\alpha = 0.05$.

Solution: The differences between the two departments for each of the fifteen weeks are given in column (4); the absolute values of the differences are given in (5), and ranked from 1 to 15 in column (6); in column (7) the signs are restored. The value of $T(+)$ is

$$3 + 12 + \ldots + 7.5 = 106.$$

From Appendix Table 16, we will reject the null hypothesis of "no difference" if $T(+) \geqslant 95$ or $T(+) \leqslant 25$. Since $T(+) \geqslant 95$, we reject the null hypothesis, and conclude that the absentee rates in the two departments do differ.

18.4. WILCOXON RANK-SUM TEST

If we wish to test the null hypothesis that two independent samples came from populations with the same distributions, then we may use a nonparametric procedure called the Wilcoxon rank-sum test. This procedure is also known as the Wilcoxon test for independent samples, the rank-sum test, and the Mann–Whitney test.

Suppose that we have samples of sizes N_1 and N_2. Combine the observations, and arrange the entire group according to size. Give the smallest observation the rank 1 and the largest the rank $N_1 + N_2$. (Ties are taken care of by giving them the average of the ranks that would have been assigned if no ties were present.) The test statistic, called $T(1)$, is the sum of the ranks for the sample of size N_1.

Example 5 Two groups of technician trainees are taught to perform a particular task in two different ways. Then four members of the first group and five of the second are compared by measuring the amount of time (in minutes) taken to perform the same task. The figures are given below:

$$\text{Group A:} \quad 24, \quad 15, \quad 30, \quad 21;$$

$$\text{Group B:} \quad 33, \quad 45, \quad 40, \quad 27, \quad 31.$$

Test the null hypothesis that there is no difference between the groups, versus the alternative that group A is better (with a 0.05 level of significance).

Solution

i) H_0: The two groups take equal time to perform the task.
$\quad H_A$: Group A takes less time to perform the task.

ii) $N_1 = 4, \quad N_2 = 5$.

iii)

Group	A	A	A	B	A	B	B	B	B
Time	15	21	24	27	30	31	33	40	45
Rank	1	2	3	4	5	6	7	8	9

iv) $T(1) = 1 + 2 + 3 + 5 = 11$.

v) From Appendix Table 17, the critical value of $T(1)$ is 12. Since $T(1) = 11 < 12$, reject H_0. Conclude that group A does take less time to perform the task than group B.

Note that if we had concentrated on the other sample, then we would have had $N_1 = 5$, $N_2 = 4$, $T(1) = 34$, and the same conclusion would have been reached. The justification for the use of $T(1)$ as the test statistic is that if the first sample has values which are all smaller than those of the second sample, then $T(1)$ will be small, and the hypothesis of equality will be rejected. Similarly, if the values of the first sample are all larger than those of the second, $T(1)$ will be large.

For large values of N_1 and N_2, T is approximately normally distributed with:

$$\text{Mean} = \frac{N_1 N_2}{2} \quad \text{and} \quad \text{Variance} = \frac{N_1 N_2 (N_1 + N_2 + 1)}{12},$$

and

$$Z = \frac{T - N_1 N_2 / 2}{\sqrt{N_1 N_2 (N_1 + N_2 + 1)/12}}$$

is approximately normal, with mean zero and unit variance.

18.5. THE KRUSKAL–WALLIS TEST

The Kruskal–Wallis test, also called the H-test, is a generalization of the Wilcoxon rank-sum test for two samples, to the case of K ($K > 2$) samples. The Kruskal–Wallis test is used to test the null hypothesis that K independent samples are from the same population. Thus it is a nonparametric counterpart of one-way analysis of variance. It is an extremely useful test when the researcher does not want to assume that the populations are normally distributed and have the same variance.

Let n_i be the size of the ith sample and $N = n_1 + n_2 + \ldots + n_K$ be the total number of observations. As in the Mann–Whitney test, each of the N observations is replaced by its rank. Let R_i be the sum of ranks in the ith sample. Then the Kruskal–Wallis test is based on the following statistic.

The Kruskal–Wallis Test

$$H = \frac{12}{N(N+1)} \left[\frac{R_1^2}{n_1} + \frac{R_2^2}{n_2} + \ldots + \frac{R_K^2}{n_K} \right] - 3(N+1).$$

The statistic H given above is approximately chi-square-distributed, with $(K-1)$ degrees of freedom. If the calculated value of H is greater than the chi-square value at a level of significance α and $(K-1)$ degrees of freedom, then we reject the null hypothesis that the K population means are equal, and accept the alternative, that the means of K populations are not equal at the level of significance α. We shall illustrate the use of the Kruskal–Wallis test by an example.

Example 6 Let us consider again Example 1 of Chapter 16 (analysis of variance). Suppose the personnel department of a certain company wants to compare the abilities of job applicants from four different universities. Four, three, four, and five applicants, respectively, from universities A, B, C, and D, are given a written test which is graded out of 100. The results are as follows:

	University		
A	B	C	D
67	71	88	83
52	94	66	95
83	90	69	91
74		73	99
			92

Recall that, in applying analysis of variance to this data, the main assumption is that the samples are drawn from normal distributions with common variance. Suppose we no longer assume that these samples are drawn from normal distributions with the common variance, and we apply the Kruskal–Wallis test to decide whether there is a significant difference (based on this written test) between the four universities.

Calculations of the sum of ranks R_i are shown in the table below:

University A		University B		University C		University D	
Observation	rank	Observation	rank	Observation	rank	Observation	rank
67	3	71	5	88	10	83	8.5
52	1	94	14	66	2	95	15
83	8.5	90	11	69	4	91	12
74	7			73	6	99	16
						92	13
R_i	19.5		30		22		64.5

From Table 17.1, the Kruskal–Wallis H-statistic may be calculated as follows:

$$H = \frac{12}{N(N+1)} \left[\frac{R_1^2}{n_1} + \frac{R_2^2}{n_2} + \frac{R_3^2}{n_3} + \frac{R_4^2}{n_4} \right] - 3(N+1)$$

$$= \frac{12}{16(17)} \left[\frac{(19.5)^2}{4} + \frac{(30)^2}{3} + \frac{(22)^2}{4} + \frac{(64.5)^2}{5} \right] - 3(17)$$

$$= 59.48 - 51 = 8.48.$$

Using a 0.05 level of significance and degrees of freedom $K-1=4-1=3$, we find, from Appendix Table 7, $\chi^2_{0.95}=7.81$. Since the value $H=8.48$ of the Kruskal–Wallis Test statistic is greater than $\chi^2_{0.95}=7.81$, we reject the null hypothesis and conclude that the job applicants from the four universities do not have the same abilities (as measured by the written test). Note that the same conclusion was reached using one-factor analysis of variance.

18.6. FRIEDMAN'S TEST FOR K RELATED SAMPLES

If we extend the ideas of the Wilcoxon test for paired data to the case where each related set has K observations, then we can test the hypothesis that the K samples came from identical populations by means of the Friedman test. We assume that there are N independent sets of data, each having K observations. The data may be arranged into a two-way table with N rows and K columns. Each row is then ranked from 1 for the smallest to K for the largest. If the K populations from which the samples came were identical, then the ranks could be expected to occur in about equal numbers in each column. If, on the other hand, one population was larger (or smaller) than the others, its ranks could be expected to be higher (or lower). For N independent samples, each with K observations, representing K populations, where the data are arranged in a two-way table of N rows and K columns, and ranked one row at a time from 1 to K, and where R_j is the sum of ranks in column j, then the null hypothesis of equality of populations is rejected if the statistic

$$S = \frac{12}{NK(K+1)} \sum_{j=1}^{K} R_j^2 - 3N(K+1)$$

is large. The upper significance levels of S are given in Appendix Table 18 for moderate values of K and N. For N and K outside this range, S has an approximate chi-square distribution, with $(K-1)$ degrees of freedom.

Example 7 Incoming trainees in a large company are given a score (from 0 to 100) concerning their management potential, by three different interviewers. The personnel manager wishes to see if the three interviewers give consistent results. Five trainees are graded by the three interviewers, giving the following results:

Trainee	Interviewer		
	1	2	3
A	71	53	66
B	84	72	85
C	57	94	47
D	78	68	75
E	65	80	70

Test the hypothesis that the three interviewers give consistent results, with a 0.05 level of significance.

Solution

i) The data are replaced by their ranks as follows:

		Interviewer	
Trainee	1	2	3
1	3	1	2
2	2	1	3
3	2	3	1
4	3	1	2
5	1	3	2
Total of ranks	11	9	10

$$R_1 = 11, \; R_2 = 9, \; R_3 = 10, \; N = 5, \; K = 3$$

ii) $S = \dfrac{12}{(5)(3)(4)} \{11^2 + 9^2 + 10^2\} - (3)(5)(4) = 0.4.$

iii) From Appendix Table 18, S must be greater than 5.2 in order to reject the null hypothesis. Thus we accept the null hypothesis, and conclude that the scores by the three interviewers are consistent with each other.

Example 8 Six quality-control laboratories are asked to analyze five chemical substances in order to see if all laboratories are performing the analyses in the same manner. Use Friedman's test, with $\alpha = 0.05$, to see if any laboratories are significantly different from any others, if the ranked data are as follows:

	Lab. 1	Lab. 2	Lab. 3	Lab. 4	Lab. 5	Lab. 6
Chemical A	1	5	3	2	4	6
Chemical B	3	2	1	4	6	5
Chemical C	3	4	2	5	1	6
Chemical D	1	4	6	3	2	5
Chemical E	4	5	1	2	3	6
Total of ranks	12	20	13	16	16	28

Solution

i) Here laboratories play the role of populations, and chemicals play the role of observations. Thus $N = 5$, $K = 6$.

ii) $R_1 = 12$, $R_2 = 20$, $R_3 = 13$, $R_4 = 16$, $R_5 = 16$, $R_6 = 28$.

iii)

$$S = \frac{12}{(5)(6)(7)} \{ 12^2 + 20^2 + 13^2 + 16^2 + 16^2 + 28^2 \} - (3)(5)(7)$$

$$= 114.8 - 105 = 9.8.$$

iv) From Appendix Table 7, with $(K-1) = 5$ degrees of freedom, $\chi^2_{5, 0.95} = 11.07$. Thus we cannot reject H_0. We cannot conclude that the laboratories are different.

18.7. RANK CORRELATION

There are many occasions in problems of business and industry when it is not possible to measure the variable under consideration quantitatively because of lack of time, lack of money, or inadequate units for measurements. For example, it may be possible for the two judges to rank (by preference) four different brands of orange juice in terms of taste, whereas it may be difficult to give them a *numerical* grade in terms of taste. Similarly, it may be possible (even easy) for two foremen to rank their ten employees for promotion, whereas it may be difficult to assign them a numerical grade. In all such cases, it is not possible to measure the relationship between the two attributes by the coefficient of correlation discussed in Chapter 12. If a set of items (or individuals) is ranked according to two different attributes, then the coefficient of correlation between the two attributes is called the *coefficient of rank correlation*, and can be measured by several different methods. We present here the best-known method, called the Spearman coefficient of rank correlation, r_S. Recall that the methods developed in Chapter 12 on regression and correlation assume normality. However, the coefficient of rank correlation r_S is nonparametric, and may be used if the data shows departures from normality, as is fairly common in problems of business and economics. The coefficient of rank correlation r_S is given by the following formula:

Coefficient of Rank Correlation

$$r_S = 1 - \frac{6 \Sigma d^2}{N(N^2 - 1)},$$

where $N =$ number of pairs in the data, and $d =$ differences between the ranks of x's and y's.

Example 9 The sales department of a certain company wishes to compare the selling abilities of its six salesmen. The tests designed to measure the selling ability are given to the salesmen, the average monthly sales and the test scores being given below:

Salesmen	Average monthly sales (in thousands of dollars) Y	Test score, X
A	15	81
B	10	63
C	12	75
D	17	85
E	13	60
F	19	87

Calculate the coefficient of rank correlation.

Solution: Calculation of coefficient of rank correlation is shown in Table 18.3.

TABLE 18.3

Salesmen	Test score, X	Average monthly sales, Y	d	d^2
A	4	4	0	0
B	2	1	+1	1
C	3	2	+1	1
D	5	5	0	0
E	1	3	−2	4
F	6	6	0	0
				$\Sigma d^2 = 6$

Substituting $N=6$, $\Sigma d^2 = 6$ into the formula for r_S, we get

$$r_S = 1 - \frac{6\Sigma d^2}{N(N^2-1)} = 1 - \frac{6(6)}{6(36-1)}$$

$$= 1 - \frac{36}{210} = 1 - 0.17$$

$$= 0.83.$$

Thus $r_S = 0.83$ indicates that there is a high relationship between the average monthly sales and the test scores of six salesmen.

SUMMARY

When the assumption concerning the normality of the population is not justified, one can use the techniques of nonparametric statistics for hypothesis
testing. These techniques are widely used in marketing, organizational behavior,
and similar areas where ranked data are the best that one can expect. Those
discussed are (i) the runs test, to test for randomness; (ii) Wilcoxon's test for
paired data, to compare the medians of two populations; (iii) the Wilcoxon
rank-sum test, to compare two populations on the basis of independent samples;
(iv) the Kruskal–Wallis test, an extension of the Wilcoxon rank-sum test from
two to K populations; (v) Friedman's test, an extension of Wilcoxon's test for
paired data from two to K populations; and (vi) rank correlation, a measure of
the relationship between two ranked variables.

Words to Remember

Runs test	Wilcoxon test (paired data)
Wilcoxon's rank-sum test	Kruskal–Wallis test
Friedman's test	Spearman's rank-correlation coefficient

Symbols to Remember

$T(+)$	The Wilcoxon test statistic
$T(1)$	The Wilcoxon rank-sum test statistic
H	Kruskal–Wallis test statistic
R_i	Sum of ranks in the ith sample
S	Friedman's test statistic
r_S	Spearman's rank-correlation coefficient

EXERCISES

1. Consider the diameter (in inches) of 11 bolts made by a machine

   $$1.3, \quad 1.6, \quad 1.5, \quad 1.2, \quad 1.4, \quad 1.1, \quad 1.7, \quad 1.8, \quad 1.8, \quad 1.0, \quad 1.1.$$

   Use the method of runs above and below the median to test whether the fluctuation
   from one bolt to another is random or whether a trend is present. Use $\alpha = 0.05$.

2. The following is a sequence of defective items, D, and nondefective items, N,
   produced by a machine:

   $$NN \quad D \quad NNN \quad DD \quad NNNNN \quad DDD \quad NNNN.$$

   Use the method of runs to test whether the defectives are occurring at random or
   not. Use $\alpha = 0.05$.

3. The following is the order in which TV sets are tested and classified as defective set, D, and nondefective set, N, as they come off an assembly line:

$$\text{NNN} \quad \text{D} \quad \text{NN} \quad \text{DD} \quad \text{NNNNN} \quad \text{DD.}$$

Test for randomness at a level of significance of 0.05.

4. A speed-reading course is given to fourteen executives (in the hope of making them more efficient readers) by measuring the amount of time required to read a passage before and after the course. With a 0.05 significance level, use Wilcoxon's test to see whether the course was effective.

Before	10	9	14	16	11	17	11	11	9	15	12	14	17	13
After	9	9	9	12	8	18	13	7	7	10	10	8	14	14

5. The effort devoted to research by 12 large companies, in 1960 and 1970, is given below as a percent of their operating budgets. Test, with $\alpha = 0.05$, to see whether the percentages dropped in 1970.

Company	A	B	C	D	E	F	G	H	I	J	K	L
1960	14	9	12	4	8	12	12	20	5	8	3	14
1970	13	6	4	3	10	17	2	15	7	14	1	19

6. In order to study the effects of refrigeration on its salad dressing, a food producer tests 10 jars each day. A bacteria count is made in the morning, the jars are refrigerated for six hours, and another count is made in the afternoon. The bacteria counts are judged on a scale of 1 to 10, to keep the figures simple. Use a 0.05-level test to see whether there is a higher level of bacteria after refrigeration.

Jar	1	2	3	4	5	6	7	8	9	10
Before ref.	2	6	4	7	8	3	2	5	7	4
After ref.	6	3	8	2	7	5	3	4	5	5

7. To study the retail prices of a certain commodity in two cities, samplers obtained 10 observations in one city and 12 observations in the other. The figures (in dollars) are given below. Is there a difference between the prices for the two cities (use $\alpha = 0.05$)?

First city	43	47	52	45	41	39	42	48	45	47		
Second city	38	37	46	45	42	47	43	40	39	44	42	50

8. In Chapter 16, Exercise 2, compare the weekly earnings of the workers in industries A and B, using a Wilcoxon rank-sum test, with $\alpha = 0.05$.

9. Workers are randomly assigned to one of two machines (machines A and B), and told to perform a particular task. The times taken to perform the task are given below (in minutes).

Machine A	Machine B
17.3	18.3
15.7	14.8
12.8	12.7
14.2	19.2
20.9	21.3
24.0	18.7
16.7	16.4
9.2	15.0
	17.1

Can we say that there is any difference between the machines in terms of the length of time needed to perform the task?

10. In an attempt to assess the effects of socioeconomic circumstances on the weight of new-born babies, an experimenter noted that 10 women from high-income families and 8 women from low-income families gave birth to babies with the following weights (in pounds).

High income	Low income
6.7	7.2
7.1	6.2
5.9	9.8
6.5	5.3
9.2	6.3
6.8	8.7
6.0	7.2
8.3	6.0
9.0	
6.4	

Test the hypothesis that socioeconomic status has no effect on the birth of new-born babies.

11. A pharmaceutical firm puts forward two new drugs as comparable analgesics, and it is desired to test them. The variable measured is the number of hours before relief begins to take place. The following results are obtained.

Drug 1	Drug 2
0.6	1.2
1.1	1.0
1.0	0.9
0.7	0.9
0.9	1.2
1.3	1.0
1.8	2.1

Use the Wilcoxon rank-sum test to test for a difference between the analgesics.

12. Refer to the data of Exercise 1, Chapter 16. Use the Kruskal–Wallis H-test to test the hypothesis that there is no difference in the retail prices of four different cities. Use a 5-percent level of significance.

13. Refer to the data of Exercise 2, Chapter 16. Use the Kruskal–Wallis H-test to test the hypothesis that there is no difference in the earnings of workers of the three industries. Use a 5-percent level of significance.

14. Refer to the data of Exercise 5, Chapter 16. Use the Kruskal–Wallis H-test to test the hypothesis that there is no difference in the diameters of the parts supplied by the three manufacturers. Use a 5-percent level of significance.

15. Refer to the data of Exercise 7, Chapter 16. Use the Kruskal–Wallis H-test to test the hypothesis that there is no difference between the mean breaking strength of cables manufactured by the three firms. Use a 1-percent level of significance.

16. Seven workers perform the same task on three different machines (in random order), since it is suspected that some machines are slower than others. The times (in minutes) taken to perform the tasks are given below.

Workers	Machine 1	Machine 2	Machine 3
1	24	22	29
2	28	30	35
3	26	21	27
4	21	25	31
5	27	22	21
6	25	28	26
7	20	24	30

Test the hypothesis that the machines are all the same in terms of time required to perform the task.

17. Four management trainees spend a month in each of seven sections of a large oil company, at the end of which time they are given a rating by the manager in question on a 1–10 scale. The results are as follows:

Department	Mr. Smith	Mr. Allan	Mr. King	Mr. Lock
Production	7	5	9	6
Accounting	8	6	9	7
Personnel	6	9	8	7
Sales	9	7	10	3
Advertising	3	4	8	9
Research	8	3	7	2
Development	8	6	5	7

With significance level 0.05, can we conclude that in overall performance, all candidates are equally good?

18. In Chapter 16, Exercise 9, use Friedman's test to see whether there is any difference between daily production for the four machines.

19. In Chapter 16, Exercise 9, use Friedman's test to see whether there is any difference between the daily production of the four employees.

20. Refer to the data of Exercise 1, Chapter 12. Calculate the coefficient of rank correlation between the aptitude test scores and weekly sales of 10 salesmen. Compare it with the coefficient of correlation.

21. Refer to the data of Exercise 2, Chapter 12. Calculate the coefficient of rank correlation between the age of the machine and repair costs. Compare it with the coefficient of correlation.

22. Refer to the data of Exercise 4, Chapter 12. Calculate the coefficient of rank correlation between disposable personal income and amount spent on furniture.

23. The following table gives ranks of scores of 10 junior executives in skill and aptitude tests. Is there any relationship between the skill and aptitude test scores?

Skill test	10	2	7	5	4	3	9	1	6	8
Aptitude test	9	1	6	10	2	5	4	3	8	7

24. The following table gives taste ranks (preferences) assigned by two judges to six brands of orange juice. Are the judges in agreement in preferences?

Judge A	6	1	3	2	4	5
Judge B	3	4	5	1	2	6

DECISION-MAKING UNDER UNCERTAINTY 19

19.1. INTRODUCTION

We have emphasized several times that business decisions must be made under conditions of uncertainty, because it is impossible for the businessman to have available to him every single fact that he may need at a given moment. The businessman who has to make decisions under uncertainty must, in effect, gamble: He makes his decisions *hoping* that he will be correct, but knowing that he may be wrong. Even if the business decisions are made with all the relevant facts present, they are usually complex. Those made under uncertainty are even more complex.

We give below some examples of problems where business decisions must be made under uncertainty.

Example 1 A newsstand owner buys papers at $0.05 each and sells them at $0.10 each. Since old (out-of-date) newspapers are valueless, any papers left over at the end of the day mean a loss for the owner. His problem is to know how many newspapers to buy: enough to make a good profit, but not too many, to avoid loss due to wastage.

Example 2 Consider again Example 2 of Chapter 1, where a company wishes to know whether an increase in the amount of flavoring in a toothpaste will increase the sales. If the flavoring is increased and sales do not increase appreciably, the company will lose money due to increased costs. If it decides not to increase the flavoring when in fact it should have, it will have lost potential sales. The company is faced with a decision problem in the face of uncertainty.

Example 3 An oil company must keep its gas stations stocked up with gas. If a station runs out, a loss is incurred due to unfilled orders. If too much gas is left at each station, the inventory levels are high, tying up resources. The decision faced here is to find some optimum levels of inventories.

Example 4 In Example 4 of Chapter 1, the quality-control expert must make one of two decisions. He must judge each day's run as acceptable or not acceptable, based on the examination of 200 similar parts. If he accepts a day's run when it is in fact of inferior quality, the company will risk losing customers through dissatisfaction. If he rejects a day's run when it is in fact acceptable, he will be incurring costs due to wastage. His information is incomplete because he measures only 200 parts and not the whole day's run. Thus his decision must be made under conditions of uncertainty.

Example 5 A car owner contemplating buying collision insurance is faced with a decision problem. If he buys the insurance and has no collision, then he pays the premiums for nothing. If he doesn't buy the insurance and has a collision, he is faced with the cost of fixing his car.

Example 6 A manager who is considering putting in a new piece of equipment has a similar problem. Should he postpone installation costs till later, at the risk of the old equipment breaking down, or should he incur costs now for new equipment and reduce the chances of breakdown?

Example 7 Whether or not to change the price of a product is a decision problem. If the price is increased, perhaps fewer items will be sold. If the price is reduced and the number of items sold does not increase, there will be losses incurred. The decision problem involves choosing an optimum price level.

We could go on giving many other examples of business problems involving decision-making; it will help the reader if he considers some problems in his own area of interest where decisions under uncertainty must be made. The salient characteristics of all of the above examples are as follows:

a) One must make a choice between several possible different acts (or actions).
b) Whichever act he chooses, there will be some definite profit or loss (we shall denote a loss as a *negative profit*). We usually call this a payoff.
c) The amount of this profit is unknown because it depends on random variables, such as customers' demand, which cannot be predicted exactly. The actual demand is called an *event*.

19.2. THE PAYOFF TABLE

In order to analyze a decision-making problem, one should first write down all the possible acts and their corresponding payoffs. We shall do this for the following example on inventories.

Example 8 A retailer has space for four units of a particular perishable product per week. The cost of a unit is $3.00, the selling price is $5.00, and the profit is $2.00. Any units not sold at the end of the week are wasted. The retailer does not know the demand, but must nevertheless decide on the number of units to stock. The payoffs are shown in Table 19.1.

TABLE 19.1 The payoff table

Event (No. of units demanded)	Act (No. of units stocked)				
	0	1	2	3	4
0	$0	$−3	$−6	$−9	$−12
1	0	2	−1	−4	−7
2	0	2	4	1	−2
3	0	2	4	6	3
4	0	2	4	6	8

Let us calculate the figures for Table 19.1 corresponding to the act "the retailer stocks two units."

Event (Demand)	Cost of two units stocked	Selling price	Profit
0	6	0	−6
1	6	5	−1
2	6	10	4
3	6	10	4
4	6	10	4

The profit figures are recorded in the fourth column of Table 19.1. Figures corresponding to other actions are similarly calculated.

The figures of Table 19.1 are sometimes called "conditional values" or "conditional profits," because their values depend on (or are conditional on) the event that occurs and the action which is taken.

If we compare any two acts in Table 19.1, we see that one of the two will be more profitable if certain events occur, while the other will be more profitable if other events occur. For example, compare the acts of stocking two or three units. If the demand is for two units, then stocking two is more profitable than stocking three. This profit picture is reversed, however, if the demand is for three or four units. Thus the "best" act is conditional on the unknown value of the random variable, demand. The decision to be made is whether to stock 0, 1, 2, 3, or 4 units. It is not obvious how many units to stock. We see that the solution to this even relatively simple problem is complex, and furthermore, we cannot even lean on intuition to guide us. We shall consider several possible criteria for helping us make a decision.

For general notation, we shall denote the event i by E_i and action j by A_j. The payoff for taking action j when event i takes place will be denoted by R_{ij}. For example, in Table 19.1, we will write E_0 when there is zero demand, E_1 when one unit is demanded, etc. We will also write A_0 when action zero takes place, that is, when the retailer stocks zero units, A_1 when he stocks one unit, etc. The payoff (say) when event 2 takes place and action 3 is taken is denoted by R_{23}. Thus $R_{23} = 1$, $R_{02} = -6$, etc.

19.3. THE LOSS TABLE

In addition to the payoff table showing conditional profits, one could construct a table showing opportunity losses (or regrets), due to imperfect information.

Opportunity Loss (or Regret)

The difference between the profit actually obtained for a certain decision and the profit which would have been obtained if the decision had been the best one for the event that actually occurred, is called the *opportunity loss*.

When there is no ambiguity, we shall use the word "loss" to indicate "opportunity loss."

Example 9 Suppose that the retailer in Example 8 had stocked two units, but that four had been demanded. His actual profit from selling the two units which he had is $4. If he had had four units he would have made $8 profit. Thus his *opportunity loss* is $8 - $4, or $4.00.

We next show how to construct the loss table from the payoff table. Table 19.2 is the payoff table for Example 8, and is identical to Table 19.1.

TABLE 19.2 The payoff table

Event (Demand)	Act (No. of units stocked)				
	0	1	2	3	4
0	$0*	$-3	$-6	$-9	$-12
1	0	2*	-1	-4	-7
2	0	2	4*	1	-2
3	0	2	4	6*	3
4	0	2	4	6	8*

We derive the loss table with the following two-step procedure:

a) The greatest possible profit for each event is marked with a star (*) in Table 19.2.

b) The loss table, Table 19.3, is constructed, row by row, by subtracting each value in Table 19.2 from the starred value in its row.

Thus for a demand of three units in Table 19.2, the starred value in the row is 6. If we subtract the profits 0, 2, 4, 6, 3 from 6, we get 6, 4, 2, 0, 3, respectively. Other rows are found similarly and entered in Table 19.3. Note that *loss can never be negative*. The loss for the best possible act is zero. For all other acts the loss is positive. Table 19.3 is often called the *conditional loss table*, since all values in it are conditional on the act taken and the event occurring.

TABLE 19.3 The loss table

Event (Demand)	Act (No. of units stocked)				
	0	1	2	3	4
0	$0	$3	$6	$9	$12
1	2	0	3	6	9
2	4	2	0	3	6
3	6	4	2	0	3
4	8	6	4	2	0

19.4. EXPECTED MONETARY VALUE

The conditional payoff and loss tables are helpful in analyzing a problem, but they do not contain enough information to help one make an optimum decision. The retailer in Example 8 would be helped in his decision-making if he knew which event was going to occur. He cannot know this of course, but often he has some idea of the *probability* of each event's occurrence.

This knowledge concerning the probability of an event may be based on past data from which the relative frequency of the event in question can be calculated; we might call such a figure an *objective probability*. There are many business situations where objective probabilities are not possible to obtain. For instance, what is the probability of selling x units of a new product, or what is the probability of success for a gas station at a new intersection? There are many similar kinds of problems where the action being considered is to be taken only once, and the concept of obtaining estimates from repeated data is meaningless. In such cases, it is often possible for the experienced businessman to give reasonable *subjective* estimates of the probabilities concerned. These estimates reflect the businessman's degree of belief in the event in question. Such probabilities are called *subjective probabilities*. Although, in general, objective probabilities would be preferred, one can obtain reliable probabilities from a subjective point of view. By reliable we mean that two experts would be expected to assign similar probabilities to the same event.

When the concept of a repeated number of observations is meaningless, when no prior information about an event exists, or when one cannot assume that the event in question is similar to events in the past, then a subjective probability may be appropriate. Such a probability is personal and represents the individual's belief (or betting odds) that the event will occur. The conclusions drawn from calculations based on subjective probabilities are the responsibility of the individual who assigned them. If his figures are poor, he will suffer as a result.

We now continue Example 8, assuming that demand data for the past 250 weeks exist. These are given in Table 19.4:

TABLE 19.4 Historical frequency of demand

Event (Demand)	Frequency of occurrence
0	25
1	25
2	63
3	125
4	12

If the businessman assigns probabilities based on the relative frequency concept, and if he believes that the present and future demand will remain as it was in the past, then he may say that the probability of a demand for zero units is 25/250, or 0.1. The probability of a demand for (say) two units is 63/250, or 0.25. Other probabilities are similarly assigned to the events, and are shown in Table 19.5.

TABLE 19.5 Probabilities of demand

Event (Demand)	Probability of event
0	0.10
1	0.10
2	0.25
3	0.50
4	0.05

If the retailer believes that the demand pattern has *changed* from the historical pattern, he should modify his probabilities according to the discussion on subjective probabilities. If these probabilities are to have any meaning, however, they must be arrived at in a reasonable manner.

Once the probabilities of the events have been arrived at, we next wish to bring them to bear on the analysis. This is done by weighting the conditional values of an event, in the conditional payoff and conditional loss tables, by the probability of the event occurring, and summing these weighted values. The resulting value is called an *expected value* (either expected payoff or expected loss, depending on which table one is using). We illustrate the calculation of expected payoff in the following example.

Example 10 Let us continue with Example 8, where the retailer must stock 0, 1, 2, 3, or 4 units of a product, where the demand for 0, 1, 2, 3, or 4 units is given in Table 19.5, and where the payoffs are given in Table 19.1. We shall now calculate the expected payoff of each act, taking into account the probabilities of demand. Consider the act of stocking two units. The respective payoffs from Table 19.1 are -6, -1, 4, 4, 4. The respective probabilities are (from Table 19.5) 0.10, 0.10, 0.25, 0.50, 0.05. These are shown more clearly in Table 19.6.

TABLE 19.6 Calculation of expected payoff for the act of stocking two units

Event (Demand)	Probability (Table 19.5)	Conditional payoff (Table 19.1)	Expected payoff (Conditional payoff times probability)
0	0.10	$-6	$-0.60
1	0.10	-1	-0.10
2	0.25	4	1.00
3	0.50	4	2.00
4	0.05	4	0.20

Expected payoff = $2.50

The expected payoffs for the other acts are similarly calculated and are shown in Table 19.7.

From Table 19.7, we see that the act of stocking two units of the product has the highest expected payoff. In that sense it is the *optimum act*. We summarize the calculation of the expected payoff as follows.

TABLE 19.7 Summary of expected payoffs

Act (No. of units stocked)	Expected payoffs
0	$0.00
1	1.50
2	2.50
3	2.25
4	−0.50

a) Construct a conditional payoff table, similar to Table 19.1.

b) Assign probabilities to the various events, making sure that the events exhaust all possible alternatives (that is, that the probabilities *add to one*), as in Table 19.5.

c) Calculate the expected payoffs for each act by multiplying (weighting) the conditional payoff values by the probabilities and adding the resulting products, as in Table 19.6.

d) Choose, as the optimum act, that act with the highest expected payoff.

If the retailer prefers to look at the *loss* figures, an alternative procedure is for him to calculate the *expected loss* for each act. The procedure is similar to that for the expected payoff, except that the conditional loss figures are used instead of the conditional payoff figures.

Where there are several alternative courses of action, and we choose one which maximizes the expected payoff or minimizes the expected loss, then this is called the Bayes strategy, and the act which achieves this is called the *Bayes act*.

Example 11 Suppose that the probabilities of demand for the retailing problem of Example 8 are given in Table 19.5. The conditional loss figures are given in Table 19.3. Consider the act of stocking two units. The respective losses are 6, 3, 0, 2 and 4 dollars, while the respective probabilities are 0.10, 0.10, 0.25, 0.50 and 0.05. The calculation of expected loss is given in Table 19.8. The expected losses for the other acts are similarly calculated and given in Table 19.9.

TABLE 19.8 Calculation of expected loss for the act of stocking two units

Event (Demand)	Probability (Table 19.5)	Conditional loss (Table 19.3)	Expected loss (Conditional loss times probability)
0	0.10	$6	$0.60
1	0.10	3	0.30
2	0.25	0	0.00
3	0.50	2	1.00
4	0.05	4	0.20

Expected loss = $2.10

TABLE 19.9 Summary of expected losses

Act (No. of units stocked)	Expected loss
0	$4.60
1	3.10
2	2.10
3	2.35
4	5.10

From Table 19.9, we see that the act of stocking two units has the smallest expected loss. Thus the Bayes act is to stock two units.

In order to make the idea of expected payoff and expected loss clearer, the student should recall the meaning of the word "expectation" discussed in Chapter 6 and, if necessary, reread that chapter. Simply said, expectation means the long-term average value. From this we conclude that the expected payoff is the long-term average payoff, or the payoff which he would expect if this procedure were carried on again and again. From Table 19.7, if the retailer takes the action of (say) stocking one unit again and again, he will realize a long-term payoff of $1.50 per week.

19.5. OTHER CRITERIA FOR DECISION-MAKING

We have mentioned one criterion for decision-making, namely the Bayes strategy. There are several other criteria which could be used in decision-making, and we mention some of them below.

a) We could decide to take that action which maximizes the maximum payoff (profit). This is called the *maximax* solution. The maximax solution is an optimistic one, since it assumes that a profit *will* be made.

Example 12 In Example 8 we see, from Table 19.1, that the maximum payoff is realized if the retailer stocks four units and sells all four of them. Thus the decision which maximizes the maximum profit is to stock four units. We see that this is not a very reasonable decision in light of the fact that there is a probability of only 0.05 that four units will be demanded.

b) We could take that action which minimizes the maximum loss or maximizes the minimum payoff. This is called the *minimax* solution. The minimax strategy is a pessimistic strategy, since it assumes that a *loss* will be incurred.

Example 13 The minimax action for Table 19.1 is to stock zero units because then the loss is zero dollars. The losses (negative profits) for buying 1, 2, 3, or 4 units are 3, 6, 9, and 12 dollars, respectively, from Table 19.1. In general, unless it is *impossible* to lose money, the minimax solution leads one to *do nothing*. In business terms, it means that a company would be unwilling to take chances and would become stagnant.

c) We could assume that each event was equally likely. If there are n possible events, then each has a probability of occurrence of $1/n$. The expected payoff for a particular act then becomes simply the sum of the payoffs for that act, divided by n.

Example 14 In Example 8, there are five possible events. If each is assumed equally likely, then each has a probability of occurrence of $\frac{1}{5}$. The conditional payoffs for (say) the act of stocking two units are, from Table 19.1, -6, -1, 4, 4, and 4 dollars. The expected payoff is thus

$$\tfrac{1}{5}(-6-1+4+4+4)=\tfrac{5}{5}=\$1.$$

The expected payoffs for the other acts are found similarly, and are recorded in Table 19.10.

TABLE 19.10 Summary of expected
payoffs for equally likely events

Act (No. of units stocked)	Expected payoff
0	$0.00
1	1.00
2	1.00
3	0.00
4	−2.00

From Table 19.10, it is apparent that stocking one or two units is optimum. It is clear, however, that if the probabilities of the events are known (as in Table 19.5), then it would be foolish to assume equally likely events. It seems that the assumption of equally likely events is best made when the decision-maker is in complete ignorance about the events.

19.6. BAYES' RULE AND DECISION-MAKING

In Chapter 5, we discussed Bayes' rule, whereby a manager could update the probabilities of an event by bringing other information to bear on the problem. For instance, he might start out with subjective probabilities and wish to update them once repeated information becomes available. We restate Bayes' rule here for clarity.

Bayes' Rule

If A and B are two mutually exclusive events and if D is an event such that $P(D) \neq 0$, then

$$P(A|D) = \frac{P(A \text{ and } D)}{P(A \text{ and } D) + P(B \text{ and } D)}$$

and

$$P(B|D) = \frac{P(B \text{ and } D)}{P(A \text{ and } D) + P(B \text{ and } D)}.$$

For three mutually exclusive events A, B, and C and the event D for which $P(D) \neq 0$, we have

$$P(A|D) = \frac{P(A \text{ and } D)}{P(A \text{ and } D) + P(B \text{ and } D) + P(C \text{ and } D)}.$$

$P(B|D)$ and $P(C|D)$ are found similarly, and the extension to more than three events is obvious. Bayes' rule is extremely useful in decision-making processes because it enables the decision-maker to update his prior probabilities in light of other information and thus obtain more realistic posterior probabilities. At this stage, it will probably be beneficial to the student to reread Section 5.7 if his knowledge of Bayes' rule, prior probabilities, and posterior probabilities is not clear. We illustrate this updating of probabilities in decision-making with the following example.

Example 15 The General Automobile Corporation interviews prospective candidates (generally those who have just finished their apprenticeship) for jobs as skilled automobile technicians. On the basis of each applicant's record, a panel of twenty people, from personnel, general management and the technical departments, judges him to be either an "excellent," "good," "fair," or "poor" prospect for the job. The evaluation of each applicant takes place as follows: For every one of the twenty people on the panel who judges him excellent he gets three points, for every one who judges him good he gets two points and for every one who judges him fair he gets one point. (Those judged as poor get no points.) Thus Mr. X, who received the judgments shown in Table 19.11, gets 47 points.

TABLE 19.11 Panel's opinion of Mr. X

Rating	Number of people	Score
Excellent	11	33
Good	5	10
Fair	4	4
Poor	0	0
Total	20	47

If an applicant gets an aggregate score of 55 or more, he is considered excellent. If he gets a score between 40 and 54, he is considered good. If he gets a score between 30 and 39, he is considered fair. Anything below 30 is poor. Mr. X is thus considered as *good*. (Those applicants judged as poor are not considered further.) Those applicants judged as excellent, good, or fair are put into a three-month training course within the company, where they rotate among several types of jobs.

At the end of the three months, each applicant is judged as superior, very good, good, or fair by his superiors in the technical departments. In the past, those judged as excellent originally and superior after their training have been offered positions in the company. Those judged as fair originally and fair after their training were generally let go. The problems arose with those who were judged as (say) excellent originally and as fair after their training, or those who were judged fair originally and superior after their training. A possible way out of the dilemma is to treat the original judgments as prior judgments, use the training period information to update the prior judgments and obtain posterior judgments; this procedure incorporates both the original opinions and the results of the training period.

Suppose that the records of the past 500 applicants were reviewed at this point and classified as to their original judgment (excellent, good, fair) and their after-training evaluation (superior, very good, good, fair). The results are given in Table 19.12.

TABLE 19.12 Classification of 500 past records

After-training evaluation	Original judgment		
	Excellent	Good	Fair
Superior	150	15	5
Very good	50	60	15
Good	40	45	30
Fair	10	30	50
Total	250	150	100

Of the 250 people judged as excellent originally, 150 (or 60 percent) were judged superior after training. Similarly, 20 percent were judged very good, 16 percent were judged good, and 4 percent were judged fair. Turning these relative frequencies into probabilities, we may say that the probability of being judged superior after training, given that one gets an original judgment of excellent, is 0.60. The other probabilities are arrived at similarly, and recorded in Table 19.13.

TABLE 19.13 Classification results as probabilities

After-training evaluation	Original judgment		
	Excellent	Good	Fair
Superior	0.60	0.10	0.05
Very good	0.20	0.40	0.15
Good	0.16	0.30	0.30
Fair	0.04	0.20	0.50
Total	1.00	1.00	1.00

We are now in a position to update or reevaluate the prior probabilities given in the original judgment of an applicant, based on what has happened to other similar applicants in the past. Reconsider the panel's opinion of Mr. X, given in Table 19.11 and reproduced for clarity in Table 19.14.

TABLE 19.14 The panel's opinion of Mr. X

Rating	No. of people	Probability of rating
Excellent	11	$\frac{11}{20} = 0.55$
Good	5	$\frac{5}{20} = 0.25$
Fair	4	$\frac{4}{20} = 0.20$
Total	20	1.00

The figures in Table 19.14 are the prior probabilities for Mr. X. Let us now consider Mr. X's after-training evaluation, which is "superior." We see, from Table 19.14, that he is given only a 0.55 probability of being excellent before training. We now might ask "In light of his superior evaluation after training, what is an updated probability of his being excellent?" Let us introduce the following notation:

A means an applicant is excellent,

B means an applicant is good,

C means an applicant is fair;

D_1 means an applicant is rated superior after training,

D_2 means an applicant is rated very good after training,

D_3 means an applicant is rated good after training,

D_4 means an applicant is rated fair after training.

Thus we are asking to find $P(A|D_1)$. Now, from the definition of Bayes' rule,

$$P(A|D_1) = \frac{P(A \text{ and } D_1)}{P(A \text{ and } D_1) + P(B \text{ and } D_1) + P(C \text{ and } D_1)}.$$

But

$$P(A) = 0.55,$$
$$P(B) = 0.25, \left.\right\} \text{from Table 19.14;}$$
$$P(C) = 0.20,$$

$$P(D_1|A) = 0.60,$$
$$P(D_1|B) = 0.10, \left.\right\} \text{from Table 19.13.}$$
$$P(D_1|C) = 0.05,$$

Thus,

$$P(A \text{ and } D_1) = P(A)P(D_1|A) = (0.55)(0.60) = 0.330,$$

$$P(B \text{ and } D_1) = P(B)P(D_1|B) = (0.25)(0.10) = 0.025,$$

$$P(C \text{ and } D_1) = P(C)P(D_1|C) = (0.20)(0.05) = 0.010.$$

So

$$P(A|D_1) = \frac{0.330}{0.330 + 0.025 + 0.010} = \frac{0.330}{0.365} = 0.904.$$

Thus, based on his superior after-training evaluation, Mr. X's original rating of excellent (0.55) can be evaluated upwards to 0.904, indicating that he is a better prospect than was first thought. The student should now update his original rating for good (0.25) and show that the updated rating is $P(B|D_1) = 0.068$. Similarly, $P(C|D_1) = 0.028$. Mr. X's prior and posterior ratings are summarized in Table 19.15.

TABLE 19.15 Mr. X's prior and posterior ratings

Ratings	Prior probability	Posterior probability
Excellent	0.55	0.904
Good	0.25	0.068
Fair	0.20	0.028
Total	1.00	1.000

In short, an original set of prior probabilities can be updated, using additional information and Bayes' rule, to obtain the set of posterior probabilities. Example 15 is an important illustration of the use of Bayes' rule because, in many business situations, the businessman will have access to other more recent (or peripheral) information which can be brought to bear on a problem and which may make the results more realistic.

SUMMARY

Because a businessman must make decisions under conditions of uncertainty, he is essentially gambling, and he needs some criteria to help him make an *optimum decision*. We may reduce a complex problem to one whose salient feature is that a choice must be made between several acts; for each act, there will be some profit or loss, and this profit depends on the outcome of a random variable called the *event*.

For any problem one constructs a payoff table and/or the opportunity loss table. The figures in these tables are called *conditional values*. If the probabilities of the various events are known, one can calculate expected payoffs and expected losses for any act, by weighting the payoffs and losses by the probability of each event. In this case, an optimum decision is one which *maximizes expected payoff*, or minimizes expected loss. A few other criteria of optimization are given. These include the *maximax principle*, the *minimax principle*, and the principle that events are equally likely to occur.

Bayes' rule is an important probabilistic technique which can profitably be applied to business problems when prior probabilities must be updated, in light of other information, to obtain posterior probabilities.

Words to Remember

Act (or action)	Conditional values
Event	Expected values
Payoff	Minimax solution
Opportunity loss	Maximax solution
Equally likely	Bayes' rule
Prior probability	Posterior probability
Objective probability	Subjective probability

EXERCISES

1. Write out the payoff table for Example 8 if the retailer has space for *five* units; each one costs him $4.00, and the selling price is $7.00.

2. A magazine seller pays $0.12 per magazine to the publisher. He charges $0.35 for each one sold. Those left over at the end of the week are thrown away. If he never stocks more than eight magazines, write out the payoff table.

3. For Exercise 1, write out the loss table.

4. For Exercise 2, write out the loss table.

5. Suppose that an analysis of past records, for Exercise 1, yielded the following figures.

Demand	Frequency of occurrence
0	50
1	50
2	150
3	250
4	350
5	50
Total	900

If demand patterns for the past can be expected to be maintained in the future, calculate the expected payoff for each act in Exercise 1.

6. The magazine seller in Exercise 2 knows that the probabilities of selling zero through eight magazines are as follows:

Demand	Probability
0	0.04
1	0.09
2	0.13
3	0.15
4	0.18
5	0.17
6	0.11
7	0.10
8	0.03
Total	1.00

Calculate the expected payoff for the act of stocking (a) three magazines, (b) five magazines.

7. Using the figures of Exercise 5, what are the expected losses for Exercise 3, for each act?

8. Using the figures of Exercise 6, what are the expected losses for Exercise 4, for each act?

9. In Exercise 5, explain what the expected payoff figures mean.
10. In Exercise 6, interpret the expected payoff figures.
11. In Exercise 5, which act maximizes the expected payoff?
12. In Exercise 6, which act maximizes the expected payoff?
13. In Exercise 5, which act:
    a) maximizes the maximum payoff?
    b) minimizes the maximum loss?
    c) maximizes the expected payoff if events are assumed equally likely?
14. In Exercise 6, which act:
    a) maximizes the maximum payoff?
    b) minimizes the maximum loss?
    c) maximizes the expected payoff if events are assumed equally likely?
15. A retailer of a perishable product stocks on a weekly basis to meet next week's demand. He buys the product at $3 per case and makes $4 profit on each case sold. The unsold product at the end of the week will represent a loss to the retailer. Write out the payoff table for the acts of stocking:
    a) 15 cases,   b) 16 cases.   c) 17 cases,   d) 18 cases.
16. Suppose the probability distribution of next week's demand for the perishable product of Exercise 15 is given in the following table.

Demand	Probability
15 cases	0.15
16 cases	0.25
17 cases	0.45
18 cases	0.15
Total	1.00

    a) Find the expected monetary value for each act in Exercise 15.
    b) Which act maximizes the expected payoff?

17. Extend Bayes' rule to four mutually exclusive events A, B, C, and D, and another event E for which $P(E) \neq 0$.
18. In Example 15, the panel's opinion of Mr. Z (parallel to Table 19.11) is as follows:

Rating	No. of people	Probability of rating
Excellent	10	0.5
Good	6	0.3
Fair	4	0.2

Using the probabilities of Table 19.13, what are the updated probabilities of Mr. Z's being rated:

a) excellent,   b) good,   c) fair,

if his after-training evaluation is "very good"?

19. In Example 15, if the analysis of past records had yielded the following results, calculate the updated probabilities for Mr. X.

After-training evaluation	Original judgment		
	Excellent	Good	Fair
Superior	0.75	0.05	0.05
Very good	0.15	0.35	0.10
Good	0.05	0.45	0.20
Fair	0.05	0.15	0.65
Total	1.00	1.00	1.00

20. Using the figures of Exercise 19, update the probabilities for Mr. Z in Exercise 18.

A LIST OF
SELECTED FORMULAS

A LIST OF
SELECTED FORMULAS

Below is a list of selected formulas which the authors think would be beneficial for computational purposes. Each formula is listed under the chapter in which it first appears.

CHAPTER 3

For ungrouped data, the mean of X is $\bar{X} = \dfrac{\Sigma X_i}{N}$.

33

For grouped data, the mean of Y is $\bar{Y} = \dfrac{\Sigma f_i Y_i}{N}$.

33

Combined means of two sets of data,

$$\bar{X} = \frac{N_1 \bar{X}_1 + N_2 \bar{X}_2}{N_1 + N_2}.$$

35

For grouped data, median $= a + \dfrac{((N/2) - N_1)(\Delta)}{N_2 - N_1}$.

37

Geometric mean, $G = (X_1 X_2 \ldots X_N)^{1/N}$.

39

CHAPTER 4

For ungrouped data, the mean deviation $= \frac{1}{N}\Sigma|X_i - \overline{X}|$.   50

For grouped data, the mean deviation $= \frac{1}{N}\Sigma f_i|Y_i - \overline{Y}|$.   51

For ungrouped data, $s^2 = \dfrac{\Sigma X_i^2 - N\overline{X}^2}{N-1}$.   53

For grouped data, $s^2 = \dfrac{\Sigma f_i Y_i^2 - N\overline{Y}^2}{N-1}$.   52

For ungrouped data, $s = \sqrt{\dfrac{\Sigma X_i^2 - N\overline{X}^2}{N-1}}$.   54

For grouped data, $s = \sqrt{\dfrac{\Sigma f_i Y_i^2 - N\overline{Y}^2}{N-1}}$.   54

Pooled variance of two sets of data,

$$s^2 = \frac{(M-1)s_1^2 + (N-1)s_2^2}{M+N-2}.$$ 56

Coefficient of variation, $\text{C.V.} = (s/\overline{X})100$.   57

CHAPTER 5

Combinations of n objects taken r at a time, $\dbinom{n}{r} = \dfrac{n!}{r!(n-r)!}$.   75

$P(A \text{ or } B) = P(A) + P(B) - P(A \text{ and } B)$.   82

$P(A \text{ or } B) = P(A) + P(B)$, if A and B are mutually exclusive.   83

$P(\text{Not } A) = 1 - P(A)$.   84

$$P(A|B) = \frac{P(A \text{ and } B)}{P(B)}.$$ 87

$$P(A \text{ and } B) = P(A|B)P(B) = P(B|A)P(A).$$ 88

$$P(A \text{ and } B) = P(A)P(B), \quad \text{if and only if } A \text{ and } B \text{ are independent.}$$ 89

Bayes' Rule: $$P(A|D) = \frac{P(A \text{ and } D)}{P(A \text{ and } D) + P(B \text{ and } D)};$$

$$P(B|D) = \frac{P(B \text{ and } D)}{P(A \text{ and } D) + P(B \text{ and } D)},$$ 92

if A and B are mutually exclusive and exhaustive and $P(D) \neq 0$.

CHAPTER 6

Population mean (mean of probability distribution; expected value of X) is $\mu = E(X) = \Sigma X f(X)$. 110

Population variance is $\sigma^2 = E(X^2) - [E(X)]^2$. 113

CHAPTER 7

Binomial distribution: $f(X) = \binom{n}{X} p^X q^{n-X}.$ 124

Mean of the binomial distribution, $\mu = np$. 128

Variance of the binomial distribution, $\sigma^2 = npq$. 130

Poisson distribution: $f(X) = \frac{e^{-\lambda} \lambda^X}{X!}.$ 131

Mean of the Poisson distribution, $\mu = \lambda$. 133

Variance of the Poisson distribution, $\sigma^2 = \lambda$. 133

CHAPTER 8

Standardization of the normal distribution, $Z = \dfrac{X - \mu}{\sigma}$. 142

Normal approximation to the binomial distribution,

$$Z = \frac{X - np}{\sqrt{npq}}.$$ 151

CHAPTER 10

$$\sigma_{\bar{X}}^2 = \frac{\sigma^2}{N}.$$ 171

Variance $(\bar{X}_1 \pm \bar{X}_2) = \dfrac{\sigma_1^2}{N_1} + \dfrac{\sigma_2^2}{N_2}.$ 174

95% confidence interval for μ, large N (variance known),

$$\left(\bar{X} - 1.96 \frac{\sigma}{\sqrt{N}} , \bar{X} + 1.96 \frac{\sigma}{\sqrt{N}} \right).$$ 177

99% confidence interval for μ, large N (variance known),

$$\left(\bar{X} - 2.58 \frac{\sigma}{\sqrt{N}} , \bar{X} + 2.58 \frac{\sigma}{\sqrt{N}} \right).$$ 177

95% confidence interval for μ, variance unknown,

$$\left(\bar{X} - t_{0.025} \frac{s}{\sqrt{N}} , \bar{X} + t_{0.025} \frac{s}{\sqrt{N}} \right).$$ 179

99% confidence interval for μ, variance unknown,

$$\left(\bar{X} - t_{0.005} \frac{s}{\sqrt{N}} , \bar{X} + t_{0.005} \frac{s}{\sqrt{N}} \right).$$ 180

Student's t-statistic, $t_{N-1} = \dfrac{\bar{X} - \mu}{s/\sqrt{N}}.$ 178

χ^2-statistic, $\chi^2_{N-1} = \dfrac{(N-1)s^2}{\sigma^2}$.   181

95% confidence interval for σ^2,

$$\left(\frac{(N-1)s^2}{\chi^2_{0.975}} , \frac{(N-1)s^2}{\chi^2_{0.025}} \right).$$ 182

99% confidence interval for σ^2,

$$\left(\frac{(N-1)s^2}{\chi^2_{0.995}} , \frac{(N-1)s^2}{\chi^2_{0.005}} \right).$$ (cf. page 182)

95% confidence interval for proportion, p,

$$\left(\hat{p} - 1.96 \sqrt{\frac{\hat{p}(1-\hat{p})}{n}} , \hat{p} + 1.96 \sqrt{\frac{\hat{p}(1-\hat{p})}{n}} \right).$$ 185

99% confidence interval for proportion, p,

$$\left(\hat{p} - 2.58 \sqrt{\frac{\hat{p}(1-\hat{p})}{n}} , \hat{p} + 2.58 \sqrt{\frac{\hat{p}(1-\hat{p})}{n}} \right).$$ (cf. page 185)

CHAPTER 11

Test statistic for the mean of a normal population (known variance),

$$Z = \frac{\overline{X} - \mu_0}{\sigma/\sqrt{N}}.$$ 199

Test statistic for the mean of a normal population (unknown variance),

$$t = \frac{\overline{X} - \mu_0}{s/\sqrt{N}}.$$ 201

Testing the equality of means of two normal populations,

$$Z = \frac{\overline{X}_1 - \overline{X}_2}{\sqrt{\dfrac{\sigma_1^2}{N_1} + \dfrac{\sigma_2^2}{N_2}}} .$$

204

Test statistic for the variance of a normal population,

$$\chi^2 = \frac{(N-1)s^2}{\sigma_0^2} .$$

206

Test statistic for a proportion,

$$Z = \frac{\hat{p} - p_0}{\sqrt{\dfrac{p_0(1 - p_0)}{n}}} \text{ where } \hat{p} = \frac{x}{n} .$$

207

Testing the equality of two proportions,

$$Z = \frac{\hat{p}_1 - \hat{p}_2}{\sqrt{\hat{p}(1 - \hat{p})\left(\dfrac{1}{n_1} + \dfrac{1}{n_2}\right)}} \text{ where } \hat{p} = \frac{X_1 + X_2}{n_1 + n_2} .$$

209

CHAPTER 12

Formulas for the constants a and b in the regression equation (prediction equation) $\hat{Y} = a + bX$

$$b = \frac{\Sigma XY - N\overline{X}\overline{Y}}{\Sigma X^2 - N(\overline{X})^2}, \qquad a = \overline{Y} - b\overline{X}.$$

222

Formulas for the constants c and d in the regression equation of
X on Y (that is, $\hat{X} = c + dY$):

$$d = \frac{\Sigma XY - N\overline{X}\overline{Y}}{\Sigma Y^2 - N(\overline{Y})^2}, \qquad c = \overline{X} - d\overline{Y}.$$ 225

Standard error of estimate, $s_{Y \cdot X} = \sqrt{\dfrac{\Sigma(Y - \hat{Y})^2}{N - 2}}$.   229

Variance of $a = s_a^2 = \dfrac{s_{Y \cdot X}^2(\Sigma X^2)}{N\Sigma(X - \overline{X})^2}$.   230

Variance of $b = s_b^2 = \dfrac{s_{Y \cdot X}^2}{\Sigma(X - \overline{X})^2}$.   230

Standard error of $\hat{Y}$ for a given value of X

$$= s_{Y \cdot X}\sqrt{\frac{1}{N} + \frac{(X - \overline{X})^2}{\Sigma(X - \overline{X})^2}}$$ 234

Standard error for an individual forecast for Y

$$= s_{Y \cdot X}\sqrt{1 + \frac{1}{N} + \frac{(X - \overline{X})^2}{\Sigma(X - \overline{X})^2}}\ .$$ 237

Correlation coefficient, $r = \dfrac{\Sigma XY - N\overline{X}\overline{Y}}{\sqrt{\Sigma X^2 - N(\overline{X})^2}\ \sqrt{\Sigma Y^2 - N(\overline{Y})^2}}$.   240

Test statistic for the correlation coefficient,

$$t = \frac{r}{\sqrt{\dfrac{1 - r^2}{N - 2}}}\ .$$ 244

CHAPTER 15

The chi-square statistic, $\chi^2 = \sum \dfrac{(o-e)^2}{e}$.

299

CHAPTER 16

For the one-factor analysis of variance: 324

$$\text{SST} = \sum\sum \left(X_{ij} - \bar{X}\right)^2, \quad \text{SSB} = \sum\left(\dfrac{T_i^2}{N_i}\right) - \dfrac{T^2}{N}, \quad \text{SSW} = \sum\sum X_{ij}^2 - \sum\left(\dfrac{T_i^2}{N_i}\right).$$

F-distribution, $F = \text{MSB}/\text{MSW}$ 327

For the two-factor analysis of variance,

$$\text{SST} = \sum\sum X_{ij}^2 - \dfrac{T^2}{rc} \; ;$$

331

$$\text{SSR} = \dfrac{\sum T_{i.}^2}{c} - \dfrac{T^2}{rc} , \quad \text{SSC} = \dfrac{\sum T_{j}^2}{r} - \dfrac{T^2}{rc} \; ;$$

332

$$\text{SSE} = \text{SST} - \text{SSR} - \text{SSC}$$

332

CHAPTER 17

$$\left.\begin{array}{l} \text{UCL} = \bar{\bar{X}} + A\bar{R} \\ \text{LCL} = \bar{\bar{X}} - A\bar{R} \end{array}\right\} \quad \bar{X}\text{-chart;}$$

346

$$\left.\begin{array}{l} \text{UCL} = B_2\bar{R} \\ \text{LCL} = B_1\bar{R} \end{array}\right\} \quad R\text{-chart.}$$

348

$$\left.\begin{array}{l} \text{UCL} = p + 3\sqrt{\dfrac{p(1-p)}{N}} \\[2em] \text{LCL} = p - 3\sqrt{\dfrac{p(1-p)}{N}} \end{array}\right\} \quad P\text{-charts.}$$

350

CHAPTER 18

Kruskal–Wallis test,

$$H = \frac{12}{N(N+1)} \left[\frac{R_1^2}{N_1} + \cdots + \frac{R_K^2}{N_K} \right] - 3(N+1).$$ 370

Friedman's test,

$$S = \frac{12}{NK(K+1)} \left(\sum R_j^2 \right) - 3N(K+1).$$ 372

Rank correlation coefficient, $r_S = 1 - \dfrac{6 \sum d^2}{N(N^2-1)}.$ 374

SELECTED
BIBLIOGRAPHY

A. SOME GENERAL TEXTS ON BUSINESS STATISTICS

1. Chou, Y., *Applied Business and Economic Statistics*. New York: Holt, Rinehart, and Winston, 1963.
2. Croxton, F. E., and D. J. Cowden, *Practical Business Statistics*, 3rd ed. Englewood Cliffs, N. J.: Prentice-Hall, Inc., 1960.
3. Freund, J. E., and F. J. Williams, *Elementary Business Statistics: The Modern Approach*. Englewood Cliffs, N. J.: Prentice-Hall Inc., 1973.
4. Hadley, G., *Introduction to Business Statistics*. San Francisco: Holden-Day, 1968.
5. Hoel, P. G., and R. J. Jessen, *Basic Statistics for Business and Economics*. New York: Wiley, 1971.
6. Kurnow, E., G. J. Glasser, and F. R. Ottman, *Statistics for Business Decisions*. Homewood, Ill.: Richard D. Irwin, 1959.
7. Mason, R. D., *Statistical Techniques in Business and Economics*. Homewood, Ill.: Richard D. Irwin, 1974.
8. Mendenhall, W., and J. E. Reinmuth, *Statistics for Management and Economics*. North Scituate, Mass.: Duxbury Press, 1971.
9. Neter, J. and W. Wasserman, *Fundamental Statistics for Business and Economics*, 3rd ed. Boston: Allyn and Bacon, 1973.
10. Spurr, W. A., and C. P. Bonini, *Statistical Analysis for Business Decisions*. Homewood, Ill.: Richard D. Irwin, 1973.

B. SOME TEXTS ON TIME SERIES ANALYSIS AND INDEX NUMBERS

1. Brown, R. G., *Smoothing, Forecasting and Prediction of Discrete Time Series*. Englewood Cliffs, N. J.: Prentice-Hall, 1962.
2. Butler, W. F., and R. A. Kavesh (eds.), *How Business Economists Forecast*. Englewood Cliffs, N. J.: Prentice-Hall, Inc., 1966.
3. Davis, H. T., *The Analysis of Economic Time Series*. Bloomington, Ind.: Principia Press, 1941.

4. Ferber, R., and P. J. Verdoorn, *Research Methods in Economics and Business*. New York: The Macmillan Company, 1962.
5. Fox, K. A., *Intermediate Economic Statistics*. New York: John Wiley & Sons, Inc., 1968.
6. Fisher, I., *The Making of Index Numbers*. Boston: Houghton Mifflin, 1923.
7. Mitchell, W. C., *The Making and Using of Index Numbers*. Bulletin No. 284, Washington D. C. Bureau of Labor Statistics, U. S. Dept. of Labor, 1921.
8. Mudgett, B. D., *Index Numbers*. New York: John Wiley & Sons, Inc., 1951.

C. SOME TEXTS ON QUALITY CONTROL

1. Cowden, D. J., *Statistical Methods in Quality Control*. Englewood Cliffs, N. J.: Prentice-Hall, Inc., 1957.
2. Duncan, A. J., *Quality Control and Industrial Statistics*, 4th ed. Homewood, Ill.: Richard D. Irwin, 1974.
3. Grant, E. L., and R. S. Leavenworth, *Statistical Quality Control*, 4th ed. New York: McGraw-Hill Book Company, 1972.

D. SOME TEXTS ON NONPARAMETRIC STATISTICS

1. Conover, W. J., *Practical Nonparametric Statistics*. New York: John Wiley & Sons, Inc., 1971.
2. Noether, G. E., *Introduction to Statistics*. Boston: Houghton Mifflin, 1974.
3. Siegel, S., *Nonparametric Statistics*. New York: McGraw-Hill, 1956.

E. SOME TEXTS ON APPLIED STATISTICAL DECISION THEORY

1. Morgan, B. W., *An Introduction to Bayesian Statistical Decision Process*. Englewood Cliffs, N. J.: Prentice-Hall, 1968.
2. Sasaki, K., *Statistics for Modern Business Decision Making*. Belmont, Calif.: Wadsworth Publishing Co., 1968.
3. Schlaifer, R., *Introduction to Statistics for Business Decisions*. New York: McGraw-Hill, 1961.
4. Schlaifer, R., *Analysis of Decisions under Uncertainty*. New York: McGraw-Hill, 1969.

F. SOME STATISTICAL TABLES

1. Beyer, W. H. (ed.), *Handbook of Tables for Probability and Statistics*. 2nd ed. Cleveland, Ohio: The Chemical Rubber Co., 1968.
2. Fisher, R. A., and F. Yates, *Statistical Tables*. New York: Hafner Publishing Co., 1949.
3. Owen, D. B., *Handbook for Statistical Tables*. Reading, Mass.: Addison-Wesley, 1962.
4. Pearson, E. S., and H. O. Hartley, *Biometrika Tables for Statisticians*. New York: Cambridge University Press, Vol. 1.

APPENDIX

TABLE 1 Squares and Square Roots

N	N²	√N	√10N	N	N²	√N	√10N
1.00	1.0000	1.00000	3.16228	**1.40**	1.9600	1.18322	3.74166
1.01	1.0201	1.00499	3.17805	1.41	1.9881	1.18743	3.75500
1.02	1.0404	1.00995	3.19374	1.42	2.0164	1.19164	3.76829
1.03	1.0609	1.01489	3.20936	1.43	2.0449	1.19583	3.78153
1.04	1.0816	1.01980	3.22490	1.44	2.0736	1.20000	3.79473
1.05	1.1025	1.02470	3.24037	1.45	2.1025	1.20416	3.80789
1.06	1.1236	1.02956	3.25576	1.46	2.1316	1.20830	3.82099
1.07	1.1449	1.03441	3.27109	1.47	2.1609	1.21244	3.83406
1.08	1.1664	1.03923	3.28634	1.48	2.1904	1.21655	3.84708
1.09	1.1881	1.04403	3.30151	1.49	2.2201	1.22066	3.86005
1.10	1.2100	1.04881	3.31662	**1.50**	2.2500	1.22474	3.87298
1.11	1.2321	1.05357	3.33167	1.51	2.2801	1.22882	3.88587
1.12	1.2544	1.05830	3.34664	1.52	2.3104	1.23288	3.89872
1.13	1.2769	1.06301	3.36155	1.53	2.3409	1.23693	3.91152
1.14	1.2996	1.06771	3.37639	1.54	2.3716	1.24097	3.92428
1.15	1.3225	1.07238	3.39116	1.55	2.4025	1.24499	3.93700
1.16	1.3456	1.07703	3.40588	1.56	2.4336	1.24900	3.94968
1.17	1.3689	1.08167	3.42053	1.57	2.4649	1.25300	3.96232
1.18	1.3924	1.08628	3.43511	1.58	2.4964	1.25698	3.97492
1.19	1.4161	1.09087	3.44964	1.59	2.5281	1.26095	3.98748
1.20	1.4400	1.09545	3.46410	**1.60**	2.5600	1.26491	4.00000
1.21	1.4641	1.10000	3.47851	1.61	2.5921	1.26886	4.01248
1.22	1.4884	1.10454	3.49285	1.62	2.6244	1.27279	4.02492
1.23	1.5129	1.10905	3.50714	1.63	2.6569	1.27671	4.03733
1.24	1.5376	1.11355	3.52136	1.64	2.6896	1.28062	4.04969
1.25	1.5625	1.11803	3.53553	1.65	2.7225	1.28452	4.06202
1.26	1.5876	1.12250	3.54965	1.66	2.7556	1.28841	4.07431
1.27	1.6129	1.12694	3.56371	1.67	2.7889	1.29228	4.08656
1.28	1.6384	1.13137	3.57771	1.68	2.8224	1.29615	4.09878
1.29	1.6641	1.13578	3.59166	1.69	2.8561	1.30000	4.11096
1.30	1.6900	1.14018	3.60555	**1.70**	2.8900	1.30384	4.12311
1.31	1.7161	1.14455	3.61939	1.71	2.9241	1.30767	4.13521
1.32	1.7424	1.14891	3.63318	1.72	2.9584	1.31149	4.14729
1.33	1.7689	1.15326	3.64692	1.73	2.9929	1.31529	4.15933
1.34	1.7956	1.15758	3.66060	1.74	3.0276	1.31909	4.17133
1.35	1.8225	1.16190	3.67423	1.75	3.0625	1.32288	4.18330
1.36	1.8496	1.16619	3.68782	1.76	3.0976	1.32665	4.19524
1.37	1.8769	1.17047	3.70135	1.77	3.1329	1.33041	4.20714
1.38	1.9044	1.17473	3.71484	1.78	3.1684	1.33417	4.21900
1.39	1.9321	1.17898	3.72827	1.79	3.2041	1.33791	4.23084

(Continued)

TABLE 1 (Continued)

N	N²	√N	√10N	N	N²	√N	√10N
1.80	3.2400	1.34164	4.24264	**2.20**	4.8400	1.48324	4.69042
1.81	3.2761	1.34536	4.25441	2.21	4.8841	1.48661	4.70106
1.82	3.3124	1.34907	4.26615	2.22	4.9284	1.48997	4.71169
1.83	3.3489	1.35277	4.27785	2.23	4.9729	1.49332	4.72229
1.84	3.3856	1.35647	4.28952	2.24	5.0176	1.49666	4.73286
1.85	3.4225	1.36015	4.30116	2.25	5.0625	1.50000	4.74342
1.86	3.4596	1.36382	4.31277	2.26	5.1076	1.50333	4.75395
1.87	3.4969	1.36748	4.32435	2.27	5.1529	1.50665	4.76445
1.88	3.5344	1.37113	4.33590	2.28	5.1984	1.50997	4.77493
1.89	3.5721	1.37477	4.34741	2.29	5.2441	1.51327	4.78539
1.90	3.6100	1.37840	4.35890	**2.30**	5.2900	1.51658	4.79583
1.91	3.6481	1.38203	4.37035	2.31	5.3361	1.51987	4.80625
1.92	3.6864	1.38564	4.38178	2.32	5.3824	1.52315	4.81664
1.93	3.7249	1.38924	4.39318	2.33	5.4289	1.52643	4.82701
1.94	3.7636	1.39284	4.40454	2.34	5.4756	1.52971	4.83735
1.95	3.8025	1.39642	4.41588	2.35	5.5225	1.53297	4.84768
1.96	3.8416	1.40000	4.42719	2.36	5.5696	1.53623	4.85798
1.97	3.8809	1.40357	4.43847	2.37	5.6169	1.53948	4.86826
1.98	3.9204	1.40712	4.44972	2.38	5.6644	1.54272	4.87852
1.99	3.9601	1.41067	4.46094	2.39	5.7121	1.54596	4.88876
2.00	4.0000	1.41421	4.47214	**2.40**	5.7600	1.54919	4.89898
2.01	4.0401	1.41774	4.48330	2.41	5.8081	1.55242	4.90918
2.02	4.0804	1.42127	4.49444	2.42	5.8564	1.55563	4.91935
2.03	4.1209	1.42478	4.50555	2.43	5.9049	1.55885	4.92950
2.04	4.1616	1.42829	4.51664	2.44	5.9536	1.56205	4.93964
2.05	4.2025	1.43178	4.52769	2.45	6.0025	1.56525	4.94975
2.06	4.2436	1.43527	4.53872	2.46	6.0516	1.56844	5.95984
2.07	4.2849	1.43875	4.54973	2.47	6.1009	1.57162	4.96991
2.08	4.3264	1.44222	4.56070	2.48	6.1504	1.57480	4.97996
2.09	4.3681	1.44568	4.57165	2.49	6.2001	1.57797	4.98999
2.10	4.4100	1.44914	4.58258	**2.50**	6.2500	1.58114	5.00000
2.11	4.4521	1.45258	4.59347	2.51	6.3001	1.58430	5.00999
2.12	4.4944	1.45602	4.60435	2.52	6.3504	1.58745	5.01996
2.13	4.5369	1.45945	4.61519	2.53	6.4009	1.59060	5.02991
2.14	4.5796	1.46287	4.62601	2.54	6.4516	1.59374	5.03984
2.15	4.6225	1.46629	4.63681	2.55	6.5025	1.59687	5.04975
2.16	4.6656	1.46969	4.64758	2.56	6.5536	1.60000	5.05964
2.17	4.7089	1.47309	4.65833	2.57	6.6049	1.60312	5.06952
2.18	4.7524	1.47648	4.66905	2.58	6.6564	1.60624	5.07937
2.19	4.7961	1.47986	4.67974	2.59	6.7081	1.60935	5.08920

TABLE 1 (Continued)

N	N²	√N	√10N	N	N²	√N	√10N
2.60	6.7600	1.61245	5.09902	**3.00**	9.0000	1.73205	5.47723
2.61	6.8121	1.61555	5.10882	3.01	9.0601	1.73494	5.48635
2.62	6.8644	1.61864	5.11859	3.02	9.1204	1.73781	5.49545
2.63	6.9169	1.62173	5.12835	3.03	9.1809	1.74069	5.50454
2.64	6.9696	1.62481	5.13809	3.04	9.2416	1.74356	5.51362
2.65	7.0225	1.62788	5.14782	3.05	9.3025	1.74642	5.52268
2.66	7.0756	1.63095	5.15752	3.06	9.3636	1.74929	5.53173
2.67	7.1289	1.63401	5.16720	3.07	9.4249	1.75214	5.54076
2.68	7.1824	1.63707	5.17687	3.08	9.4864	1.75499	5.54977
2.69	7.2361	1.64012	5.18652	3.09	9.5481	1.75784	5.55878
2.70	7.2900	1.64317	5.19615	**3.10**	9.6100	1.76068	5.56776
2.71	7.3441	1.64621	5.20577	3.11	9.6721	1.76352	5.57674
2.72	7.3984	1.64924	5.21536	3.12	9.7344	1.76636	5.58570
2.73	7.4529	1.65227	5.22494	3.13	9.7969	1.76918	5.59464
2.74	7.5076	1.65529	5.23450	3.14	9.8596	1.77200	5.60357
2.75	7.5625	1.65831	5.24404	3.15	9.9225	1.77482	5.61249
2.76	7.6176	1.66132	5.25357	3.16	9.9856	1.77764	5.62139
2.77	7.6729	1.66433	5.26308	3.17	10.0489	1.78045	5.63028
2.78	7.7284	1.66733	5.27257	3.18	10.1124	1.78326	5.63915
2.79	7.7841	1.67033	5.28205	3.19	10.1761	1.78606	5.64801
2.80	7.8400	1.67332	5.29150	**3.20**	10.2400	1.78885	5.65685
2.81	7.8961	1.67631	5.30094	3.21	10.3041	1.79165	5.66569
2.82	7.9524	1.67929	5.31037	3.22	10.3684	1.79444	5.67450
2.83	8.0089	1.68226	5.31977	3.23	10.4329	1.79722	5.68331
2.84	8.0656	1.68523	5.32917	3.24	10.4976	1.80000	5.69210
2.85	8.1225	1.68819	5.33854	3.25	10.5625	1.80278	5.70088
2.86	8.1796	1.69115	5.34790	3.26	10.6276	1.80555	5.70964
2.87	8.2369	1.69411	5.35724	3.27	10.6929	1.80831	5.71839
2.88	8.2944	1.69706	5.36656	3.28	10.7584	1.81108	5.72713
2.89	8.3521	1.70000	5.37587	3.29	10.8241	1.81384	5.73585
2.90	8.4100	1.70294	5.38516	**3.30**	10.8900	1.81659	5.74456
2.91	8.4681	1.70587	5.39444	3.31	10.9561	1.81934	5.75326
2.92	8.5264	1.70880	5.40370	3.32	11.0224	1.82209	5.76194
2.93	8.5849	1.71172	5.41295	3.33	11.0889	1.82483	5.77062
2.94	8.6436	1.71464	5.42218	3.34	11.1556	1.82757	5.77927
2.95	8.7025	1.71756	5.43139	3.35	11.2225	1.83030	5.78792
2.96	8.7616	1.72047	5.44059	3.36	11.2896	1.83303	5.79655
2.97	8.8209	1.72337	5.44977	3.37	11.3569	1.83576	5.80517
2.98	8.8804	1.72627	5.45894	3.38	11.4244	1.83848	5.81378
2.99	8.9401	1.72916	5.46809	3.39	11.4921	1.84120	5.82237

(Continued)

TABLE 1 (Continued)

N	N²	√N	√10N	N	N²	√N	√10N
3.40	11.5600	1.84391	5.83095	**3.80**	14.4400	1.94936	6.16441
3.41	11.6281	1.84662	5.83952	3.81	14.5161	1.95192	6.17252
3.42	11.6964	1.84932	5.84808	3.82	14.5924	1.95448	6.18061
3.43	11.7649	1.85203	5.85662	3.83	14.6689	1.95704	6.18870
3.44	11.8336	1.85472	5.86515	3.84	14.7456	1.95959	6.19677
3.45	11.9025	1.85742	5.87367	3.85	14.8225	1.96214	6.20484
3.46	11.9716	1.86011	5.88218	3.86	14.8996	1.96469	6.21289
3.47	12.0409	1.86279	5.89067	3.87	14.9769	1.96732	6.22093
3.48	12.1104	1.86548	5.89915	3.88	15.0544	1.96977	6.22896
3.49	12.1801	1.86815	5.90762	3.89	15.1321	1.97231	6.23699
3.50	12.2500	1.87083	5.91608	**3.90**	15.2100	1.97484	6.24500
3.51	12.3201	1.87350	5.92453	3.91	15.2881	1.97737	6.25300
3.52	12.3904	1.87617	5.93296	3.92	15.3664	1.97990	6.26099
3.53	12.4609	1.87883	5.94138	3.93	15.4449	1.98242	6.26897
3.54	12.5316	1.88149	5.94979	3.94	15.5236	1.98494	6.27694
3.55	12.6025	1.88414	5.95819	3.95	15.6025	1.98746	6.28490
3.56	12.6736	1.88680	5.96657	3.96	15.6816	1.98997	6.29285
3.57	12.7449	1.88944	5.97495	3.97	15.7609	1.99249	6.30079
3.58	12.8164	1.89209	5.98331	3.98	15.8404	1.99499	6.30872
3.59	12.8881	1.89473	5.99166	3.99	15.9201	1.99750	6.31664
3.60	12.9600	1.89737	6.00000	**4.00**	16.0000	2.00000	6.32456
3.61	13.0321	1.90000	6.00833	4.01	16.0801	2.00250	6.33246
3.62	13.1044	1.90263	6.01664	4.02	16.1604	2.00499	6.34035
3.63	13.1769	1.90526	6.02495	4.03	16.2409	2.00749	6.34823
3.64	13.2496	1.90788	6.03324	4.04	16.3216	2.00998	6.35610
3.65	13.3225	1.91050	6.04152	4.05	16.4025	2.01246	6.36396
3.66	13.3956	1.91311	6.04979	4.06	16.4836	2.01494	6.37181
3.67	13.4689	1.91572	6.05805	4.07	16.5649	2.01742	6.37966
3.68	13.5424	1.91833	6.06630	4.08	16.6464	2.01990	6.38749
3.69	13.6161	1.92094	6.07454	4.09	16.7281	2.02237	6.39531
3.70	13.6900	1.92354	6.08276	**4.10**	16.8100	2.02485	6.40312
3.71	13.7641	1.92614	6.09098	4.11	16.8921	2.02731	6.41093
3.72	13.8384	1.92873	6.09918	4.12	16.9744	2.02978	6.41872
3.73	13.9129	1.93132	6.10737	4.13	17.0569	2.03224	6.42651
3.74	13.9876	1.93391	6.11555	4.14	17.1396	2.03470	6.43428
3.75	14.0625	1.93649	6.12372	4.15	17.2225	2.03715	6.44205
3.76	14.1376	1.93907	6.13188	4.16	17.3056	2.03961	6.44981
3.77	14.2129	1.94165	6.14003	4.17	17.3889	2.04206	6.45755
3.78	14.2884	1.94422	6.14817	4.18	17.4724	2.04450	6.46529
3.79	14.3641	1.94679	6.15630	4.19	17.5561	2.04695	6.47302

TABLE 1 (Continued)

N	N²	√N	√10N	N	N²	√N	√10N
4.20	17.6400	2.04939	6.48074	**4.60**	21.1600	2.14476	6.78233
4.21	17.7241	2.05183	6.48845	4.61	21.2521	2.14709	6.78970
4.22	17.8084	2.05426	6.49615	4.62	21.3444	2.14942	6.79706
4.23	17.8929	2.05670	6.50384	4.63	21.4369	2.15174	6.80441
4.24	17.9776	2.05913	6.51153	4.64	21.5296	2.15407	6.81175
4.25	18.0625	2.06155	6.51920	4.65	21.6225	2.15639	6.81909
4.26	18.1476	2.06398	6.52687	4.66	21.7156	2.15870	6.82642
4.27	18.2329	2.06640	6.53452	4.67	21.8089	2.16102	6.83374
4.28	18.3184	2.06882	6.54217	4.68	21.9024	2.16333	6.84105
4.29	18.4041	2.07123	6.54981	4.69	21.9961	2.16564	6.84836
4.30	18.4900	2.07364	6.55744	**4.70**	22.0900	2.16795	6.85565
4.31	18.5761	2.07605	6.56506	4.71	22.1841	2.17025	6.86294
4.32	18.6624	2.07846	6.57267	4.72	22.2784	2.17256	6.87023
4.33	18.7489	2.08087	6.58027	4.73	22.3729	2.17486	6.87750
4.34	18.8356	2.08327	6.58787	4.74	22.4676	2.17715	6.88477
4.35	18.9225	2.08567	6.59545	4.75	22.5625	2.17945	6.89202
4.36	19.0096	2.08806	6.60303	4.76	22.6576	2.18174	6.89928
4.37	19.0969	2.09045	6.61060	4.77	22.7529	2.18403	6.90652
4.38	19.1844	2.09284	6.61816	4.78	22.8484	2.18632	6.91375
4.39	19.2721	2.09523	6.62571	4.79	22.9441	2.18861	6.92098
4.40	19.3600	2.09762	6.63325	**4.80**	23.0400	2.19089	6.92820
4.41	19.4481	2.10000	6.64078	4.81	23.1361	2.19317	6.93542
4.42	19.5364	2.10238	6.64831	4.82	23.2324	2.19545	6.94262
4.43	19.6249	2.10476	6.65582	4.83	23.3289	2.19773	6.94982
4.44	19.7136	2.10713	6.66333	4.84	23.4256	2.20000	6.95701
4.45	19.8025	2.10950	6.67083	4.85	23.5225	2.20227	6.96419
4.46	19.8916	2.11187	6.67832	4.86	23.6196	2.20454	6.97137
4.47	19.9809	2.11424	6.68581	4.87	23.7169	2.20681	6.97854
4.48	20.0704	2.11660	6.69328	4.88	23.8144	2.20907	6.98570
4.49	20.1601	2.11896	6.70075	4.89	23.9121	2.21133	6.99285
4.50	20.2500	2.12132	6.70820	**4.90**	24.0100	2.21359	7.00000
4.51	20.3401	2.12368	6.71565	4.91	24.1081	2.21585	7.00714
4.52	20.4304	2.12603	6.72309	4.92	24.2064	2.21811	7.01427
4.53	20.5209	2.12838	6.73053	4.93	24.3049	2.22036	7.02140
4.54	20.6116	2.13073	6.73795	4.94	24.4036	2.22261	7.02851
4.55	20.7025	2.13307	6.74537	4.95	24.5025	2.22486	7.03562
4.56	20.7936	2.13542	6.75278	4.96	24.6016	2.22711	7.04273
4.57	20.8849	2.13776	6.76018	4.97	24.7009	2.22935	7.04982
5.48	20.9764	2.14009	6.76757	4.98	24.8004	2.23159	7.05691
4.59	21.0681	2.14243	6.77495	4.99	24.9901	2.23383	7.06399

(Continued)

TABLE 1 (Continued)

N	N²	√N	√10N	N	N²	√N	√10N
5.00	25.0000	2.23607	7.07107	**5.40**	29.1600	2.32379	7.34847
5.01	25.1001	2.23830	7.07814	5.41	29.2681	2.32594	7.35527
5.02	25.2004	2.24054	7.08520	5.42	29.3764	2.32809	7.36206
5.03	25.3009	2.24277	7.09225	5.43	29.4849	2.33024	7.36885
5.04	25.4016	2.24499	7.09930	5.44	29.5936	2.33238	7.37564
5.05	25.5025	2.24722	7.10634	5.45	29.7025	2.33452	7.38241
5.06	25.6036	2.24944	7.11337	5.46	29.8116	2.33666	7.38918
5.07	25.7049	2.25167	7.12039	5.47	29.9209	2.33880	7.39594
5.08	25.8064	2.25389	7.12741	5.48	30.0304	2.34094	7.40270
5.09	25.9081	2.25610	7.13442	5.49	30.1401	2.34307	7.40945
5.10	26.0100	2.25832	7.14143	**5.50**	30.2500	2.34521	7.41620
5.11	26.1121	2.26053	7.14843	5.51	30.3601	2.34734	7.42294
5.12	26.2144	2.26274	7.15542	5.52	30.4704	2.34947	7.42967
5.13	26.3169	2.26495	7.16240	5.53	30.5809	2.35160	7.43640
5.14	26.4196	2.26716	7.16938	5.54	30.6916	2.35372	7.44312
5.15	26.5225	2.26936	7.17635	5.55	30.8025	2.35584	7.44983
5.16	26.6256	2.27156	7.18331	5.56	30.9136	2.35797	7.45654
5.17	26.7289	2.27376	7.19027	5.57	31.0249	2.36008	7.46324
5.18	26.8324	2.27596	7.19722	5.58	21.1364	2.36220	7.46994
5.19	26.9361	2.27816	7.20417	5.59	31.2481	2.36432	7.47663
5.20	27.0400	2.28035	7.21110	**5.60**	31.3600	2.36643	7.48331
5.21	27.1441	2.28254	7.21803	5.61	31.4721	2.26854	7.48999
5.22	27.2484	2.28473	7.22496	5.62	31.5844	2.37065	7.49667
5.23	27.3529	2.28692	7.23187	5.63	31.6969	2.37276	7.50333
5.24	27.4576	2.28910	7.23878	5.64	31.8096	2.37487	7.50999
5.25	27.5625	2.29129	7.24569	5.65	31.9225	2.37697	7.51665
5.26	27.6676	2.29347	7.25259	5.66	32.0356	2.37908	7.52330
5.27	27.7729	2.29565	7.25948	5.67	32.1489	2.38118	7.52994
5.28	27.8784	2.29783	7.26636	5.68	32.2624	2.38328	7.53658
5.29	27.9841	2.30000	7.27324	5.69	32.3761	2.38537	7.54321
5.30	28.0900	2.30217	7.28011	**5.70**	32.4900	2.38747	7.54983
5.31	28.1961	2.30434	7.28697	5.71	32.6041	2.38956	7.55645
5.32	28.3024	2.30651	7.29383	5.72	32.7184	2.39165	7.56307
5.33	28.4089	2.30868	7.30068	5.73	32.8329	2.39374	7.56968
5.34	28.5156	2.31084	7.30753	5.74	32.9476	2.39583	7.57628
5.35	28.6225	2.31301	7.31437	5.75	33.0625	2.39792	7.58288
5.36	28.7296	2.31517	7.32120	5.76	33.1776	2.40000	7.58947
5.37	28.8369	2.31733	7.32803	5.77	33.2929	2.40208	7.59605
5.38	28.9444	2.31948	7.33485	5.78	33.4048	2.40416	7.60263
5.39	29.0521	2.32164	7.34166	5.79	33.5241	2.40624	7.60920

TABLE 1 (Continued)

N	N²	√N	√10N	N	N²	√N	√10N
5.80	33.6400	2.40832	7.61577	**6.20**	38.4400	2.48998	7.87401
5.81	33.7561	2.41039	7.62234	6.21	38.5641	2.49199	7.88036
5.82	33.8724	2.41247	7.62889	6.22	38.6884	2.49399	7.88670
5.83	33.9889	2.41454	7.63544	6.23	38.8129	2.49600	7.89303
5.84	34.1056	2.41661	7.64199	6.24	38.9376	2.49800	7.89937
5.85	34.2225	2.41868	7.64853	6.25	39.0625	2.50000	7.90569
5.86	34.3396	2.42074	7.65506	6.26	39.1876	2.50200	7.91202
5.87	34.4569	2.42281	7.66159	6.27	39.3129	2.50400	7.91833
5.88	34.5744	2.42487	7.66812	6.28	39.4384	2.50599	7.92465
5.89	34.6921	2.42693	7.67463	6.29	39.5641	2.50799	7.93095
5.90	34.8100	2.42899	7.68115	**6.30**	39.6900	2.50998	7.93725
5.91	34.9281	2.43105	7.68765	6.31	39.8161	2.51197	7.94355
5.92	35.0464	2.43311	7.69415	6.32	39.9424	2.51396	7.94984
5.93	35.1649	2.43516	7.70065	6.33	40.0689	2.51595	7.95613
5.94	35.2836	2.43721	7.70714	6.34	40.1956	2.51794	7.96241
5.95	35.4025	2.43926	7.71362	6.35	40.3225	2.51992	7.96869
5.96	35.5216	2.44131	7.72010	6.36	40.4496	2.52190	7.97496
5.97	35.6409	2.44336	7.72658	6.37	40.5769	2.52389	7.98123
5.98	35.7604	2.44540	7.73305	6.38	40.7044	2.52587	7.98749
5.99	35.8801	2.44745	7.73951	6.39	40.8321	2.52784	7.99375
6.00	36.0000	2.44949	7.74597	**6.40**	40.9600	2.52982	8.00000
6.01	36.1201	2.45153	7.75242	6.41	41.0881	2.53180	8.00625
6.02	36.2404	2.45357	7.75887	6.42	41.2164	2.53377	8.01249
6.03	36.3609	2.45561	7.76531	6.43	41.3449	2.53574	8.01873
6.04	36.4816	2.45764	7.77174	6.44	41.4736	2.53772	8.02496
6.05	36.6025	2.45967	7.77817	6.45	41.6025	2.53969	8.03119
6.06	36.7236	2.46171	7.78460	6.46	41.7316	2.54165	8.03741
6.07	36.8449	2.46374	7.79102	6.47	41.8609	2.54362	8.04363
6.08	36.9664	2.46577	7.79744	6.48	41.9904	2.54558	8.04984
6.09	37.0881	2.46779	7.80385	6.49	42.1201	2.54755	8.05605
6.10	37.2100	2.46982	7.81025	**6.50**	42.2500	2.54951	8.06226
6.11	37.3321	2.47184	7.81665	6.51	42.3801	2.55147	8.06846
6.12	37.4544	2.47386	7.82304	6.52	42.5104	2.55343	8.07465
6.13	37.5769	2.47588	7.82943	6.53	42.6409	2.55539	8.08084
6.14	37.6996	2.47790	7.83582	6.54	42.7716	2.55734	8.08703
6.15	37.8225	2.47992	7.84219	6.55	42.9025	2.55930	8.09321
6.16	37.9456	2.48193	7.84857	5.56	43.0336	2.56125	8.09938
6.17	38.0689	2.48395	7.85493	6.57	43.1649	2.56320	8.10555
6.18	38.1924	2.48596	7.86130	6.58	43.2964	2.56515	8.11172
6.19	38.3161	2.48797	7.86766	6.59	43.4281	2.56710	8.11788

(Continued)

TABLE 1 (Continued)

N	N²	√N	√10N	N	N²	√N	√10N
6.60	43.5600	2.56905	8.12404	**7.00**	49.0000	2.64575	8.36660
6.61	43.6921	2.57099	8.13019	7.01	49.1401	2.64764	8.37257
6.62	43.8244	2.57294	8.13634	7.02	49.2804	2.64953	8.37854
6.63	43.9569	2.57488	8.14248	7.03	49.4209	2.65141	8.38451
6.64	44.0896	2.57682	8.14862	7.04	49.5616	2.65330	8.39047
6.65	44.2225	2.57876	8.15475	7.05	49.7025	2.65518	8.39643
6.66	44.3556	2.58070	8.16088	7.06	49.8436	2.65707	8.40238
6.67	44.4889	2.58263	8.16701	7.07	49.9849	2.65895	8.40833
6.68	44.6224	2.58457	8.17313	7.08	50.1264	2.66083	8.41427
6.69	44.7561	2.58650	8.17924	7.09	50.2681	2.66271	8.42021
6.70	44.8900	2.58844	8.18535	**7.10**	50.4100	2.66458	8.42615
6.71	45.0241	2.59037	8.19146	7.11	50.5521	2.66646	8.43208
6.72	45.1584	2.59230	8.19756	7.12	50.6944	2.66833	8.43801
6.73	45.2929	2.59422	8.20366	7.13	50.8369	2.67021	8.44393
6.74	45.4276	2.59615	8.20975	7.14	50.9796	2.67208	8.44985
6.75	45.5625	2.59808	8.21584	7.15	51.1225	2.67395	8.45577
6.76	45.6976	2.60000	8.22192	7.16	51.2656	2.67582	8.46168
6.77	45.8329	2.60192	8.22800	7.17	51.4089	2.67769	8.46759
6.78	45.9684	2.60384	8.23408	7.18	51.5524	2.67955	8.47349
6.79	46.1041	2.60576	8.24015	7.19	51.6961	2.68142	8.47939
6.80	46.2400	2.60768	8.24621	**7.20**	51.8400	2.68328	8.48528
6.81	46.3761	2.60960	8.25227	7.21	51.9841	2.68514	8.49117
6.82	46.5124	2.61151	8.25833	7.22	52.1284	2.68701	8.49706
6.83	46.6489	2.61343	8.26438	7.23	52.2729	2.68887	8.50294
6.84	46.7856	2.61534	8.27043	7.24	52.4176	2.69072	8.50882
6.85	46.9225	2.61725	8.27647	7.25	52.5625	2.69258	8.51469
6.86	47.0596	2.61916	8.28251	7.26	52.7076	2.69444	8.52056
6.87	47.1969	2.62107	8.28855	7.27	52.8529	2.69629	8.52643
6.88	47.3344	2.62298	8.29458	7.28	52.9984	2.69815	8.53229
6.89	47.4721	2.62488	8.30060	7.29	53.1441	2.70000	8.53815
6.90	47.6100	2.62679	8.30662	**7.30**	53.2900	2.70185	8.54400
6.91	47.7481	2.62869	8.31264	7.31	53.4361	2.70370	8.54985
6.92	47.8864	2.63059	8.31865	7.32	53.5824	2.70555	8.55570
6.93	48.0249	2.63249	8.32466	7.33	53.7289	2.70740	8.56154
6.94	48.1636	2.63439	8.33067	7.34	53.8756	2.70924	8.56738
6.95	48.3025	2.63629	8.33667	7.35	54.0225	2.71109	8.57321
6.96	48.4416	2.63818	8.34266	7.36	54.1696	2.71293	8.57904
6.97	48.5809	2.64008	8.34865	7.37	54.3169	2.71477	8.58487
6.98	48.7204	2.64197	8.35464	7.38	54.4644	2.71662	8.59069
6.99	48.8601	2.64386	8.36062	7.39	54.6121	2.71846	8.59651

TABLE 1 (Continued)

N	N²	√N̄	√10N̄	N	N²	√N̄	√10N̄
7.40	54.7600	2.72029	8.60233	**7.80**	60.8400	2.79285	8.83176
7.41	54.9081	2.72213	8.60814	7.81	60.9961	2.79464	8.83742
7.42	55.0564	2.72397	8.61394	7.82	61.1524	2.79643	8.84308
7.43	55.2049	2.72580	8.61974	7.83	61.3089	2.79821	8.84873
7.44	55.3536	2.72764	8.62554	7.84	61.4656	2.80000	8.85438
7.45	55.5025	2.72947	8.63134	7.85	61.6225	2.80179	8.86002
7.46	55.6516	2.73130	8.63713	7.86	61.7796	2.80357	8.86566
7.47	55.8009	2.73313	8.64292	7.87	61.9369	2.80535	8.87130
7.48	55.9504	2.73496	8.64870	7.88	62.0944	2.80713	8.87694
7.49	56.1001	2.73679	8.65448	7.89	62.2521	2.80891	8.88257
7.50	56.2500	2.73861	8.66025	**7.90**	62.4100	2.81069	8.88819
7.51	56.4001	2.74044	8.66603	7.91	62.5681	2.81247	8.98382
7.52	56.5504	2.74226	8.67179	7.92	62.7264	2.81425	8.89944
7.53	56.7009	2.74408	8.67756	7.93	62.8849	2.81603	8.90505
7.54	56.8516	2.74591	8.68332	7.94	63.0436	2.81780	8.91067
7.55	57.0025	2.74773	8.68907	7.95	63.2025	2.81957	8.91628
7.56	57.1536	2.74955	8.69483	7.96	63.3616	2.82135	8.92188
7.57	57.3049	2.75136	8.70057	7.97	63.5209	2.82312	8.92749
7.58	57.4564	2.75318	8.70632	7.98	63.6804	2.82489	8.93308
7.59	57.6081	2.75500	8.71206	7.99	63.8401	2.82666	8.93868
7.60	57.7600	2.75681	8.71780	**8.00**	64.0000	2.82843	8.94427
7.61	57.9121	2.75862	8.72353	8.01	64.1601	2.83019	8.94986
7.62	58.0644	2.76043	8.72926	8.02	64.3204	2.83196	8.95545
7.63	58.2169	2.76225	8.73499	8.03	64.4809	2.83373	8.96103
7.64	58.3696	2.76405	8.74071	8.04	64.6416	2.83549	8.96660
7.65	58.5225	2.76586	8.74643	8.05	64.8025	2.83725	8.97218
7.66	58.6756	2.76767	8.75214	8.06	64.9636	2.83901	8.97775
7.67	58.8289	2.76948	8.75785	8.07	65.1249	2.84077	8.98332
7.68	58.9824	2.77128	8.76356	8.08	65.2864	2.84253	8.98888
7.69	59.1361	2.77308	8.76926	8.09	65.4481	2.84429	8.99444
7.70	59.2900	2.77489	8.77496	**8.10**	65.6100	2.84605	9.00000
7.71	59.4441	2.77669	8.78066	8.11	65.7721	2.84781	9.00555
7.72	59.5984	2.77849	8.78635	8.12	65.9344	2.84956	9.01110
7.73	59.7529	2.78029	8.79204	8.13	66.0969	2.85132	9.01665
7.74	59.9076	2.78209	8.79773	8.14	66.2596	2.85307	9.02219
7.75	60.0625	2.78388	8.80341	8.15	66.4225	2.85482	9.02774
7.76	60.2176	2.78568	8.80909	8.16	66.5856	2.85657	9.03327
7.77	60.3729	2.78747	8.81476	8.17	66.7489	2.85832	9.03881
7.78	60.5284	2.78927	8.82043	8.18	66.9124	2.86007	9.04434
7.79	60.6841	2.79106	8.82610	8.19	67.0761	2.86182	9.04986

(Continued)

TABLE 1 (Continued)

N	N²	√N	√10N	N	N²	√N	√10N
8.20	67.2400	2.86356	9.05539	**8.60**	73.9600	2.93258	9.27362
8.21	67.4041	2.86531	9.06091	8.61	74.1321	2.93428	9.27901
8.22	67.5684	2.86705	9.06642	8.62	74.3044	2.93598	9.28440
8.23	67.7329	2.86880	9.07193	8.63	74.4769	2.93769	9.28978
8.24	67.8976	2.87054	9.07744	8.64	74.6496	2.93939	9.29516
8.25	68.0625	2.87228	9.08295	8.65	74.8225	2.94109	9.30054
8.26	68.2276	2.87402	9.08845	8.66	74.9956	2.94279	9.30591
8.27	68.3929	2.87576	9.09395	8.67	75.1689	2.94449	9.31128
8.28	68.5584	2.87750	9.09945	8.68	75.3424	2.94618	9.31665
8.29	68.7241	2.87924	9.10494	8.69	75.5161	2.94788	9.32202
8.30	68.8900	2.88097	9.11043	**8.70**	75.6900	2.94958	9.32738
8.31	69.0561	2.88271	9.11592	8.71	75.8641	2.95127	9.33274
8.32	69.2224	2.88444	9.12140	8.72	76.0384	2.95296	9.33809
8.33	69.3889	2.88617	9.12688	8.73	76.2129	2.95466	9.34345
8.34	69.5556	2.88791	9.13236	8.74	76.3876	2.95635	9.34880
8.35	69.7225	2.88964	9.13783	8.75	76.5625	2.95804	9.35414
8.36	69.8896	2.89137	9.14330	8.76	76.7376	2.95973	9.35949
8.37	70.0569	2.89310	9.14877	8.77	76.9129	2.96142	9.36483
8.38	70.2244	2.89482	9.15423	8.78	77.0884	2.96311	9.37017
8.39	70.3921	2.89655	9.15969	8.79	77.2641	2.96479	9.37550
8.40	70.5600	2.89828	9.16515	**8.80**	77.4400	2.96648	9.38083
8.41	70.7281	2.90000	9.17061	8.81	77.6161	2.96816	9.38616
8.42	70.8964	2.90172	9.17606	8.82	77.7924	2.96985	9.39149
8.43	71.0649	2.90345	9.18150	8.83	77.9689	2.97153	9.39681
8.44	71.2336	2.90517	9.18695	8.84	78.1456	2.97321	9.40213
8.45	71.4025	2.90689	9.19239	8.85	78.3225	2.97489	9.40744
8.46	71.5716	2.90861	9.19783	8.86	78.4996	2.97658	9.41276
8.47	71.7490	2.91033	9.20326	8.87	78.6769	2.97825	9.41807
8.48	71.9104	2.91204	9.20869	8.88	78.8544	2.97993	9.42338
8.49	72.0801	2.91376	9.21412	8.89	79.0321	2.98161	9.42868
8.50	72.2500	2.91548	9.21954	**8.90**	79.2100	2.98329	9.43398
8.51	72.4201	2.91719	9.22497	8.91	79.3881	2.98496	9.43928
8.52	72.5904	2.91890	9.23038	8.92	79.5664	2.98664	9.44458
8.53	72.7609	2.92062	9.23580	8.93	79.7449	2.98831	9.44987
8.54	72.9316	2.92233	9.24121	8.94	79.9236	2.98998	9.45516
8.55	73.1025	2.92404	9.24662	8.95	80.1025	2.99166	9.46044
8.56	73.2736	2.92575	9.25203	8.96	80.2816	2.99333	9.46573
8.57	73.4449	2.92746	9.25743	8.97	80.4609	2.99500	9.47101
8.58	73.6164	2.92916	9.26283	8.98	80.6404	2.99666	9.47629
8.59	73.7881	2.93087	9.26823	8.99	80.8201	2.99833	9.48156

TABLE 1 (Continued)

N	N²	√N̄	√10N̄	N	N²	√N̄	√10N̄
9.00	81.0000	3.00000	9.48683	**9.40**	88.3600	3.06594	9.69536
9.01	81.1801	3.00167	9.49210	9.41	88.5481	3.06757	9.70052
9.02	81.3604	3.00333	9.49737	9.42	88.7364	3.06920	9.70567
9.03	81.5409	3.00500	9.50263	9.43	88.9249	3.07083	9.71082
9.04	81.7216	3.00666	9.50789	9.44	89.1136	3.07246	9.71597
9.05	81.9025	3.00832	9.51315	9.45	89.3025	3.07409	9.72111
9.06	82.0836	3.00998	9.51840	9.46	89.4916	3.07571	9.72625
9.07	82.2649	3.01164	9.52365	9.47	89.6809	3.07734	9.73139
9.08	82.4464	3.01330	9.52890	9.48	89.8704	3.07896	9.73653
9.09	82.6281	3.01496	9.53415	9.49	90.0601	3.08058	9.74166
9.10	82.8100	3.01662	9.53939	**9.50**	90.2500	3.08221	9.74679
9.11	82.9921	3.01828	9.54463	9.51	90.4401	3.08383	9.75192
9.12	83.1744	3.01993	9.54987	9.52	90.6304	3.08545	9.75705
9.13	83.3569	3.02159	9.55510	9.53	90.8209	3.08707	9.76217
9.14	83.5396	3.02324	9.56033	9.54	91.0116	3.08869	9.76729
9.15	83.7225	3.02490	9.56556	9.55	91.2025	3.09031	9.77241
9.16	83.9056	3.02655	9.57079	9.56	91.3936	3.09192	9.77753
9.17	84.0889	3.02820	9.57601	9.57	91.5849	3.09354	9.78264
9.18	84.2724	3.02985	9.58123	9.58	91.7764	3.09516	9.78775
9.19	84.4561	3.03150	9.58645	9.59	91.9681	3.09677	9.79285
9.20	84.6400	3.03315	9.59166	**9.60**	92.1600	3.09839	9.79796
9.21	84.8241	3.03480	9.59687	9.61	92.3521	3.10000	9.80306
9.22	85.0084	3.03645	9.60208	9.62	92.5444	3.10161	9.80816
9.23	85.1929	3.03809	9.60729	9.63	92.7369	3.10322	9.81326
9.24	85.3776	3.03974	9.61249	9.64	92.9296	3.10483	9.81835
9.25	85.5625	3.04138	9.61769	9.65	93.1225	3.10644	9.82344
9.26	85.7476	3.04302	9.62289	9.66	93.3156	3.10805	9.82853
9.27	85.9329	3.04467	9.62808	9.67	93.5089	3.10966	9.83362
9.28	86.1184	3.04631	9.63328	9.68	93.7024	3.11127	9.83870
9.29	86.3041	3.04795	9.63846	9.69	93.8961	3.11288	9.84378
9.30	86.4900	3.04959	9.64365	**9.70**	94.0900	3.11448	9.84886
9.31	86.6761	3.05123	9.64883	9.71	94.2841	3.11609	9.85393
9.32	86.8624	3.05287	9.65401	9.72	94.4784	3.11769	9.85901
9.33	87.0489	3.05450	9.65919	9.73	94.6729	3.11929	9.86408
9.34	87.2356	3.05614	9.66437	9.74	94.8676	3.12090	9.86914
9.35	87.4225	3.05778	9.66954	9.75	95.0625	3.12250	9.87421
9.36	87.6096	3.05941	9.67471	9.76	95.2576	3.12410	9.87927
9.37	87.7969	3.06105	9.67988	9.77	95.4529	3.12570	9.88433
9.38	87.9844	3.06268	9.68504	9.78	95.6484	3.12730	9.88939
9.39	88.1721	3.06431	9.69020	9.79	95.8441	3.12890	9.89444

(Continued)

TABLE 1 (Continued)

N	N²	√N	√10N	N	N²	√N	√10N
9.80	96.0400	3.13050	9.89949	**9.90**	98.0100	3.14643	9.94987
9.81	96.2361	3.13209	9.90454	9.91	98.2081	3.14802	9.95490
9.82	96.4324	3.13369	9.90959	9.92	98.4064	3.14960	9.95992
9.83	96.6289	3.13528	9.91464	9.93	98.6049	3.15119	9.96494
9.84	96.8256	3.13688	9.91968	9.94	98.8036	3.15278	9.96995
9.85	97.0225	3.13847	9.92472	9.95	99.0025	3.15436	9.97497
9.86	97.2196	3.14006	9.92975	9.96	99.2016	3.15595	9.97998
9.87	97.4169	3.14166	9.93479	9.97	99.4009	3.15753	9.98499
9.88	97.6144	3.14325	9.93982	9.98	99.6004	3.15911	9.98999
9.89	97.8121	3.14484	9.94485	9.99	99.8001	3.16070	9.99500
9.90	98.0100	3.14643	9.94987	**10.00**	100.0000	3.16228	10.00000

TABLE 2 Random Numbers

93108	77033	68325	10160	38667	62441	87023	94372	06164	30700
28271	08589	83279	48838	60935	70541	53814	95588	05832	80235
21841	35545	11148	34775	17308	88034	97765	35959	52843	44895
22025	79554	19698	25255	50283	94037	57463	92925	12042	91414
09210	20779	02994	02258	86978	85092	54052	18354	20914	28460
90552	71129	03621	20517	16908	06668	29916	51537	93658	29525
01130	06995	20258	10351	99248	51660	38861	49668	74742	47181
22604	56719	21784	68788	38358	59827	19270	99287	81193	43366
06690	01800	34272	65497	94891	14537	91358	21587	95765	72605
59809	69982	71809	64984	48709	43991	24987	69246	86400	29559
56475	02726	58511	95405	70293	84971	06676	44075	32338	31980
02730	34870	83209	03138	07715	31557	55242	61308	26507	06186
74482	33990	13509	92588	10462	76546	46097	01825	20153	36271
19793	22487	94238	81054	95488	23617	15539	94335	73822	93481
19020	27856	60526	24144	98021	60564	46373	86928	52135	74919
69565	60635	65709	77887	42766	86698	14004	94577	27936	47220
69274	23208	61035	84263	15034	28717	76146	22021	23779	98562
83658	14204	09445	41081	49630	34215	89806	40930	97194	21747
78612	51102	66826	40430	54072	62164	68977	95583	11765	81072
14980	74158	78216	38985	60838	82836	42777	85321	90463	11813
63172	28010	29405	91554	75195	51183	65805	87525	35952	83204
71167	37984	52737	06869	38122	95322	41356	19391	96787	64410
78530	56410	19195	34434	83712	50397	80920	15464	81350	18673
98324	03774	07573	67864	06497	20758	83454	22756	83959	96347
55793	30055	08373	32652	02654	75980	02095	87545	88815	80086
05674	34471	61967	91266	38814	44728	32455	17057	08339	93997
15643	22245	07592	22078	73628	60902	41561	54608	41023	98345
66750	19609	70358	03622	64898	82220	69304	46235	97332	64539
42320	74314	50222	82339	51564	42885	50482	98501	02245	88990
73752	73818	15470	04914	24936	65514	56633	72030	30856	85183
97546	02188	46373	21486	28221	08155	23486	66134	88799	49496
32569	52162	38444	42004	78011	16909	94194	79732	47114	23919
36048	93973	82596	28739	86985	58144	65007	08786	14826	04896
40455	36702	38965	56042	80023	28169	04174	65533	52718	55255
33597	47071	55618	51796	71027	46690	08002	45066	02870	60012
22828	96380	35883	15910	17211	42358	14056	55438	98148	35384
00631	95925	19324	31497	88118	06283	84596	72091	53987	01477
75722	36478	07634	63114	27164	15467	03983	09141	60562	65725
80577	01771	61510	17099	28731	41426	18853	41523	14914	76661
10524	20900	65463	83680	05005	11611	64426	59065	06758	02892
93815	69446	75253	51915	97839	75427	90685	60352	96288	34248
81867	97119	93446	20862	46591	97677	42704	13718	44975	67145
64649	07689	16711	12169	15238	74106	60655	56289	74166	78561
55768	09210	52439	33355	57884	36791	00853	49969	74814	09270
38080	49460	48137	61589	42742	92035	21766	19435	92579	27683
22360	16332	05343	34613	24013	98831	17157	44089	07366	66196
40521	09057	00239	51284	71556	22605	41293	54854	39736	05113
19292	69862	59951	49644	53486	28244	20714	56030	39292	45166
79504	40078	06838	05509	68581	39400	85615	52314	83202	40313
64138	27983	84048	42631	58658	62243	82572	45211	37060	15017

* Abstracted with permission from *A Million Random Digits with 100,000 Normal Deviates* by The Rand Corporation, The Free Press of Glencoe, New York (1955).

TABLE 3 Binomial Coefficients $\dbinom{n}{x} = \dfrac{n!}{x!(n-x)!}$

n	0	1	2	3	4	5	6	7	8	9	10
0	1										
1	1	1									
2	1	2	1								
3	1	3	3	1							
4	1	4	6	4	1						
5	1	5	10	10	5	1					
6	1	6	15	20	15	6	1				
7	1	7	21	35	35	21	7	1			
8	1	8	28	56	70	56	28	8	1		
9	1	9	36	84	126	126	84	36	9	1	
10	1	10	45	120	210	252	210	120	45	10	1
11	1	11	55	165	330	462	462	330	165	55	11
12	1	12	66	220	495	792	924	792	495	220	66
13	1	13	78	286	715	1287	1716	1716	1287	715	286
14	1	14	91	364	1001	2002	3003	3432	3003	2002	1001
15	1	15	105	455	1365	3003	5005	6435	6435	5005	3003
16	1	16	120	560	1820	4368	8008	11440	12870	11440	8008
17	1	17	136	680	2380	6188	12376	19448	24310	24310	19448
18	1	18	153	816	3060	8568	18564	31824	43758	48620	43758
19	1	19	171	969	3876	11628	27132	50388	75582	92378	92378
20	1	20	190	1140	4845	15504	38760	77520	125970	167960	184756

TABLE 4 Values of Negative Exponential Function e^{-x}.

x	e^{-x}	x	e^{-x}	x	e^{-x}	x	e^{-x}	x	e^{-x}
0.0	1.000	2.0	0.135	4.0	0.018	6.0	0.0025	8.0	0.00034
0.1	0.905	2.1	0.122	4.1	0.017	6.1	0.0022	8.1	0.00030
0.2	0.819	2.2	0.111	4.2	0.015	6.2	0.0020	8.2	0.00028
0.3	0.741	2.3	0.100	4.3	0.014	6.3	0.0018	8.3	0.00025
0.4	0.670	2.4	0.091	4.4	0.012	6.4	0.0017	8.4	0.00023
0.5	0.607	2.5	0.082	4.5	0.011	6.5	0.0015	8.5	0.00020
0.6	0.549	2.6	0.074	4.6	0.010	6.6	0.0014	8.6	0.00018
0.7	0.497	2.7	0.067	4.7	0.009	6.7	0.0012	8.7	0.00017
0.8	0.449	2.8	0.061	4.8	0.008	6.8	0.0011	8.8	0.00015
0.9	0.407	2.9	0.055	4.9	0.007	6.9	0.0010	8.9	0.00014
1.0	0.368	3.0	0.050	5.0	0.0067	7.0	0.0009	9.0	0.00012
1.1	0.333	3.1	0.045	5.1	0.0061	7.1	0.0008	9.1	0.00011
1.2	0.301	3.2	0.041	5.2	0.0055	7.2	0.0007	9.2	0.00010
1.3	0.273	3.3	0.037	5.3	0.0050	7.3	0.0007	9.3	0.00009
1.4	0.247	3.4	0.033	5.4	0.0045	7.4	0.0006	9.4	0.00008
1.5	0.223	3.5	0.030	5.5	0.0041	7.5	0.00055	9.5	0.00008
1.6	0.202	3.6	0.027	5.6	0.0037	7.6	0.00050	9.6	0.00007
1.7	0.183	3.7	0.025	5.7	0.0033	7.7	0.00045	9.7	0.00006
1.8	0.165	3.8	0.022	5.8	0.0030	7.8	0.00041	9.8	0.00006
1.9	0.150	3.9	0.020	5.9	0.0027	7.9	0.00037	9.9	0.00005

TABLE 5 Areas Under the Normal Probability Curve

A denotes the area between the line of symmetry (i.e. $Z = 0$) and the given Z-value.

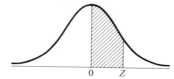

Z	A	Z	A	Z	A	Z	A
0.00	0.0000	0.47	0.1808	0.94	0.3264	1.41	0.4207
.01	.0040	.48	.1844	.95	.3289	1.42	.4222
.02	.0080	.49	.1879	.96	.3315	1.43	.4236
.03	.0120	.50	.1915	.97	.3340	1.44	.4251
.04	.0160	.51	.1950	.98	.3365	1.45	.4265
.05	.0199	.52	.1985	.99	.3389	1.46	.4279
.06	.0239	.53	.2019	1.00	.3413	1.47	.4292
.07	.0279	.54	.2054	1.01	.3438	1.48	.4306
.08	.0319	.55	.2088	1.02	.3461	1.49	.4319
.09	.0359	.56	.2123	1.03	.3485	1.50	.4332
.10	.0398	.57	.2157	1.04	.3508	1.51	.4345
.11	.0438	.58	.2190	1.05	.3531	1.52	.4357
.12	.0478	.59	.2224	1.06	.3554	1.53	.4370
.13	.0517	.60	.2258	1.07	.3577	1.54	.4382
.14	.0557	.61	.2291	1.08	.3599	1.55	.4394
.15	.0596	.62	.2324	1.09	.3621	1.56	.4406
.16	.0636	.63	.2357	1.10	.3643	1.57	.4418
.17	.0675	.64	.2389	1.11	.3665	1.58	.4430
.18	.0714	.65	.2422	1.12	.3686	1.59	.4441
.19	.0754	.66	.2454	1.13	.3708	1.60	.4452
.20	.0793	.67	.2486	1.14	.3729	1.61	.4463
.21	.0832	.68	.2518	1.15	.3749	1.62	.4474
.22	.0871	.69	.2549	1.16	.3770	1.63	.4485
.23	.0910	.70	.2580	1.17	.3790	1.64	.4495
.24	.0948	.71	.2612	1.18	.3810	1.65	.4505
.25	.0987	.72	.2642	1.19	.3830	1.66	.4515
.26	.1026	.73	.2673	1.20	.3849	1.67	.4525
.27	.1064	.74	.2704	1.21	.3869	1.68	.4535
.28	.1103	.75	.2734	1.22	.3888	1.69	.4545
.29	.1141	.76	.2764	1.23	.3907	1.70	.4554
30	.1179	.77	.2794	1.24	.3925	1.71	.4564
.31	.1217	.78	2823	1.25	.3944	1.72	.4573
.32	.1255	.79	.2852	1.26	.3962	1.73	.4582
.33	.1293	.80	.2881	1.27	.3980	1.74	.4591
.34	.1331	.81	.2910	1.28	.3997	1.75	.4599
.35	.1368	.82	.2939	1.29	.4015	1.76	4608
.36	.1406	.83	.2967	1.30	.4032	1.77	.4616
.37	.1443	.84	.2996	1.31	.4049	1.78	.4625
.38	.1480	.85	.3023	1.32	.4066	1.79	.4633
.39	.1517	.86	.3051	1.33	.4082	1.80	.4641
.40	.1554	.87	.3079	1.34	.4099	1.81	.4649
.41	.1591	.88	.3106	1.35	.4115	1.82	.4656
.42	.1628	.89	.3133	1.36	.4131	1.83	.4664
.43	.1664	.90	.3159	1.37	.4147	1.84	.4671
.44	.1700	.91	.3186	1.38	.4162	1.85	.4678
.45	.1736	.92	.3212	1.39	.4177	1.86	.4686
.46	.1772	.93	3238	1.40	.4192	1.87	.4693

TABLE 5 (Continued)

Z	A	Z	A	Z	A	Z	A
1.88	0.4700	2.41	0.4920	2.94	0.4984	3.47	0.4997
1.89	.4706	2.42	.4922	2.95	.4984	3.48	.4998
1.90	.4713	2.43	.4925	2.96	.4985	3.49	.4998
1.91	.4719	2.44	.4927	2.97	.4985	3.50	.4998
1.92	.4726	2.45	.4929	2.98	.4986	3.51	.4998
1.93	.4732	2.46	.4931	2.99	.4986	3.52	.4998
1.94	.4738	2.47	.4932	3.00	.4987	3.53	.4998
1.95	.4744	2.48	.4934	3.01	.4987	3.54	.4998
1.96	.4750	2.49	.4936	3.02	.4987	3.55	.4998
1.97	.4756	2.50	.4938	3.03	.4988	3.56	.4998
1.98	.4762	2.51	.4940	3.04	.4988	3.57	.4998
1.99	.4767	2.52	.4941	3.05	.4989	3.58	.4998
2.00	.4773	2.53	.4943	3.06	.4989	3.59	.4998
2.01	.4778	2.54	.4945	3.07	.4989	3.60	.4998
2.02	.4783	2.55	.4946	3.08	.4990	3.61	.4999
2.03	.4788	2.56	.4948	3.09	.4990	3.62	.4999
2.04	.4793	2.57	.4949	3.10	.4990	3.63	.4999
2.05	.4798	2.58	.4951	3.11	.4991	3.64	.4999
2.06	.4803	2.59	.4952	3.12	.4991	3.65	.4999
2.07	.4808	2.60	.4953	3.13	.4991	3.66	.4999
2.08	.4812	2.61	.4955	3.14	.4992	3.67	.4999
2.09	.4817	2.62	.4956	3.15	.4992	3.68	.4999
2.10	.4821	2.63	.4957	3.16	.4992	3.69	.4999
2.11	.4826	2.64	.4959	3.17	.4992	3.70	.4999
2.12	.4830	2.65	.4960	3.18	.4993	3.71	.4999
2.13	.4834	2.66	.4961	3.19	.4993	3.72	.4999
2.14	.4838	2.67	.4962	3.20	.4993	3.73	.4999
2.15	.4842	2.68	.4963	3.21	.4993	3.74	.4999
2.16	.4846	2.69	.4964	3.22	.4994	3.75	.4999
2.17	.4850	2.70	.4965	3.23	.4994	3.76	.4999
2.18	.4854	2.71	.4966	3.24	.4994	3.77	.4999
2.19	.4857	2.72	.4967	3.25	.4994	3.78	.4999
2.20	.4861	2.73	.4968	3.26	.4994	3.79	.4999
2.21	.4865	2.74	.4969	3.27	.4995	3.80	.4999
2.22	.4868	2.75	.4970	3.28	.4995	3.81	.4999
2.23	.4871	2.76	.4971	3.29	.4995	3.82	.4999
2.24	.4875	2.77	.4972	3.30	.4995	3.83	.4999
2.25	.4878	2.78	.4973	3.31	.4995	3.84	.4999
2.26	.4881	2.79	.4974	3.32	.4996	3.85	.4999
2.27	.4884	2.80	.4974	3.33	.4996	3.86	.4999
2.28	.4887	2.81	.4975	3.34	.4996	3.87	.5000
2.29	.4890	2.82	.4976	3.35	.4996	3.88	.5000
2.30	.4893	2.83	.4977	3.36	.4996	3.89	.5000
2.31	.4896	2.84	.4977	3.37	.4996		
2.32	.4898	2.85	.4978	3.38	.4996		
2.33	.4901	2.86	.4979	3.39	.4997		
2.34	.4904	2.87	.4980	3.40	.4997		
2.35	.4906	2.88	.4980	3.41	.4997		
2.36	.4909	2.89	.4981	3.42	.4997		
2.37	.4911	2.90	.4981	3.43	.4997		
2.38	.4910	2.91	.4982	3.44	.4997		
2.39	.4916	2.92	.4983	3.45	.4997		
2.40	.4918	2.93	.4983	3.46	.4997		

TABLE 6 Percentage Points of the *t*-distribution

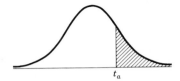

t_a

d. f.	$t_{.40}$	$t_{.30}$	$t_{.20}$	$t_{.10}$	$t_{.05}$	$t_{.025}$	$t_{.01}$	$t_{.005}$	$t_{.0005}$
1	0.3250	0.7270	1.376	3.078	6.3138	12.706	31.821	63.657	636.619
2	.2885	.6172	1.061	1.886	2.9200	4.3027	6.965	9.9248	31.598
3	.2766	.5840	.978	1.638	2.3534	3.1825	4.541	5.8409	12.924
4	.2707	.5692	.941	1.533	2.1318	2.7764	3.747	4.6041	8.610
5	.2672	.5598	.920	1.476	2.0150	2.5706	3.365	4.0321	6.869
6	.2648	.5536	.906	1.440	1.9432	2.4469	3.143	3.7074	5.959
7	.2632	.5493	.896	1.415	1.8946	2.3646	2.998	3.4995	5.408
8	.2619	.5461	.889	1.397	1.8595	2.3060	2.896	3.3554	5.041
9	.2610	.5436	.883	1.383	1.8331	2.2622	2.821	3.2498	4.781
10	.2602	.5416	.879	1.372	1.8125	2.2281	2.764	3.1693	4.587
11	.2596	.5400	.876	1.363	1.7939	2.2010	2.718	3.1058	4.437
12	.2590	.5387	.873	1.356	1.7823	2.1788	2.681	3.0545	4.318
13	.2586	.5375	.870	1.350	1.7709	2.1604	2.650	3.0123	4.221
14	.2582	.5366	.868	1.345	1.7613	2.1448	2.624	2.9768	4.140
15	.2579	.5358	.866	1.341	1.7530	2.1315	2.602	2.9467	4.073
16	.2576	.5351	.865	1.337	1.7459	2.1199	2.583	2.9208	4.015
17	.2574	.5344	.863	1.333	1.7396	2.1098	2.567	2.8982	3.965
18	.2571	.5338	.862	1.330	1.7341	2.1009	2.552	2.8784	3.922
19	.2569	.5333	.861	1.328	1.7291	2.0930	2.539	2.8609	3.883
20	.2567	.5329	.860	1.325	1.7247	2.0860	2.528	2.8453	3.850
21	.2566	.5325	.859	1.323	1.7207	2.0796	2.518	2.8314	3.819
22	.2564	.5321	.858	1.321	1.7171	2.0739	2.508	2.8188	3.792
23	.2563	.5318	.858	1.319	1.7139	2.0687	2.500	2.9073	3.767
24	.2562	.5315	.857	1.318	1.7109	2.0639	2.492	2.7969	3.745
25	.2561	.5312	.856	1.316	1.7081	2.0595	2.485	2.7874	3.725
26	.2560	.5309	.856	1.315	1.7056	2.0555	2.479	2.7787	3.707
27	.2559	.5307	.855	1.314	1.7033	2.0518	2.473	2.7707	3.690
28	.2558	.5304	.855	1.313	1.7011	2.0484	2.467	2.7633	3.674
29	.2557	.5302	.854	1.311	1.6991	2.0452	2.462	2.7564	3.659
30	.2556	.5300	.854	1.310	1.6973	2.0423	2.457	2.7500	3.616
35	.2553	.5292	.8521	1.3062	1.6896	2.0301	2.438	2.7239	3.5919
40	.2550	.5286	.8507	1.3031	1.6839	2.0211	2.423	2.7045	3.5511
45	.2549	.5281	.8497	1.3007	1.6794	2.0141	2.412	2.6896	3.5207
50	.2547	.5278	.8489	1.2987	1.6759	2.0086	2.403	2.6778	3.4965
60	.2545	.5272	.8477	1.2959	1.6707	2.0003	2.390	2.6603	3.4606
70	.2543	.5268	.8468	1.2938	1.6669	1.9945	2.381	2.6480	3.4355
80	.2542	.5265	.8462	1.2922	1.6641	1.9901	2.374	2.6388	3.4169
90	.2541	.5263	.8457	1.2910	1.6620	1.9867	2.368	2.6316	3.4022
100	.2540	.5261	.8452	1.2901	1.6602	1.9840	2.364	2.6260	3.3909
120	2539	.5258	.8446	1.2887	1.6577	1.9799	2.358	2.6175	3.3736
140	.2538	.5256	.8442	1.2876	1.6558	1.9771	2.353	2.6114	3.3615
160	.2538	.5255	.8439	1.2869	1.6545	1.9749	2.350	2.6070	3.3527
180	.2537	.5253	.8436	1.2863	1.6534	1.9733	2.347	2.6035	3.3456
200	.2537	.5252	.8434	1.2858	1.6525	1.9719	2.345	2.6006	3.3400
∞	.2533	.5244	.8416	1.2816	1.6449	1.9600	2.326	2.5758	3.2905

* Reproduced from *Documenta Geigy Scientific Tables,* 7th edition, by permission of CIBA-GEIGY Limited, Basle, Switzerland.

TABLE 7 Percentage Points of the Chi-Square distribution

$\chi^2 \cdot 1 - \alpha$

χ^2

# of classes
-1

d.f.	$\chi^2_{.005}$	$\chi^2_{.010}$	$\chi^2_{.025}$	$\chi^2_{.050}$	$\chi^2_{.100}$	$\chi^2_{.250}$	$\chi^2_{.500}$
1	392704.10^{-10}	157088.10^{-9}	982069.10^{-9}	393214.10^{-8}	0·0157908	0·1015308	0·454936
2	0·0100251	0·0201007	0·0506356	0·102587	0·210721	0·575364	1·38629
3	0·0717218	0·114832	0·215795	0·351846	0·584374	1·212534	2·36597
4	0·206989	0·297109	0·484419	0·710723	1·063623	1·92256	3·35669
5	0·411742	0·554298	0·831212	1·145476	1·61031	2·67460	4·35146
6	0·675727	0·872090	1·23734	1·63538	2·20413	3·45460	5·34812
7	0·989256	1·239043	1·68987	2·16735	2·83311	4·25485	6·34581
8	1·34441	1·64650	2·17973	2·73264	3·48954	5·07064	7·34412
9	1·73493	2·08790	2·70039	3·32511	4·16816	5·89883	8·34283
10	2·15586	2·55821	3·24697	3·94030	4·86518	6·73720	9·34182
11	2·60322	3·05348	3·81575	4·57481	5·57778	7·58414	10·3410
12	3·07382	3·57057	4·40379	5·22603	6·30380	8·43842	11·3403
13	3·56503	4·10692	5·00875	5·89186	7·04150	9·29907	12·3398
14	4·07467	4·66043	5·62873	6·57063	7·78953	10·1653	13·3393
15	4·60092	5·22935	6·26214	7·26094	8·54676	11·0365	14·3389
16	5·14221	5·81221	6·90766	7·96165	9·31224	11·9122	15·3385
17	5·69722	6·40776	7·56419	8·67176	10·0852	12·7919	16·3382
18	6·26480	7·01491	8·23075	9·39046	10·8649	13·6753	17·3379
19	6·84397	7·63273	8·90652	10·1170	11·6509	14·5620	18·3377
20	7·43384	8·26040	9·59078	10·8508	12·4426	15·4518	19·3374
21	8·03365	8·89720	10·28293	11·5913	13·2396	16·3444	20·3372
22	8·64272	9·54249	10·9823	12·3380	14·0415	17·2396	21·3370
23	9·26043	10·19567	11·6886	13·0905	14·8480	18·1373	22·3369
24	9·88623	10·8564	12·4012	13·8484	15·6587	19·0373	23·3367
25	10·5197	11·5240	13·1197	14·6114	16·4734	19·9393	24·3366
26	11·1602	12·1981	13·8439	15·3792	17·2919	20·8434	25·3365
27	11·8076	12·8785	14·5734	16·1514	18·1139	21·7494	26·3363
28	12·4613	13·5647	15·3079	16·9279	18·9392	22·6572	27·3362
29	13·1211	14·2565	16·0471	17·7084	19·7677	23·5666	28·3361
30	13·7867	14·9535	16·7908	18·4927	20·5992	24·4776	29·3360
40	20·7065	22·1643	24·4330	26·5093	29·0505	33·6603	39·3353
50	27·9907	29·7067	32·3574	34·7643	37·6886	42·9421	49·3349
60	35·5345	37·4849	40·4817	43·1880	46·4589	52·2938	59·3347
70	43·2752	45·4417	48·7576	51·7393	55·3289	61·6983	69·3345
80	51·1719	53·5401	57·1532	60·3915	64·2778	71·1445	79·3343
90	59·1963	61·7541	65·6466	69·1260	73·2911	80·6247	89·3342
100	67·3276	70·0649	74·2219	77·9295	82·3581	90·1332	99·3341
X	− 2·5758	− 2·3263	− 1·9600	− 1·6449	− 1·2816	− 0·6745	0·0000

* Abridged with permission from *Biometrika Tables for Statisticians*, Vol. I, Edited by E. S. Pearson and H. O. Hartley, Cambridge University Press (1966).

TABLE 7 (Continued)

$d.f.$	$\chi^2_{.750}$	$\chi^2_{.900}$	$\chi^2_{.950}$	$\chi^2_{.975}$	$\chi^2_{.990}$	$\chi^2_{.995}$	$\chi^2_{.999}$
1	1·32330	2·70554	3·84146	5·02389	6·63490	7·87944	10·828
2	2·77259	4·60517	5·99146	7·37776	9·21034	10·5966	13·816
3	4·10834	6·25139	7·81473	9·34840	11·3449	12·8382	16·266
4	5·38527	7·77944	9·48773	11·1433	13·2767	14·8603	18·467
5	6·62568	9·23636	11·0705	12·8325	15·0863	16·7496	20·515
6	7·84080	10·6446	12·5916	14·4494	16·8119	18·5476	22·458
7	9·03715	12·0170	14·0671	16·0128	18·4753	20·2777	24·322
8	10·2189	13·3616	15·5073	17·5345	20·0902	21·9550	26·125
9	11·3888	14·6837	16·9190	19·0228	21·6660	23·5894	27·877
10	12·5489	15·9872	18·3070	20·4832	23·2093	25·1882	29·588
11	13·7007	17·2750	19·6751	21·9200	24·7250	26·7568	31·264
12	14·8454	18·5493	21·0261	23·3367	26·2170	28·2995	32·909
13	15·9839	19·8119	22·3620	24·7356	27·6882	29·8195	34·528
14	17·1169	21·0641	23·6848	26·1189	29·1412	31·3194	36·123
15	18·2451	22·3071	24·9958	27·4884	30·5779	32·8013	37·697
16	19·3689	23·5418	26·2962	28·8454	31·9999	34·2672	39·252
17	20·4887	24·7690	27·5871	30·1910	33·4087	35·7185	40·790
18	21·6049	25·9894	28·8693	31·5264	34·8053	37·1565	42·312
19	22·7178	27·2036	30·1435	32·8523	36·1909	38·5823	43·820
20	23·8277	28·4120	31·4104	34·1696	37·5662	39·9968	45·315
21	24·9348	29·6151	32·6706	35·4789	38·9322	41·4011	46·797
22	26·0393	30·8133	33·9244	36·7807	40·2894	42·7957	48·268
23	27·1413	32·0069	35·1725	38·0756	41·6384	44·1813	49·728
24	28·2412	33·1962	36·4150	39·3641	42·9798	45·5585	51·179
25	29·3389	34·3816	37·6525	40·6465	44·3141	46·9279	52·618
26	30·4346	35·5632	38·8851	41·9232	45·6417	48·2899	54·052
27	31·5284	36·7412	40·1133	43·1945	46·9629	49·6449	55·476
28	32·6205	37·9159	41·3371	44·4608	48·2782	50·9934	56·892
29	33·7109	39·0875	42·5570	45·7223	49·5879	52·3356	58·301
30	34·7997	40·2560	43·7730	46·9792	50·8922	53·6720	59·703
40	45·6160	51·8051	55·7585	59·3417	63·6907	66·7660	73·402
50	56·3336	63·1671	67·5048	71·4202	76·1539	79·4900	86·661
60	66·9815	74·3970	79·0819	83·2977	88·3794	91·9517	99·607
70	77·5767	85·5270	90·5312	95·0232	100·425	104·215	112·317
80	88·1303	96·5782	101·879	106·629	112·329	116·321	124·839
90	98·6499	107·565	113·145	118·136	124·116	128·299	137·208
100	109·141	118·498	124·342	129·561	135·807	140·169	149·449
X	+0·6745	+1·2816	+1·6449	+1·9600	+2·3263	+2·5758	+3·0902

TABLE 8(a)   Binomial Confidence Intervals with $1 - \alpha = 0.95$

The numbers printed along the curves indicate the sample size n. If for a given value of the abscissa $\hat{p}$, p_A and p_B are the ordinates read from (or interpolated between) the appropriate lower and upper curves, then

$$Pr\{p_A \leqslant p \leqslant p_B\} \leqslant 1 - 2\alpha.$$

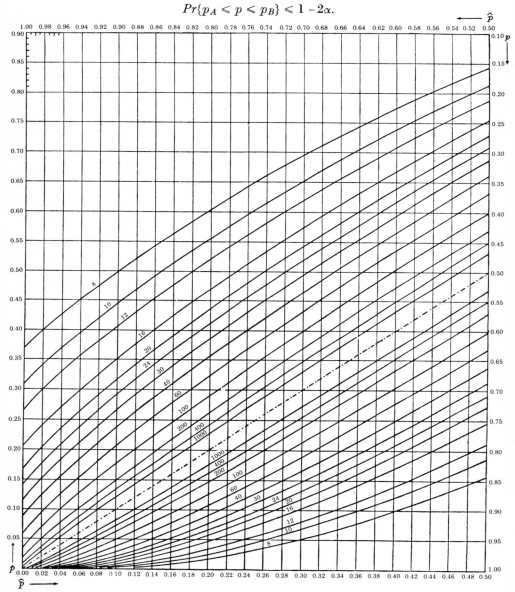

* Reproduced with permission from *Biometrika Tables for Statisticians*, Vol. I, Edited by E. S. Pearson and H. O. Hartley, Cambridge University Press (1966).

TABLE 8(b) Binomial Confidence Intervals with $1 - \alpha = 0.99$

The numbers printed along the curves indicate the sample size n.

Note: the process of reading from the curves can be simplified with the help of the right-angled corner of a loose sheet of paper or thin card, along the edges of which are marked off the scales shown in the top left-hand corner of each chart.

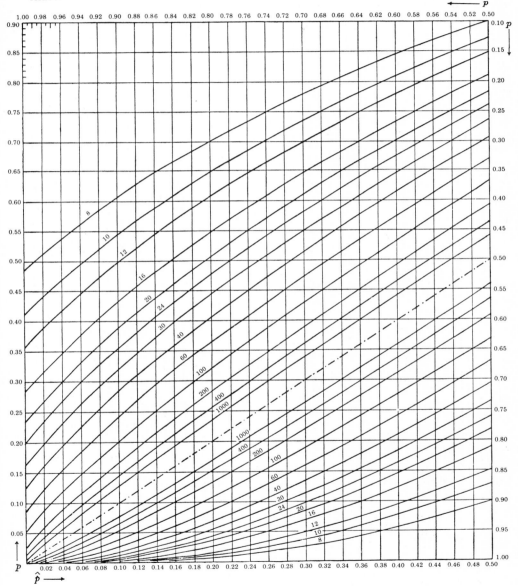

TABLE 9(a)   95% Significance points of the F-distribution

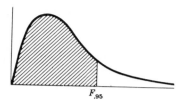

$F_{.95}$

ν_1 / ν_2	1	2	3	4	5	6	7	8	9
1	161·4	199·5	215·7	224·6	230·2	234·0	236·8	238·9	240·5
2	18·51	19·00	19·16	19·25	19·30	19·33	19·35	19·37	19·38
3	10·13	9·55	9·28	9·12	9·01	8·94	8·89	8·85	8·81
4	7·71	6·94	6·59	6·39	6·26	6·16	6·09	6·04	6·00
5	6·61	5·79	5·41	5·19	5·05	4·95	4·88	4·82	4·77
6	5·99	5·14	4·76	4·53	4·39	4·28	4·21	4·15	4·10
7	5·59	4·74	4·35	4·12	3·97	3·87	3·79	3·73	3·68
8	5·32	4·46	4·07	3·84	3·69	3·58	3·50	3·44	3·39
9	5·12	4·26	3·86	3·63	3·48	3·37	3·29	3·23	3·18
10	4·96	4·10	3·71	3·48	3·33	3·22	3·14	3·07	3·02
11	4·84	3·98	3·59	3·36	3·20	3·09	3·01	2·95	2·90
12	4·75	3·89	3·49	3·26	3·11	3·00	2·91	2·85	2·80
13	4·67	3·81	3·41	3·18	3·03	2·92	2·83	2·77	2·71
14	4·60	3·74	3·34	3·11	2·96	2·85	2·76	2·70	2·65
15	4·54	3·68	3·29	3·06	2·90	2·79	2·71	2·64	2·59
16	4·49	3·63	3·24	3·01	2·85	2·74	2·66	2·59	2·54
17	4·45	3·59	3·20	2·96	2·81	2·70	2·61	2·55	2·49
18	4·41	3·55	3·16	2·93	2·77	2·66	2·58	2·51	2·46
19	4·38	3·52	3·13	2·90	2·74	2·63	2·54	2·48	2·42
20	4·35	3·49	3·10	2·87	2·71	2·60	2·51	2·45	2·39
21	4·32	3·47	3·07	2·84	2·68	2·57	2·49	2·42	2·37
22	4·30	3·44	3·05	2·82	2·66	2·55	2·46	2·40	2·34
23	4·28	3·42	3·03	2·80	2·64	2·53	2·44	2·37	2·32
24	4·26	3·40	3·01	2·78	2·62	2·51	2·42	2·36	2·30
25	4·24	3·39	2·99	2·76	2·60	2·49	2·40	2·34	2·28
26	4·23	3·37	2·98	2·74	2·59	2·47	2·39	2·32	2·27
27	4·21	3·35	2·96	2·73	2·57	2·46	2·37	2·31	2·25
28	4·20	3·34	2·95	2·71	2·56	2·45	2·36	2·29	2·24
29	4·18	3·33	2·93	2·70	2·55	2·43	2·35	2·28	2·22
30	4·17	3·32	2·92	2·69	2·53	2·42	2·33	2·27	2·21
40	4·08	3·23	2·84	2·61	2·45	2·34	2·25	2·18	2·12
60	4·00	3·15	2·76	2·53	2·37	2·25	2·17	2·10	2·04
120	3·92	3·07	2·68	2·45	2·29	2·17	2·09	2·02	1·96
∞	3·84	3·00	2·60	2·37	2·21	2·10	2·01	1·94	1·88

* Abridged with permission from *Biometrika Tables for Statisticians*, Vol. I, Edited by E. S. Pearson and H. O. Hartley, Cambridge University Press (1966).

TABLE 9 (a) (Continued)

ν_2 \ ν_1	10	12	15	20	24	30	40	60	120	∞
1	241·9	243·9	245·9	248·0	249·1	250·1	251·1	252·2	253·3	254·3
2	19·40	19·41	19·43	19·45	19·45	19·46	19·47	19·48	19·49	19·50
3	8·79	8·74	8·70	8·66	8·64	8·62	8·59	8·57	8·55	8·53
4	5·96	5·91	5·86	5·80	5·77	5·75	5·72	5·69	5·66	5·63
5	4·74	4·68	4·62	4·56	4·53	4·50	4·46	4·43	4·40	4·36
6	4·06	4·00	3·94	3·87	3·84	3·81	3·77	3·74	3·70	3·67
7	3·64	3·57	3·51	3·44	3·41	3·38	3·34	3·30	3·27	3·23
8	3·35	3·28	3·22	3·15	3·12	3·08	3·04	3·01	2·97	2·93
9	3·14	3·07	3·01	2·94	2·90	2·86	2·83	2·79	2·75	2·71
10	2·98	2·91	2·85	2·77	2·74	2·70	2·66	2·62	2·58	2·54
11	2·85	2·79	2·72	2·65	2·61	2·57	2·53	2·49	2·45	2·40
12	2·75	2·69	2·62	2·54	2·51	2·47	2·43	2·38	2·34	2·30
13	2·67	2·60	2·53	2·46	2·42	2·38	2·34	2·30	2·25	2·21
14	2·60	2·53	2·46	2·39	2·35	2·31	2·27	2·22	2·18	2·13
15	2·54	2·48	2·40	2·33	2·29	2·25	2·20	2·16	2·11	2·07
16	2·49	2·42	2·35	2·28	2·24	2·19	2·15	2·11	2·06	2·01
17	2·45	2·38	2·31	2·23	2·19	2·15	2·10	2·06	2·01	1·96
18	2·41	2·34	2·27	2·19	2·15	2·11	2·06	2·02	1·97	1·92
19	2·38	2·31	2·23	2·16	2·11	2·07	2·03	1·98	1·93	1·88
20	2·35	2·28	2·20	2·12	2·08	2·04	1·99	1·95	1·90	1·84
21	2·32	2·25	2·18	2·10	2·05	2·01	1·96	1·92	1·87	1·81
22	2·30	2·23	2·15	2·07	2·03	1·98	1·94	1·89	1·84	1·78
23	2·27	2·20	2·13	2·05	2·01	1·96	1·91	1·86	1·81	1·76
24	2·25	2·18	2·11	2·03	1·98	1·94	1·89	1·84	1·79	1·73
25	2·24	2·16	2·09	2·01	1·96	1·92	1·87	1·82	1·77	1·71
26	2·22	2·15	2·07	1·99	1·95	1·90	1·85	1·80	1·75	1·69
27	2·20	2·13	2·06	1·97	1·93	1·88	1·84	1·79	1·73	1·67
28	2·19	2·12	2·04	1·96	1·91	1·87	1·82	1·77	1·71	1·65
29	2·18	2·10	2·03	1·94	1·90	1·85	1·81	1·75	1·70	1·64
30	2·16	2·09	2·01	1·93	1·89	1·84	1·79	1·74	1·68	1·62
40	2·08	2·00	1·92	1·84	1·79	1·74	1·69	1·64	1·58	1·51
60	1·99	1·92	1·84	1·75	1·70	1·65	1·59	1·53	1·47	1·39
120	1·91	1·83	1·75	1·66	1·61	1·55	1·50	1·43	1·35	1·25
∞	1·83	1·75	1·67	1·57	1·52	1·46	1·39	1·32	1·22	1·00

TABLE 9(b) 99% Significance points of the F-distribution

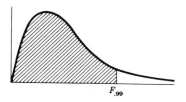

$F_{.99}$

ν_2 \ ν_1	1	2	3	4	5	6	7	8	9
1	4052	4999·5	5403	5625	5764	5859	5928	5981	6022
2	98·50	99·00	99·17	99·25	99·30	99·33	99·36	99·37	99·39
3	34·12	30·82	29·46	28·71	28·24	27·91	27·67	27·49	27·35
4	21·20	18·00	16·69	15·98	15·52	15·21	14·98	14·80	14·66
5	16·26	13·27	12·06	11·39	10·97	10·67	10·46	10·29	10·16
6	13·75	10·92	9·78	9·15	8·75	8·47	8·26	8·10	7·98
7	12·25	9·55	8·45	7·85	7·46	7·19	6·99	6·84	6·72
8	11·26	8·65	7·59	7·01	6·63	6·37	6·18	6·03	5·91
9	10·56	8·02	6·99	6·42	6·06	5·80	5·61	5·47	5·35
10	10·04	7·56	6·55	5·99	5·64	5·39	5·20	5·06	4·94
11	9·65	7·21	6·22	5·67	5·32	5·07	4·89	4·74	4·63
12	9·33	6·93	5·95	5·41	5·06	4·82	4·64	4·50	4·39
13	9·07	6·70	5·74	5·21	4·86	4·62	4·44	4·30	4·19
14	8·86	6·51	5·56	5·04	4·69	4·46	4·28	4·14	4·03
15	8·68	6·36	5·42	4·89	4·56	4·32	4·14	4·00	3·89
16	8·53	6·23	5·29	4·77	4·44	4·20	4·03	3·89	3·78
17	8·40	6·11	5·18	4·67	4·34	4·10	3·93	3·79	3·68
18	8·29	6·01	5·09	4·58	4·25	4·01	3·84	3·71	3·60
19	8·18	5·93	5·01	4·50	4·17	3·94	3·77	3·63	3·52
20	8·10	5·85	4·94	4·43	4·10	3·87	3·70	3·56	3·46
21	8·02	5·78	4·87	4·37	4·04	3·81	3·64	3·51	3·40
22	7·95	5·72	4·82	4·31	3·99	3·76	3·59	3·45	3·35
23	7·88	5·66	4·76	4·26	3·94	3·71	3·54	3·41	3·30
24	7·82	5·61	4·72	4·22	3·90	3·67	3·50	3·36	3·26
25	7·77	5·57	4·68	4·18	3·85	3·63	3·46	3·32	3·22
26	7·72	5·53	4·64	4·14	3·82	3·59	3·42	3·29	3·18
27	7·68	5·49	4·60	4·11	3·78	3·56	3·39	3·26	3·15
28	7·64	5·45	4·57	4·07	3·75	3·53	3·36	3·23	3·12
29	7·60	5·42	4·54	4·04	3·73	3·50	3·33	3·20	3·09
30	7·56	5·39	4·51	4·02	3·70	3·47	3·30	3·17	3·07
40	7·31	5·18	4·31	3·83	3·51	3·29	3·12	2·99	2·89
60	7·08	4·98	4·13	3·65	3·34	3·12	2·95	2·82	2·72
120	6·85	4·79	3·95	3·48	3·17	2·96	2·79	2·66	2·56
∞	6·63	4·61	3·78	3·32	3·02	2·80	2·64	2·51	2·41

* Abridged with permission from *Biometrika Tables for Statisticians*, Vol. I, Edited by E. S. Pearson and H. O. Hartley, Cambridge University Press (1966).

TABLE 9 (b) (Continued)

ν_2 \ ν_1	10	12	15	20	24	30	40	60	120	∞
1	6056	6106	6157	6209	6235	6261	6287	6313	6339	6366
2	99·40	99·42	99·43	99·45	99·46	99·47	99·47	99·48	99·49	99·50
3	27·23	27·05	26·87	26·69	26·60	26·50	26·41	26·32	26·22	26·13
4	14·55	14·37	14·20	14·02	13·93	13·84	13·75	13·65	13·56	13·46
5	10·05	9·89	9·72	9·55	9·47	9·38	9·29	9·20	9·11	9·02
6	7·87	7·72	7·56	7·40	7·31	7·23	7·14	7·06	6·97	6·88
7	6·62	6·47	6·31	6·16	6·07	5·99	5·91	5·82	5·74	5·65
8	5·81	5·67	5·52	5·36	5·28	5·20	5·12	5·03	4·95	4·86
9	5·26	5·11	4·96	4·81	4·73	4·65	4·57	4·48	4·40	4·31
10	4·85	4·71	4·56	4·41	4·33	4·25	4·17	4·08	4·00	3·91
11	4·54	4·40	4·25	4·10	4·02	3·94	3·86	3·78	3·69	3·60
12	4·30	4·16	4·01	3·86	3·78	3·70	3·62	3·54	3·45	3·36
13	4·10	3·96	3·82	3·66	3·59	3·51	3·43	3·34	3·25	3·17
14	3·94	3·80	3·66	3·51	3·43	3·35	3·27	3·18	3·09	3·00
15	3·80	3·67	3·52	3·37	3·29	3·21	3·13	3·05	2·96	2·87
16	3·69	3·55	3·41	3·26	3·18	3·10	3·02	2·93	2·84	2·75
17	3·59	3·46	3·31	3·16	3·08	3·00	2·92	2·83	2·75	2·65
18	3·51	3·37	3·23	3·08	3·00	2·92	2·84	2·75	2·66	2·57
19	3·43	3·30	3·15	3·00	2·92	2·84	2·76	2·67	2·58	2·49
20	3·37	3·23	3·09	2·94	2·86	2·78	2·69	2·61	2·52	2·42
21	3·31	3·17	3·03	2·88	2·80	2·72	2·64	2·55	2·46	2·36
22	3·26	3·12	2·98	2·83	2·75	2·67	2·58	2·50	2·40	2·31
23	3·21	3·07	2·93	2·78	2·70	2·62	2·54	2·45	2·35	2·26
24	3·17	3·03	2·89	2·74	2·66	2·58	2·49	2·40	2·31	2·21
25	3·13	2·99	2·85	2·70	2·62	2·54	2·45	2·36	2·27	2·17
26	3·09	2·96	2·81	2·66	2·58	2·50	2·42	2·33	2·23	2·13
27	3·06	2·93	2·78	2·63	2·55	2·47	2·38	2·29	2·20	2·10
28	3·03	2·90	2·75	2·60	2·52	2·44	2·35	2·26	2·17	2·06
29	3·00	2·87	2·73	2·57	2·49	2·41	2·33	2·23	2·14	2·03
30	2·98	2·84	2·70	2·55	2·47	2·39	2·30	2·21	2·11	2·01
40	2·80	2·66	2·52	2·37	2·29	2·20	2·11	2·02	1·92	1·80
60	2·63	2·50	2·35	2·20	2·12	2·03	1·94	1·84	1·73	1·60
120	2·47	2·34	2·19	2·03	1·95	1·86	1·76	1·66	1·53	1·38
∞	2·32	2·18	2·04	1·88	1·79	1·70	1·59	1·47	1·32	1·00

A–27

TABLE 10 Constants for Constructing Control Charts for Averages*

Number of observations in each sample	A
2	1.880
3	1.023
4	0.729
5	0.577
6	0.483
7	0.419
8	0.373
9	0.337
10	0.308
11	0.285
12	0.266
13	0.249
14	0.235
15	0.223
16	0.212
17	0.203
18	0.194
19	0.187
20	0.180
21	0.173
22	0.167
23	0.162
24	0.157
25	0.153

*Reproduced with permission from the ASTM Manual on Quality Control of Materials (1951). Published by the American Society for Testing and Materials, Philadelphia, Pa.

TABLE 11   Constants for Constructing Control Charts for Ranges*

Number of observations in each sample	B_1	B_2
2	0	3.267
3	0	2.575
4	0	2.282
5	0	2.115
6	0	2.004
7	0.076	1.924
8	0.136	1.864
9	0.184	1.816
10	0.223	1.777
11	0.256	1.744
12	0.284	1.716
13	0.308	1.692
14	0.329	1.671
15	0.348	1.652
16	0.364	1.636
17	0.379	1.621
18	0.392	1.608
19	0.404	1.596
20	0.414	1.586
21	0.425	1.575
22	0.434	1.566
23	0.443	1.557
24	0.452	1.548
25	0.459	1.541

*Reproduced with permission from the ASTM Manual on Quality Control of Materials (1951). Published by the American Society for Testing and Materials, Philadelphia, Pa.

TABLE 12 Table for Constructing an Acceptance Sampling Plan

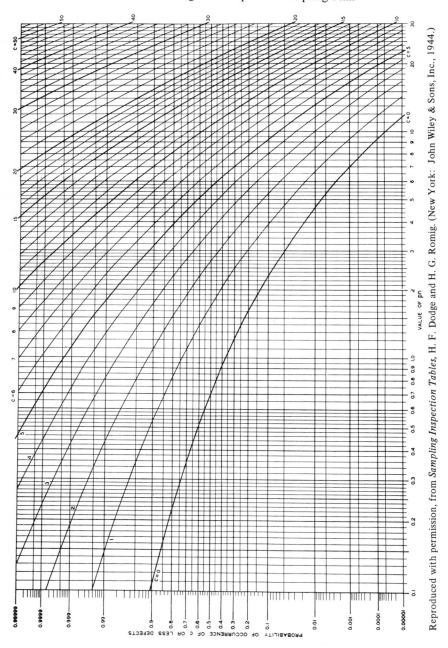

Reproduced with permission, from *Sampling Inspection Tables*, H. F. Dodge and H. G. Romig. (New York: John Wiley & Sons, Inc., 1944.)

TABLE 13(a) Sample Size Code Letters

Lot or batch size			Special inspection levels				General inspection levels		
			S-1	S-2	S-3	S-4	I	II	III
2	to	8	A	A	A	A	A	A	B
9	to	15	A	A	A	A	A	B	C
16	to	25	A	A	B	B	B	C	D
26	to	50	A	B	B	C	C	D	E
51	to	90	B	B	C	C	C	E	F
91	to	150	B	B	C	D	D	F	G
151	to	280	B	C	D	E	E	G	H
281	to	500	B	C	D	E	F	H	J
501	to	1200	C	C	E	F	G	J	K
1201	to	3200	C	D	E	G	H	K	L
3201	to	10000	C	D	F	G	J	L	M
10001	to	35000	C	D	F	H	K	M	N
35001	to	150000	D	E	G	J	L	N	P
150001	to	500000	D	E	G	J	M	P	Q
500001	and	over	D	E	H	K	N	Q	R

TABLE 13(b) Single Sampling, Normal Inspection

Acceptable Quality Levels (normal inspection)

Each AQL cell gives the pair "Ac Re". ↓ = use first sampling plan below arrow; ↑ = use first sampling plan above arrow.

Sample size code letter	Sample size	0.010	0.015	0.025	0.040	0.065	0.10	0.15	0.25	0.40	0.65	1.0	1.5	2.5	4.0	6.5	10	15	25	40	65	100	150	250	400	650	1000
A	2	↓	↓	↓	↓	↓	↓	↓	↓	↓	↓	↓	↓	↓	↓	↓	↓	0 1	1 2	2 3	3 4	5 6	7 8	10 11	14 15	21 22	30 31
B	3	↓	↓	↓	↓	↓	↓	↓	↓	↓	↓	↓	↓	↓	↓	↓	0 1	1 2	2 3	3 4	5 6	7 8	10 11	14 15	21 22	30 31	44 45
C	5	↓	↓	↓	↓	↓	↓	↓	↓	↓	↓	↓	↓	↓	↓	0 1	1 2	2 3	3 4	5 6	7 8	10 11	14 15	21 22	30 31	44 45	↑
D	8	↓	↓	↓	↓	↓	↓	↓	↓	↓	↓	↓	↓	↓	0 1	1 2	2 3	3 4	5 6	7 8	10 11	14 15	21 22	30 31	44 45	↑	↑
E	13	↓	↓	↓	↓	↓	↓	↓	↓	↓	↓	↓	↓	0 1	1 2	2 3	3 4	5 6	7 8	10 11	14 15	21 22	30 31	44 45	↑	↑	↑
F	20	↓	↓	↓	↓	↓	↓	↓	↓	↓	↓	↓	0 1	1 2	2 3	3 4	5 6	7 8	10 11	14 15	21 22	30 31	44 45	↑	↑	↑	↑
G	32	↓	↓	↓	↓	↓	↓	↓	↓	↓	↓	0 1	1 2	2 3	3 4	5 6	7 8	10 11	14 15	21 22	30 31	44 45	↑	↑	↑	↑	↑
H	50	↓	↓	↓	↓	↓	↓	↓	↓	↓	0 1	1 2	2 3	3 4	5 6	7 8	10 11	14 15	21 22	30 31	44 45	↑	↑	↑	↑	↑	↑
J	80	↓	↓	↓	↓	↓	↓	↓	↓	0 1	1 2	2 3	3 4	5 6	7 8	10 11	14 15	21 22	30 31	44 45	↑	↑	↑	↑	↑	↑	↑
K	125	↓	↓	↓	↓	↓	↓	↓	0 1	1 2	2 3	3 4	5 6	7 8	10 11	14 15	21 22	30 31	44 45	↑	↑	↑	↑	↑	↑	↑	↑
L	200	↓	↓	↓	↓	↓	↓	0 1	1 2	2 3	3 4	5 6	7 8	10 11	14 15	21 22	30 31	44 45	↑	↑	↑	↑	↑	↑	↑	↑	↑
M	315	↓	↓	↓	↓	↓	0 1	1 2	2 3	3 4	5 6	7 8	10 11	14 15	21 22	30 31	44 45	↑	↑	↑	↑	↑	↑	↑	↑	↑	↑
N	500	↓	↓	↓	↓	0 1	1 2	2 3	3 4	5 6	7 8	10 11	14 15	21 22	30 31	44 45	↑	↑	↑	↑	↑	↑	↑	↑	↑	↑	↑
P	800	↓	↓	↓	0 1	1 2	2 3	3 4	5 6	7 8	10 11	14 15	21 22	30 31	44 45	↑	↑	↑	↑	↑	↑	↑	↑	↑	↑	↑	↑
Q	1250	↓	↓	0 1	1 2	2 3	3 4	5 6	7 8	10 11	14 15	21 22	30 31	44 45	↑	↑	↑	↑	↑	↑	↑	↑	↑	↑	↑	↑	↑
R	2000	↓	0 1	1 2	2 3	3 4	5 6	7 8	10 11	14 15	21 22	30 31	44 45	↑	↑	↑	↑	↑	↑	↑	↑	↑	↑	↑	↑	↑	↑

⇩ = Use first sampling plan below arrow. If sample size equals, or exceeds, lot or batch size, do 100 percent inspection.

⇧ = Use first sampling plan above arrow.

Ac = Acceptance number.

Re = Rejection number.

TABLE 13(c) Single Sampling, Tightened Inspection

Acceptable Quality Levels (tightened inspection)

Each cell below shows the values "Ac Re" (Acceptance number, Rejection number) for the given Acceptable Quality Level. ↓ = Use first sampling plan below arrow. ↑ = Use first sampling plan above arrow.

Sample size code letter	Sample size	0.010	0.015	0.025	0.040	0.065	0.10	0.15	0.25	0.40	0.65	1.0	1.5	2.5	4.0	6.5	10	15	25	40	65	100	150	250	400	650	1000
A	2	↓	↓	↓	↓	↓	↓	↓	↓	↓	↓	↓	↓	↓	↓	↓	↓	0 1	1 2	2 3	3 4	5 6	8 9	12 13	18 19	27 28	41 42
B	3	↓	↓	↓	↓	↓	↓	↓	↓	↓	↓	↓	↓	↓	↓	↓	0 1	1 2	2 3	3 4	5 6	8 9	12 13	18 19	27 28	41 42	↑
C	5	↓	↓	↓	↓	↓	↓	↓	↓	↓	↓	↓	↓	↓	↓	0 1	1 2	2 3	3 4	5 6	8 9	12 13	18 19	27 28	41 42	↑	↑
D	8	↓	↓	↓	↓	↓	↓	↓	↓	↓	↓	↓	↓	↓	0 1	1 2	2 3	3 4	5 6	8 9	12 13	18 19	27 28	41 42	↑	↑	↑
E	13	↓	↓	↓	↓	↓	↓	↓	↓	↓	↓	↓	↓	0 1	1 2	2 3	3 4	5 6	8 9	12 13	18 19	27 28	41 42	↑	↑	↑	↑
F	20	↓	↓	↓	↓	↓	↓	↓	↓	↓	↓	↓	0 1	1 2	2 3	3 4	5 6	8 9	12 13	18 19	27 28	41 42	↑	↑	↑	↑	↑
G	32	↓	↓	↓	↓	↓	↓	↓	↓	↓	↓	0 1	1 2	2 3	3 4	5 6	8 9	12 13	18 19	27 28	41 42	↑	↑	↑	↑	↑	↑
H	50	↓	↓	↓	↓	↓	↓	↓	↓	↓	0 1	1 2	2 3	3 4	5 6	8 9	12 13	18 19	27 28	41 42	↑	↑	↑	↑	↑	↑	↑
J	80	↓	↓	↓	↓	↓	↓	↓	↓	0 1	1 2	2 3	3 4	5 6	8 9	12 13	18 19	27 28	41 42	↑	↑	↑	↑	↑	↑	↑	↑
K	125	↓	↓	↓	↓	↓	↓	↓	0 1	1 2	2 3	3 4	5 6	8 9	12 13	18 19	27 28	41 42	↑	↑	↑	↑	↑	↑	↑	↑	↑
L	200	↓	↓	↓	↓	↓	↓	0 1	1 2	2 3	3 4	5 6	8 9	12 13	18 19	27 28	41 42	↑	↑	↑	↑	↑	↑	↑	↑	↑	↑
M	315	↓	↓	↓	↓	↓	0 1	1 2	2 3	3 4	5 6	8 9	12 13	18 19	27 28	41 42	↑	↑	↑	↑	↑	↑	↑	↑	↑	↑	↑
N	500	↓	↓	↓	↓	0 1	1 2	2 3	3 4	5 6	8 9	12 13	18 19	27 28	41 42	↑	↑	↑	↑	↑	↑	↑	↑	↑	↑	↑	↑
P	800	↓	↓	↓	0 1	1 2	2 3	3 4	5 6	8 9	12 13	18 19	27 28	41 42	↑	↑	↑	↑	↑	↑	↑	↑	↑	↑	↑	↑	↑
Q	1250	↓	↓	0 1	1 2	2 3	3 4	5 6	8 9	12 13	18 19	27 28	41 42	↑	↑	↑	↑	↑	↑	↑	↑	↑	↑	↑	↑	↑	↑
R	2000	↓	0 1	1 2	2 3	3 4	5 6	8 9	12 13	18 19	27 28	41 42	↑	↑	↑	↑	↑	↑	↑	↑	↑	↑	↑	↑	↑	↑	↑
S	3150	0 1	1 2	2 3	3 4	5 6	8 9	12 13	18 19	27 28	41 42	↑	↑	↑	↑	↑	↑	↑	↑	↑	↑	↑	↑	↑	↑	↑	↑

↓ = Use first sampling plan below arrow. If sample size equals or exceeds lot or batch size, do 100 percent inspection.
↑ = Use first sampling plan above arrow.
Ac = Acceptance number.
Re = Rejection number.

A–33

TABLE 13(d) Single Sampling, Reduced Inspection

Acceptable Quality Levels (reduced inspection)†

Values shown as "Ac Re" (Acceptance number, Rejection number). ↓ = use first sampling plan below arrow; ↑ = use first sampling plan above arrow.

Sample size code letter	Sample size	0.010	0.015	0.025	0.040	0.065	0.10	0.15	0.25	0.40	0.65	1.0	1.5	2.5	4.0	6.5	10	15	25	40	65	100	150	250	400	650	1000
A	2	↓	↓	↓	↓	↓	↓	↓	↓	↓	↓	↓	↓	↓	↓	0 1	0 2	1 3	1 4	2 5	3 6	5 8	7 10	10 13	14 17	21 24	30 31
B	2	↓	↓	↓	↓	↓	↓	↓	↓	↓	↓	↓	↓	↓	0 1	0 2	1 3	1 4	2 5	3 6	5 8	7 10	10 13	14 17	21 24	30 31	↑
C	2	↓	↓	↓	↓	↓	↓	↓	↓	↓	↓	↓	↓	0 1	0 2	1 3	1 4	2 5	3 6	5 8	7 10	10 13	14 17	21 24	30 31	↑	↑
D	3	↓	↓	↓	↓	↓	↓	↓	↓	↓	↓	↓	0 1	0 2	1 3	1 4	2 5	3 6	5 8	7 10	10 13	14 17	21 24	30 31	↑	↑	↑
E	5	↓	↓	↓	↓	↓	↓	↓	↓	↓	↓	0 1	0 2	1 3	1 4	2 5	3 6	5 8	7 10	10 13	14 17	21 24	30 31	↑	↑	↑	↑
F	8	↓	↓	↓	↓	↓	↓	↓	↓	↓	0 1	0 2	1 3	1 4	2 5	3 6	5 8	7 10	10 13	14 17	21 24	30 31	↑	↑	↑	↑	↑
G	13	↓	↓	↓	↓	↓	↓	↓	↓	0 1	0 2	1 3	1 4	2 5	3 6	5 8	7 10	10 13	14 17	21 24	30 31	↑	↑	↑	↑	↑	↑
H	20	↓	↓	↓	↓	↓	↓	↓	0 1	0 2	1 3	1 4	2 5	3 6	5 8	7 10	10 13	14 17	21 24	30 31	↑	↑	↑	↑	↑	↑	↑
J	32	↓	↓	↓	↓	↓	↓	0 1	0 2	1 3	1 4	2 5	3 6	5 8	7 10	10 13	14 17	21 24	30 31	↑	↑	↑	↑	↑	↑	↑	↑
K	50	↓	↓	↓	↓	↓	0 1	0 2	1 3	1 4	2 5	3 6	5 8	7 10	10 13	14 17	21 24	30 31	↑	↑	↑	↑	↑	↑	↑	↑	↑
L	80	↓	↓	↓	↓	0 1	0 2	1 3	1 4	2 5	3 6	5 8	7 10	10 13	14 17	21 24	30 31	↑	↑	↑	↑	↑	↑	↑	↑	↑	↑
M	125	↓	↓	↓	0 1	0 2	1 3	1 4	2 5	3 6	5 8	7 10	10 13	14 17	21 24	30 31	↑	↑	↑	↑	↑	↑	↑	↑	↑	↑	↑
N	200	↓	↓	0 1	0 2	1 3	1 4	2 5	3 6	5 8	7 10	10 13	14 17	21 24	30 31	↑	↑	↑	↑	↑	↑	↑	↑	↑	↑	↑	↑
P	315	↓	0 1	0 2	1 3	1 4	2 5	3 6	5 8	7 10	10 13	14 17	21 24	30 31	↑	↑	↑	↑	↑	↑	↑	↑	↑	↑	↑	↑	↑
Q	500	0 1	0 2	1 3	1 4	2 5	3 6	5 8	7 10	10 13	14 17	21 24	30 31	↑	↑	↑	↑	↑	↑	↑	↑	↑	↑	↑	↑	↑	↑
R	800	0 2	1 3	1 4	2 5	3 6	5 8	7 10	10 13	14 17	21 24	30 31	↑	↑	↑	↑	↑	↑	↑	↑	↑	↑	↑	↑	↑	↑	↑

⇩ = Use first sampling plan below arrow. If sample size equals or exceeds lot or batch size, do 100 percent inspection.

⇧ = Use first sampling plan above arrow.

Ac = Acceptance number.

Re = Rejection number.

† = If the acceptance number has been exceeded, but the rejection number has not been reached, accept the lot, but reinstate normal inspection (see 10.1.4).

TABLE 14 Double Sampling, Normal Inspection

Acceptable Quality Levels (normal inspection)

Sample size code letter	Sample	Sample size	Cumulative sample size	0.010	0.015	0.025	0.040	0.065	0.10	0.15	0.25	0.40	0.65	1.0	1.5	2.5	4.0	6.5	10	15	25	40	65	100	150	250	400	650	1000
				Ac Re	Ac Re	Ac Re	Ac Re	Ac Re	Ac Re	Ac Re	Ac Re	Ac Re	Ac Re	Ac Re	Ac Re	Ac Re	Ac Re	Ac Re	Ac Re	Ac Re	Ac Re	Ac Re	Ac Re	Ac Re	Ac Re	Ac Re	Ac Re	Ac Re	Ac Re
A				↓	↓	↓	↓	↓	↓	↓	↓	↓	↓	↓	↓	↓	↓	↓	↓	↓	↓	↓	↓	•	↑	↑	↑	↑	↑
B First		2	2	↓	↓	↓	↓	↓	↓	↓	↓	↓	↓	↓	↓	↓	↓	↓	↓	↓	↓	•	0 2	0 3	1 4	2 5	3 7	5 9	7 11
B Second		2	4																				1 2	3 4	4 5	6 7	8 9	12 13	18 19
C First		3	3	↓	↓	↓	↓	↓	↓	↓	↓	↓	↓	↓	↓	↓	↓	↓	↓	↓	•	0 2	0 3	1 4	2 5	3 7	5 9	7 11	11 16
C Second		3	6																			1 2	3 4	4 5	6 7	8 9	12 13	18 19	26 27
D First		5	5	↓	↓	↓	↓	↓	↓	↓	↓	↓	↓	↓	↓	↓	↓	↓	↓	•	0 2	0 3	1 4	2 5	3 7	5 9	7 11	11 16	17 22
D Second		5	10																		1 2	3 4	4 5	6 7	8 9	12 13	18 19	26 27	37 38
E First		8	8	↓	↓	↓	↓	↓	↓	↓	↓	↓	↓	↓	↓	↓	↓	↓	•	0 2	0 3	1 4	2 5	3 7	5 9	7 11	11 16	17 22	25 31
E Second		8	16																	1 2	3 4	4 5	6 7	8 9	12 13	18 19	26 27	37 38	56 57
F First		13	13	↓	↓	↓	↓	↓	↓	↓	↓	↓	↓	↓	↓	↓	↓	•	0 2	0 3	1 4	2 5	3 7	5 9	7 11	11 16	17 22	25 31	↑
F Second		13	26																1 2	3 4	4 5	6 7	8 9	12 13	18 19	26 27	37 38	56 57	
G First		20	20	↓	↓	↓	↓	↓	↓	↓	↓	↓	↓	↓	↓	↓	•	0 2	0 3	1 4	2 5	3 7	5 9	7 11	11 16	17 22	25 31	↑	↑
G Second		20	40															1 2	3 4	4 5	6 7	8 9	12 13	18 19	26 27	37 38	56 57		
H First		32	32	↓	↓	↓	↓	↓	↓	↓	↓	↓	↓	↓	↓	•	0 2	0 3	1 4	2 5	3 7	5 9	7 11	11 16	17 22	25 31	↑	↑	↑
H Second		32	64														1 2	3 4	4 5	6 7	8 9	12 13	18 19	26 27	37 38	56 57			
J First		50	50	↓	↓	↓	↓	↓	↓	↓	↓	↓	↓	↓	•	0 2	0 3	1 4	2 5	3 7	5 9	7 11	11 16	17 22	25 31	↑	↑	↑	↑
J Second		50	100													1 2	3 4	4 5	6 7	8 9	12 13	18 19	26 27	37 38	56 57				
K First		80	80	↓	↓	↓	↓	↓	↓	↓	↓	↓	↓	•	0 2	0 3	1 4	2 5	3 7	5 9	7 11	11 16	17 22	25 31	↑	↑	↑	↑	↑
K Second		80	160												1 2	3 4	4 5	6 7	8 9	12 13	18 19	26 27	37 38	56 57					
L First		125	125	↓	↓	↓	↓	↓	↓	↓	↓	↓	•	0 2	0 3	1 4	2 5	3 7	5 9	7 11	11 16	17 22	25 31	↑	↑	↑	↑	↑	↑
L Second		125	250											1 2	3 4	4 5	6 7	8 9	12 13	18 19	26 27	37 38	56 57						
M First		200	200	↓	↓	↓	↓	↓	↓	↓	↓	•	0 2	0 3	1 4	2 5	3 7	5 9	7 11	11 16	17 22	25 31	↑	↑	↑	↑	↑	↑	↑
M Second		200	400										1 2	3 4	4 5	6 7	8 9	12 13	18 19	26 27	37 38	56 57							
N First		315	315	↓	↓	↓	↓	↓	↓	↓	•	0 2	0 3	1 4	2 5	3 7	5 9	7 11	11 16	17 22	25 31	↑	↑	↑	↑	↑	↑	↑	↑
N Second		315	630									1 2	3 4	4 5	6 7	8 9	12 13	18 19	26 27	37 38	56 57								
P First		500	500	↓	↓	↓	↓	↓	↓	•	0 2	0 3	1 4	2 5	3 7	5 9	7 11	11 16	17 22	25 31	↑	↑	↑	↑	↑	↑	↑	↑	↑
P Second		500	1000								1 2	3 4	4 5	6 7	8 9	12 13	18 19	26 27	37 38	56 57									
Q First		800	800	↓	↓	↓	↓	↓	•	0 2	0 3	1 4	2 5	3 7	5 9	7 11	11 16	17 22	25 31	↑	↑	↑	↑	↑	↑	↑	↑	↑	↑
Q Second		800	1600							1 2	3 4	4 5	6 7	8 9	12 13	18 19	26 27	37 38	56 57										
R First		1250	1250	↓	↓	↓	↓	•	0 2	0 3	1 4	2 5	3 7	5 9	7 11	11 16	17 22	25 31	↑	↑	↑	↑	↑	↑	↑	↑	↑	↑	↑
R Second		1250	2500						1 2	3 4	4 5	6 7	8 9	12 13	18 19	26 27	37 38	56 57											

⇩ = Use first sampling plan below arrow. If sample size equals or exceeds lot or batch size, do 100 percent inspection.

⇧ = Use first sampling plan above arrow.

Ac = Acceptance number

Re = Rejection number

• = Use corresponding single sampling plan (or alternatively, use double sampling plan below, where available).

TABLE 15 Critical Values for a 5% Two-Sided Runs Test

	5	6	7	8	9	10	11	12	13	14	15	16	17	18	19	20
2								2/6	2/6	2/6	2/6	2/6	2/6	2/6	2/6	2/6
3		2/8	2/8	2/8	2/8	2/8	2/8	2/8	2/8	2/8	3/8	3/8	3/8	3/8	3/8	3/8
4	2/9	2/9	2/10	3/10	3/10	3/10	3/10	3/10	3/10	3/10	3/10	4/10	4/10	4/10	4/10	4/10
5	2/10	3/10	3/11	3/11	3/12	3/12	4/12	4/12	4/12	4/12	4/12	4/12	4/12	5/12	5/12	5/12
6		3/11	3/12	3/12	4/13	4/13	4/13	4/13	5/14	5/14	5/14	5/14	5/14	5/14	6/14	6/14
7			3/13	4/13	4/14	5/14	5/14	5/14	5/15	5/15	6/15	6/16	6/16	6/16	6/16	6/16
8				4/14	5/14	5/15	5/15	6/16	6/16	6/16	6/16	6/17	7/17	7/17	7/17	7/17
9					5/15	5/16	6/16	6/16	6/17	7/17	7/18	7/18	7/18	8/18	8/18	8/18
10						6/16	6/17	7/17	7/18	7/18	7/18	8/19	8/19	8/19	8/20	9/20
11							7/17	7/18	7/19	8/19	8/19	8/20	9/20	9/20	9/21	9/21
12								7/19	8/19	8/20	8/20	9/21	9/21	9/21	10/22	10/22
13									8/20	9/20	9/21	9/21	10/22	10/22	10/23	10/23
14										9/21	9/22	10/22	10/23	10/23	11/23	11/24
15											10/22	10/23	11/23	11/24	11/24	12/25
16												11/23	11/24	11/25	12/25	12/25
17													11/25	12/25	12/26	13/26
18														12/26	13/26	13/27
19															13/27	13/27
20																14/28

* From C. Eisenhart and F. Swed, "Tables for Testing Randomness of Grouping in a Sequence of Alternatives," *Annals of Mathematical Statistics*, Vol. 14 (1943), p. 66. Reprinted by permission of the authors and the publisher.

TABLE 16 Critical Values for the Wilcoxon Test for Paired Data

n	$2\alpha \leq 0.10$	$2\alpha \leq 0.05$	$2\alpha \leq 0.02$	$2\alpha \leq 0.01$
5	0– 15	–	–	–
6	2– 19	0– 21	–	–
7	3– 25	2– 26	0– 28	–
8	5– 31	3– 33	1– 35	0– 36
9	8– 37	5– 40	3– 42	1– 44
10	10– 45	8– 47	5– 50	3– 52
11	13– 53	10– 56	7– 59	5– 61
12	17– 61	13– 65	9– 69	7– 71
13	21– 70	17– 74	12– 79	9– 82
14	25– 80	21– 84	15– 90	12– 93
15	30– 90	25– 95	19–101	15–105
16	35–101	29–107	23–113	19–117
17	41–112	34–119	28–125	23–130
18	47–124	40–131	32–139	27–144
19	53–137	46–144	37–153	32–158
20	60–150	52–158	43–167	37–173
21	67–164	58–173	49–182	42–189
22	75–178	66–187	55–198	48–205
23	83–193	73–203	62–214	54–222
24	91–209	81–219	69–231	61–239
25	100–225	89–236	76–249	68–257

* Reproduced from *Documenta Geigy Scientific Tables*, 7th edition, by permission of CIBA-GEIGY Limited, Basle, Switzerland.

TABLE 17 Critical Values for the Wilcoxon Rank-Sum Test

For one-sided tests $\alpha = .05$, for two-sided tests $\alpha = .10$

N_1, N_2	4	5	6	7	8	9	10	11	12	13	14	15	16	17	18	19	20	21	22	23	24	25

* Reproduced from *Documenta Geigy Scientific Tables*, 7th edition, by permission of CIBA-GEIGY Limited, Basle, Switzerland.

TABLE 17 (Continued)

For one-sided tests α = .01, for two-sided tests α = .02

N_1	4	5	6	7	8	9	10	11	12	13	14	15	16	17	18	19	20	21	22	23	24	25
N_2	T_l T_r	T_l T_r	T_l T_r	T_l T_r	T_l T_r	T_l T_r	T_l T_r	T_l T_r	T_l T_r	T_l T_r	T_l T_r	T_l T_r	T_l T_r	T_l T_r	T_l T_r	T_l T_r	T_l T_r	T_l T_r	T_l T_r	T_l T_r	T_l T_r	T_l T_r
4	—	15- 35	22- 44	29- 55	38- 66	48- 78	58- 92	70-106	83-121	96-138	111-155	127-173	143-193	161-213	180-234	199-257	220-280	242-304	264-330	288- 356	313- 383	338- 412
5	10- 30	16- 39	23- 49	31- 60	40- 72	50- 85	61- 99	73-114	86-129	99-147	114-165	131-184	148-204	166-225	185-247	205-270	226-294	248-319	271-345	295- 372	320- 400	346- 429
6	11- 33	17- 43	24- 54	32- 66	42- 78	52- 92	63-107	75-123	89-139	103-157	118-176	135-195	152-216	171-237	190-260	210-284	232-308	254-334	277-361	302- 388	327- 417	354- 446
7	11- 37	18- 47	25- 59	34- 71	43- 85	54- 99	66-114	78-131	92-148	107-166	122-186	139-206	157-227	176-248	195-273	216-297	238-322	261-348	284-376	309- 404	335- 433	361- 464
8	12- 40	19- 51	27- 63	35- 77	45- 91	56-106	68-122	81-139	95-157	111-175	127-195	144-216	162-238	181-261	201-285	222-310	244-336	267-363	291-391	316- 420	342- 450	370- 480
9	13- 43	20- 55	28- 68	37- 82	47- 97	59-112	71-129	84-147	99-165	114-185	131-205	148-227	167-249	186-273	207-297	228-323	250-350	274-377	298-406	324- 435	350- 466	378- 497
10	13- 47	21- 59	29- 73	39- 87	49-103	61-119	74-136	88-154	102-174	118-194	135-215	153-237	172-260	191-285	212-310	234-336	257-363	281-391	306-420	331- 451	358- 482	386- 514
11	14- 50	22- 63	30- 78	40- 93	51-109	63-126	77-143	91-162	106-182	122-203	139-225	157-248	177-271	197-296	218-322	240-349	263-377	288-405	313-435	339- 466	366- 498	394- 531
12	15- 53	23- 67	32- 82	42- 98	53-115	66-132	79-151	94-170	109-191	126-212	143-235	162-258	182-282	202-308	224-334	246-362	270-390	295-419	320-450	347- 481	374- 514	403- 547
13	15- 57	24- 71	33- 87	44-103	56-120	68-139	82-158	97-178	113-199	130-221	148-244	167-268	187-293	208-319	230-346	253-374	276-404	301-434	327-465	354- 497	382- 530	411- 564
14	16- 60	25- 75	34- 92	45-109	58-126	71-145	85-165	100-186	116-208	134-230	152-254	171-279	192-304	213-331	235-359	259-387	283-417	308-448	335-479	362- 512	391- 545	420- 580
15	17- 63	26- 79	36- 96	47-114	60-132	73-152	88-172	103-194	120-216	138-239	156-264	176-289	197-315	218-343	241-371	265-400	290-430	315-462	342-494	370- 527	399- 561	429- 596
16	17- 67	27- 83	37-101	49-119	62-138	76-158	91-179	107-201	124-224	142-248	161-273	181-299	202 326	224-354	247-383	271-413	296-444	322-476	349-509	377- 543	407- 577	437- 613
17	18- 70	28- 87	39-105	51-124	64-144	78-165	93-187	110-209	127-233	146-257	165-283	185-310	207-337	229-366	253-395	278-425	303-457	330-489	357-523	386- 557	416- 592	446- 629
18	19- 73	29- 91	40-110	52-130	66-150	80-171	96-194	113-217	131-242	150-268	170-295	190-323	212-351	235-380	259-410	284-441	310-473	337-506	365-540	394- 575	424- 611	455- 648
19	19- 77	30- 95	41-115	54-135	68-156	83-178	99-201	116-225	134-250	154-275	174-302	195-330	217-359	240-388	265-418	290-451	317-483	344-517	373-551	402- 587	432- 624	464- 661
20	20- 80	31- 99	43-119	56-140	70-162	85-185	101-209	119-233	138-259	157-285	178-312	200-340	222-370	246-400	271-431	297-463	323-497	351-531	380-566	410- 602	441- 639	473- 677
21	21- 83	32-103	44-124	58-145	72-168	88-191	105-215	123-240	142-266	161-294	182-322	204-351	228-380	252-411	277-443	303-476	330-510	359-544	388-580	418- 617	449- 655	482- 693
22	21- 87	33-107	45-129	59-151	74-174	90-198	108-222	126-248	146-275	165-303	187-331	209-361	233-391	257-423	283-455	310-488	337-523	366-558	395-595	426- 632	458- 670	490- 710
23	22- 90	34-111	47-133	61-156	76-180	93-204	110-230	129-256	149-283	169-312	191-341	214-371	238-402	263-434	289-467	316-501	344-536	373-572	403-609	434- 647	466- 686	499- 726
24	23- 93	35-115	48-138	63-161	78-186	95-211	113-237	132-264	152-292	173-321	196-350	219-381	243-413	269-445	295-479	322-514	351-549	380-586	411-623	442- 662	475- 701	508- 742
25	23- 97	36-119	50-1.2	64-167	81-191	98-217	116-244	135-272	156-300	177-330	200-360	224-391	248-424	274-457	301-491	329-526	358-562	388-599	418-638	450- 677	483- 717	517- 758
26	24-100	37-123	51-147	66-172	83-197	100-224	119-251	138-280	159-309	181-339	204-370	229-401	254-434	280-468	307-503	335-539	365-575	395-613	426-652	458- 692	492- 732	526- 774
27	25-103	38-127	52-152	68-177	85-203	103-230	121-259	142-287	163-317	185-348	209-379	233-412	259-445	285-480	313-515	342-551	371-589	402-627	434-666	466- 707	500- 748	535- 790
28	26-106	39-131	54-156	70-182	87-209	105-237	124-266	145-295	167-325	189-357	213-388	238-422	264-456	291-491	319-527	348-564	378-602	409-641	441-681	475- 721	509- 763	544- 806
29	26-110	40-135	55-161	71-188	89-215	108-243	127-273	148-303	170-334	193-366	218-398	243-432	269-467	297-502	325-539	355-576	385-615	417-654	449-695	483- 736	517- 779	553- 822
30	27-113	41-139	56-166	73-193	91-221	110-250	130-280	151-311	174-342	198-374	222-408	248-442	275-477	302-514	331-551	361-589	392-628	424-668	457-709	491- 751	526- 794	562- 838
31	28-116	42-143	58-170	75-198	93-227	112-257	133-287	155-318	178-350	202-383	227-417	253-452	280-488	308-525	337-563	368-601	399-641	431-682	465-723	499- 766	534- 810	571- 854
32	28-120	43-147	59-175	77-203	95-233	115-263	136-294	158-326	181-359	206-392	231-427	258-462	285-499	314-536	343-575	374-614	406-654	438-696	472-738	507- 781	543- 825	580- 870
33	29-123	44-151	61-179	78-209	97-239	117-270	139-301	161-334	185-367	210-401	236-437	262-473	290-510	319-548	349-587	381-626	413-667	446-709	480-752	515- 796	552- 840	589- 886
34	30-126	45-155	62-184	79-215	99-245	120-276	141-309	164-342	189-375	214-410	240-446	267-483	296-520	325-559	356-598	387-639	420-680	453-723	488-766	523- 811	560- 856	598- 902
35	30-130	46-159	63-189	81-220	101-251	122-283	144-316	168-349	192-384	218-419	244-456	272-493	301-531	331-570	362-610	394-651	426-694	461-736	496-780	532- 825	569- 870	608- 918
36	31-133	47-163	65-193	83-225	103-257	125-289	147-323	171-357	196-392	222-428	249-465	277-503	306-542	336-582	368-622	400-664	433-707	468-750	503-795	540- 840	577- 887	616- 934
37	32-136	48-167	66-198	84-231	105-263	127-296	150-330	174-365	199-401	225-439	253-475	282-513	311-553	342-593	374-634	407-677	440-720	475-764	511-809	548- 855	586- 902	625- 950
38	32-140	49-171	67-203	86-236	107-269	129-303	153-337	177-373	203-409	230-446	258-484	287-523	317-563	348-604	380-646	413-689	447-733	483-777	519-823	556- 870	595- 917	634- 966
39	33-143	50-175	69-207	88-241	109-275	132-309	156-344	180-382	206-417	234-455	262-494	292-533	322-574	353-616	386-658	420-701	454-746	490-791	527-837	564- 885	603- 933	643- 982
40	34-146	51-179	70-212	90-246	111-281	134-316	159-351	184-388	210-426	238-464	267-503	296-544	327-585	359-627	392-670	426-714	461-759	497-805	534-852	573- 899	612- 948	652- 998
41	34-150	52-183	72-216	91-252	113-287	137-322	161-359	187-396	214-434	242-473	271-513	301-555	333-595	365-638	398-682	433-726	468-772	505-818	542-866	581- 914	620- 964	662-1014
42	35-153	53-187	73-221	93-257	115-292	139-329	164-366	190-404	218-442	246-482	276-522	306-566	338-606	371-649	404-694	439-739	475-785	512-832	550-880	589- 929	629- 979	670-1030
43	35-157	54-191	74-226	95-262	118-298	142-335	167-373	194-411	221-451	250-491	280-532	311-574	343-617	376-661	410-706	446-751	482-798	519-846	558-894	597- 944	638- 994	679-1046
44	36-160	55-195	76-230	97-267	120-304	144-342	170-380	197-419	225-459	254-500	284-542	316-584	348-628	382-672	416-718	452-764	489-811	527-859	565-909	605- 959	646-1010	688-1062
45	37-163	56-199	77-235	98-273	122-310	147-348	173-387	200-427	229-467	258-509	289-551	321-594	354-638	388-683	423-729	459-776	496-824	534-873	573-923	614- 973	655-1025	697-1078
46	37-167	57-203	78-240	100-278	124-316	149-355	176-394	203-435	232-476	262-518	293-560	325-605	359-648	393-695	429-741	465-789	503-837	541-887	581-937	622- 988	663-1041	706-1094
47	38-170	58-207	80-244	102-283	126-322	152-361	179-401	207-442	236-484	266-527	298-570	330-615	364-660	399-706	435-753	472-801	510-850	549-900	589-951	630-1003	672-1056	715-1110
48	39-173	59-211	81-249	103-289	128-328	154-368	181-409	210-449	239-493	270-536	302-580	335-625	369-671	405-717	441-765	479-813	516-864	556-913	597-965	638-1018	681-1071	724-1126
49	39-177	60-215	82-254	105-294	130-334	157-374	184-416	213-458	243-501	274-545	307-589	340-635	375-681	410-729	447-777	485-826	524-876	563-928	604-980	646-1033	689-1087	733-1142
50	40-180	61-219	84-258	107-299	132-340	159-381	187-423	216-466	247-509	279-553	311-599	345-645	380-692	416-740	453-789	491-839	531-889	571-941	612-994	655-1047	698-1102	743-1157

TABLE 17 (Continued)

For one-sided tests $\alpha = .025$, for two-sided tests $\alpha = .05$

N_2 \ N_1 (T_l–T_r)	4	5	6	7	8	9	10	11	12	13	14	15	16	17	18	19	20	21	22	23	24	25
4	**10-26**	16-34	23-43	31-53	40-64	49-77	60-90	72-104	85-119	99-135	114-152	130-170	147-189	164-210	183-231	203-253	224-276	246-300	269-325	293-351	317-379	343-407
5	11-29	**17-38**	24-48	33-58	42-70	52-83	63-97	75-112	89-127	103-144	118-162	134-181	151-201	170-221	189-243	209-266	230-290	253-314	276-340	300-367	325-395	352-423
6	12-32	18-42	**26-52**	34-64	44-76	55-89	66-104	79-119	92-136	107-153	122-172	139-191	157-211	175-233	195-255	215-279	237-303	260-328	283-355	308-382	333-411	360-440
7	13-35	20-45	27-57	**36-69**	46-82	57-96	69-111	82-127	96-144	111-162	127-181	144-201	162-222	181-244	201-267	222-291	244-316	267-342	291-369	316-397	342-426	369-456
8	14-38	21-49	29-61	38-74	**49-87**	60-102	72-118	85-135	100-152	115-171	131-191	149-211	167-233	187-255	207-279	228-304	251-329	274-356	298-384	324-412	350-442	378-472
9	14-42	22-53	31-65	40-79	51-93	**62-109**	75-125	89-142	104-160	119-180	136-200	154-221	173-243	192-267	213-291	235-316	258-342	281-370	306-398	332-427	359-457	387-488
10	15-45	23-57	32-70	42-84	53-99	65-115	**78-132**	92-150	107-169	124-188	141-209	159-231	178-254	198-278	219-303	242-328	265-355	289-383	314-412	340-442	368-472	396-504
11	16-48	24-61	34-74	44-89	55-105	68-121	81-139	**96-157**	111-177	128-197	145-219	164-241	183-265	204-289	226-314	248-341	272-368	296-397	322-426	349-456	376-488	405-520
12	17-51	26-64	35-79	46-94	58-110	71-127	84-146	99-165	**115-185**	132-206	150-228	169-251	189-275	210-300	232-326	255-353	279-381	304-410	330-440	357-471	385-503	414-536
13	18-54	27-68	37-83	48-99	60-116	73-134	88-152	103-172	119-193	**136-215**	155-237	174-261	195-285	216-311	239-337	262-365	286-394	312-423	338-454	365-486	394-518	423-552
14	19-57	28-72	38-88	50-104	62-122	76-140	91-159	106-180	123-201	141-223	**160-246**	179-271	200-296	222-322	245-349	269-377	293-407	319-437	346-468	374-500	403-533	433-567
15	20-60	29-76	40-92	52-109	65-127	79-146	94-166	110-187	127-209	145-232	164-256	**184-281**	206-306	228-333	251-361	275-390	301-419	327-450	354-482	382-515	412-548	442-583
16	21-63	30-80	42-96	54-114	67-133	82-152	97-173	113-195	131-217	150-240	169-265	190-290	**211-317**	234-344	258-372	282-402	308-432	335-463	362-496	391-529	421-563	451-599
17	21-67	32-83	43-101	56-119	70-138	84-159	100-180	117-202	135-225	154-249	174-274	195-300	217-327	**240-355**	264-384	289-414	315-445	342-477	370-510	399-544	429-579	461-614
18	22-70	33-87	45-105	58-124	72-144	87-165	103-187	121-209	139-233	159-257	179-283	200-310	223-337	246-366	**271-395**	296-426	322-458	350-490	378-524	408-558	438-594	470-630
19	23-73	34-91	46-110	60-129	74-150	90-171	107-193	124-217	143-241	163-266	184-292	205-320	228-348	252-377	277-407	**303-438**	330-470	358-503	387-537	416-573	447-609	479-646
20	24-76	35-95	48-114	62-134	77-155	93-177	110-200	128-224	147-249	167-275	188-302	211-329	234-358	258-388	283-419	310-450	**337-483**	365-517	395-551	425-587	456-624	489-661
21	25-79	37-98	50-118	64-139	79-161	95-184	113-207	131-232	151-257	172-283	193-311	216-339	240-368	264-399	290-430	317-462	344-496	**373-530**	403-565	434-601	465-639	498-677
22	26-82	38-102	51-123	66-144	81-167	98-190	116-214	135-239	155-265	176-292	198-320	221-349	245-379	270-410	296-442	324-474	352-508	381-543	**411-579**	442-616	474-654	508-692
23	27-85	39-106	53-127	68-149	84-172	101-196	119-221	139-246	159-273	180-301	203-329	226-359	251-389	276-421	303-453	330-487	359-521	389-556	419-593	**451-630**	483-669	517-708
24	27-89	40-110	54-132	70-154	86-178	104-202	123-227	142-254	163-281	185-309	208-338	232-368	257-399	282-432	309-465	337-499	366-534	396-570	427-607	459-645	**492-684**	527-723
25	28-92	42-113	56-136	72-159	89-183	107-208	126-234	146-261	167-289	189-318	213-347	237-378	262-410	289-442	316-476	344-511	374-546	404-583	436-620	468-659	502-698	**536-739**
26	29-95	43-117	58-140	74-164	91-189	109-215	129-241	150-268	171-297	194-326	217-357	242-388	268-420	295-453	322-488	351-523	381-559	412-596	444-634	477-673	511-713	545-755
27	30-98	44-121	59-145	76-169	93-195	112-221	132-248	153-276	175-305	198-335	222-366	247-398	273-431	301-464	329-499	358-535	388-572	420-609	452-648	485-688	520-728	555-770
28	31-101	45-125	61-149	78-174	96-200	115-227	135-255	157-283	179-313	203-343	227-375	253-407	279-441	307-475	335-511	365-547	396-584	427-623	460-662	494-702	529-743	564-786
29	32-104	47-128	63-153	80-179	98-206	118-233	139-261	160-291	183-321	207-352	232-384	258-417	285-451	313-486	342-522	372-559	403-597	435-636	468-676	503-716	538-758	574-801
30	33-107	48-132	64-158	82-184	101-211	121-239	142-268	164-298	187-329	211-361	237-393	263-427	291-461	319-497	348-534	379-571	411-609	443-649	477-689	511-731	547-773	584-816
31	34-110	49-136	66-162	84-189	103-217	123-246	145-275	167-306	191-337	216-369	242-402	268-437	296-472	325-508	355-545	386-583	418-622	451-662	485-703	520-745	556-788	593-832
32	34-114	50-140	67-167	86-194	106-222	126-252	148-282	171-313	195-345	220-378	246-412	274-446	302-482	331-519	362-556	393-595	425-635	459-675	493-717	529-759	565-803	603-847
33	35-117	52-143	69-171	88-199	108-228	129-258	151-289	175-320	199-353	225-386	251-421	279-456	308-492	337-530	368-568	400-607	433-647	467-688	501-731	537-774	574-818	612-863
34	36-120	53-147	71-175	90-204	110-234	132-264	155-295	178-328	203-361	229-395	256-430	284-466	313-503	343-541	375-579	407-619	440-660	474-702	510-744	546-788	583-833	622-878
35	37-123	54-151	72-180	92-209	113-239	135-270	158-302	182-335	207-369	234-403	261-439	289-476	319-513	350-551	381-591	414-631	447-673	482-715	518-758	555-802	592-848	631-894
36	38-126	55-155	74-184	94-214	115-245	137-277	161-309	185-343	211-377	238-412	266-448	295-485	325-523	356-562	388-602	421-643	455-685	490-728	526-772	563-817	602-862	641-909
37	39-129	57-158	76-188	96-219	117-251	140-283	164-316	189-350	215-385	242-421	271-457	300-495	330-534	362-573	394-614	428-655	462-698	498-741	534-786	572-831	611-877	650-925
38	40-132	58-162	77-193	98-224	120-256	143-289	167-323	193-357	219-393	247-429	275-467	305-505	336-544	368-584	401-625	435-667	470-710	506-754	543-799	581-845	620-892	660-940
39	41-135	59-166	79-197	100-229	122-262	146-295	170-330	196-365	223-401	251-438	280-476	311-514	342-554	374-595	407-637	442-679	477-723	514-767	551-813	589-860	629-907	670-955
40	41-139	60-170	80-202	102-234	125-267	149-301	174-336	200-372	227-409	256-446	285-485	316-524	347-565	380-606	414-648	449-691	485-735	521-781	559-827	598-874	638-922	679-971
41	42-142	61-174	82-206	104-239	127-273	151-308	177-343	204-379	231-417	260-455	290-494	321-534	353-575	386-617	420-660	456-703	492-748	529-794	568-840	607-888	647-937	689-986
42	43-145	63-177	84-210	106-244	129-279	154-314	180-350	207-387	235-425	265-463	295-503	326-544	359-585	392-628	427-671	463-715	499-761	537-807	576-854	616-902	656-952	698-1002
43	44-148	64-181	85-215	108-249	132-284	157-320	183-357	211-394	239-433	269-472	300-512	332-553	365-595	399-638	434-682	470-727	507-773	545-820	584-868	624-917	666-966	708-1017
44	45-151	65-185	87-219	110-254	134-290	160-326	186-364	214-402	243-441	273-481	305-521	337-563	370-606	405-649	440-694	477-739	514-786	553-833	592-882	633-931	675-981	717-1033
45	46-154	66-189	88-224	112-259	137-295	163-332	190-370	218-409	247-449	278-489	309-531	342-573	376-616	411-660	447-705	484-751	522-798	561-846	601-895	642-945	684-996	727-1048
46	47-157	68-192	90-228	114-264	139-301	165-339	193-377	222-416	251-457	282-498	314-540	347-583	382-626	417-671	453-717	491-763	529-811	568-860	609-909	651-959	693-1011	737-1063
47	48-160	69-196	92-232	116-269	141-307	168-345	196-384	225-424	255-465	287-506	319-549	353-592	387-637	423-682	460-728	498-775	536-824	576-873	617-923	659-974	702-1026	746-1079
48	48-164	70-200	93-237	118-274	144-312	171-351	199-391	229-431	259-473	291-515	324-558	358-602	393-647	429-693	466-740	505-787	544-836	584-886	626-936	668-988	711-1041	756-1094
49	49-167	71-204	95-241	120-279	146-318	174-357	202-398	232-439	263-481	296-523	329-567	363-612	399-657	435-704	473-751	512-799	551-849	592-899	634-950	677-1002	721-1055	766-1109
50	50-170	73-207	97-245	122-284	149-323	177-363	206-404	236-446	267-489	300-532	334-576	369-621	405-667	441-715	480-762	519-811	559-861	600-912	642-964	685-1017	730-1070	775-1125

TABLE 18(a) Probabilities Associated with Obtaining Values as Large as Observed Values with Friedman's Test. $(k = 3)$

χ_r^2	p	χ_r^2	p	χ_r^2	p	χ_r^2	p
N = 2		**N = 3**		**N = 4**		**N = 5**	
0	1.000	000	1.000	.0	1.000	.0	1.000
1	.833	.667	.944	.5	.931	.4	.954
3	.500	2.000	.528	1.5	.653	1.2	.691
4	.167	2.667	.361	2.0	.431	1.6	.522
		4.667	.194	3.5	.273	2.8	.367
		6.000	028	4.5	.125	3.6	.182
				6.0	.069	4.8	.124
				6.5	.042	5.2	.093
				8.0	.0046	6.4	.039
						7.6	.024
						8.4	.0085
						10.0	.00077

χ_r^2	p	χ_r^2	p	χ_r^2	p	χ_r^2	p
N = 6		**N = 7**		**N = 8**		**N = 9**	
.00	1.000	.000	1.000	.00	1.000	.000	1.000
.33	.956	.286	.964	.25	.967	.222	.971
1.00	.740	.857	.768	.75	.794	.667	.814
1.33	.570	1.143	.620	1.00	.654	.889	.865
2.33	.430	2.000	.486	1.75	.531	1.556	.569
3.00	.252	2.571	.305	2.25	.355	2.000	.398
4.00	.184	3.429	.237	3.00	.285	2.667	.328
4.33	.142	3.714	.192	3.25	.236	2.889	.278
5.33	.072	4.571	.112	4.00	.149	3.556	.187
6.33	.052	5.429	.085	4.75	.120	4.222	.154
7.00	.029	6.000	.052	5.25	.079	4.667	.107
8.33	.012	7.143	.027	6.25	.047	5.556	.069
9.00	.0081	7.714	.021	6.75	.038	6.000	.057
9.33	.0055	8.000	.016	7.00	.038	6.222	.048
10.33	.0017	8.857	.0084	7.75	.018	6.889	.031
12.00	.00013	10.286	.0036	9.00	.0099	8.000	.019
		10.571	.0027	9.25	.0080	8.222	.016
		11.143	.0012	9.75	.0048	8.667	.010
		12.286	.00032	10.75	.0024	9.556	.0060
		14.000	.000021	12.00	.0011	10.667	.0035
				12.25	.00086	10.889	.0029
				13.00	.00026	11.556	.0013
				14.25	.000061	12.667	.00066
				16.00	.0000036	13.556	.00035
						14.000	.00020
						14.222	.000097
						14.889	.000054
						16.222	.000011
						18.000	.0000006

* Adapted from Friedman, M. "The Use of Ranks to Avoid the Assumption of Normality Implicit in the Analysis of Variance," *J. Amer. Statist. Ass.*, **32**, 688–689 (1937), with permission of the author and publisher.

TABLE 18(b) Probabilities Associated with Obtaining Values as Large as Observed Values with Friedman's Test. $(k = 4)$

χ_r^2	p	χ_r^2	p	χ_r^2	p	χ_r^2	p
N = 2		**N = 3**		**N = 4**			
.0	1.000	.2	1.000	.0	1.000	5.7	.141
.6	.958	.6	.958	.3	.992	6.0	.105
1.2	.834	1.0	.910	.6	.928	6.3	.094
1.8	.792	1.8	.727	.9	.900	6.6	.077
2.4	.625	2.2	.608	1.2	.800	6.9	068
3.0	.542	2.6	.524	1.5	.754	7.2	.054
3.6	.458	3.4	.446	1.8	.677	7.5	.052
4.2	.375	3.8	.342	2.1	.649	7.8	.036
4.8	.208	4.2	.300	2.4	.524	8.1	.033
5.4	.167	5.0	.207	2.7	.508	8.4	.019
6.0	.042	5.4	.175	3.0	.432	8.7	.014
		5.8	.148	3.3	.389	9.3	.012
		6.6	.075	3.6	.355	9.6	.0069
		7.0	054	3.9	.324	9.9	.0062
		7.4	.033	4.5	.242	10.2	.0027
		8.2	.017	4.8	.200	10.8	.0016
		9.0	.0017	5.1	.190	11.1	.00094
				5.4	.158	12.0	.000072

* Adapted from Friedman, M. "The Use of Ranks to Avoid the Assumption of Normality Implicit in the Analysis of Variance," *J. Amer. Statist. Ass.*, **32**, 688-689 (1937), with permission of the author and the publisher.

ANSWERS TO
ODD-NUMBERED EXERCISES

ANSWERS TO
ODD-NUMBERED EXERCISES

CHAPTER 2

7. b) Data for pie chart

Store	No. of degrees
A	45°
B	159°
C	35°
D	93°
E	28°
Total	360°

CHAPTER 3

1. a) $f_1 X_1^2 + f_2 X_2^2 + \cdots + f_{10} X_{10}^2$

    b) $X_6^2 + X_7^2 + X_8^2 + X_9^2$

    c) $[X_1(X_1-2) + X_2(X_2-2) + X_3(X_3-2) + X_4(X_4-2)]^2$

3. a) 8 b) 0 c) 26
5. a) 3.48 b) 3.02
7. 3.4
9. a) 21,700 b) 11,350
11. a) 43.3
15. 195.2, 121   17. 3.1641, 2.8   19. 12000, 12000
23. $G = 156.9$ 25. 123.33
27. a) 1655 b) 1647.05

CHAPTER 4

1. a) 8.0   3. a) 0.0041
7. 23.25% (ungrouped data), 23.20% (grouped data)
9. 22254.7
13. 4.206, 2.051   15. 34.267, 5.854   17. 0
19. No   21. 2.84   23. $s^2 = 325100.503$, $s = 570.176$

CHAPTER 5

1. $S = \{HH, HT, TH, TT\}$
3.

r \ b	1	2	3	4	5	6
1	$(1,1)$	$(1,2)$	$(1,3)$	$(1,4)$	$(1,5)$	$(1,6)$
2	$(2,1)$	$(2,2)$	$(2,3)$	$(2,4)$	$(2,5)$	$(2,6)$
3	$(3,1)$	$(3,2)$	$(3,3)$	$(3,4)$	$(3,5)$	$(3,6)$
4	$(4,1)$	$(4,2)$	$(4,3)$	$(4,4)$	$(4,5)$	$(4,6)$
5	$(5,1)$	$(5,2)$	$(5,3)$	$(5,4)$	$(5,5)$	$(5,6)$
6	$(6,1)$	$(6,2)$	$(6,3)$	$(6,4)$	$(6,5)$	$(6,6)$

5. $\frac{1}{2}$
7. a) $\frac{9}{14}$ b) $\frac{1}{8}$ c) $\frac{3}{8}$ d) $\frac{3}{56}$
9. a) 0.0875 b) 0.4875   11. $\frac{13}{36}$
13. a) $\frac{4}{9}$ b) $\frac{83}{192}$ 15. 0.36878
17. a) 0.86 b) 0.14   19. 0.8
21. $\frac{1}{12}$ 23. $\frac{2}{5}$, $\frac{3}{5}$ 25. 0.24
27. a) 0.2 b) 0.95 c) 0.35
29. a) $\frac{3}{10}$ b) $\frac{7}{15}$ c) $\frac{6}{10}$ 31. $\frac{30}{38}$
33. a) 0.2353 b) 0.2353 c) 0.3529

CHAPTER 6

1. a) Discrete   b) Continuous   c) Discrete   d) Continuous
3. a) $S = \{DD, DN, ND, NN\}$ b) 4
   c) $P(DD) = 0.09$, $P(ND) = 0.21$, $P(DN) = 0.21$, $P(NN) = 0.49$
5. Expected value $= 0.6$, variance $= 0.42$

7. a)   b) $E(X) = 3.1643$

Length of time	Probability
0.5	0.1698
1.5	0.1425
2.5	0.2292
3.5	0.1575
4.5	0.1104
5.5	0.0632
6.5	0.0585
7.5	0.0689

9. a) 5.64 thousand gallons   b) 4.9904 thousand gallons
11. $11,200
13. a) 125,000   b) 2,500,000,000   15. a) $175,000   b) $18,000
17. 136   19. 0.17

CHAPTER 7

1. a) 0.729   b) 0.243   c) 0.027   d) 0.001
3. $1 - (0.9)^{10} = 0.6513$
5. a) 0.5695   b) 0.4305   c) 5
7. $\frac{10}{3}$ 9. a) 0.2335   b) 0.1029
11. a) 0.2048   b) 0.32768   c) 0.00032
13. 300
15. a) $(0.8)^{15} = 0.03518$ b) $(0.2)^{15}$ c) 0.1319
17. 0.0527   19. a) 0.1353   b) 0.3711
21. 0.3711   23. 0.8008
25. a) 0.0025   b) 0.9975

CHAPTER 8

1. a) 0.2580   b) 0.1915   c) 0.7118
    d) 0.2638   e) 0.8643   f) 0.1151

3. a) 0.025   b) 0.9974   c) 0.1587
5. a) $x_0 = 12.9$ b) $x_0 = 6.9$ 7. 2.14%
9. a) 0.0227   b) 0.1587   c) 0.6826
11. 86.5   13. 86.44%
15. a) 0.5   b) 0.5   c) 0.4772
17. a) 0.0505   b) 0.9426   c) 0.0021
19. a) 0.2033   b) 0.7967   c) 0.5055
21. a) 0.0354   b) 0.4356   c) 0.0721   23. 0.9649

CHAPTER 10

7. a) $\bar{X}_A + \bar{X}_B$ is normal with mean 7, variance $\frac{1}{5}$;
    b) $\bar{X}_B - \bar{X}_A$ is normal with mean 3; variance $\frac{1}{5}$
9. 0   11. 0
13. 95% C.I.   (0.5, 2.1)
     99% C.I.   (0.04, 2.6)
15. (7.7, 7.9)   17. (34.8, 35.9)
19. (3.71, 7.05)   21. (63.7, 66.2)
23. (0.16, 0.45)
25. a) (0.16, 0.43)   b) (0.13, 0.55)
27. (8.118, 13.165)
29. (0.035, 0.22)   31. (0.66, 0.83)
33. (3.55, 7.21)   35. (0.202, 0.318)
37. (0.11, 0.29)   39. (0, 0.16)

CHAPTER 11

1. 0.5
3. $Z = -21.08$; store does have lower sales than average.
5. $\beta = 0.3613$ 7. $\beta = 0.0001$
9.

μ	β	Power
55	0.0104	0.9896
60	0	1

11. $t = 0.62$; accept $H_0 : \mu = 10$
13. $t = -2.25$; reject H_0; i.e., $\mu \neq 1550$
15. $t = 1.4161$; accept $H_0 : \mu = 4$
17. $\chi^2 = 46.1733$; $\sigma > 1.5$
19. $\chi^2 = 30.375$; accept H_0: $\sigma^2 = 16$
21. $t = 0.06243$; accept $H_0 : \mu_1 = \mu_2$
23. $t = 23.0718$; life lengths do differ.
25. $Z = 2.95$; $\mu_1 \neq \mu_2$
27. $Z = 4.08$; $p \neq 0.1$
29. $Z = 6.33$; colored appliances preferred
31. $Z = 1.29$; accept $H_0 : p = 0.1$
33. $Z = 1.59$; accept $H_0 : p_1 = p_2$
35. $t = 3.4054$; accept $H_A : \mu_A > \mu_B$

CHAPTER 12

1. a) $\hat{Y} = -1.356 + 0.091X$ b) 5.469
3. b) $\hat{Y} = 4414.7598 + 48.398X$ c) 6834.6598   d) Yes
5. a) $\hat{Y} = 54.3929 + 0.525X$ b) 83.2679
7. $\hat{\sigma}^2 = 776.6809$
9. SS $= 4317971.875$,   SSE $= 1085805.3491$,   SSR $= 3232166.52$
11. $-113.88 < A < -2.28$ 13. $20.375 < B < 76.421$
15. $-1.42 < A < 0.54$ 17. $r = 0.97$ 19. $r = 0.99$

CHAPTER 13

1. b) $\hat{Y}_A = 40.5090 + 2.6424X$ (Origin, July 1, 1961; X, 1 year units; Y, annual sales
   in millions of dollars)
   c) $\hat{Y}_B = 45.6727 + 0.6727X$ (Origin, July 1, 1961; X, 1 year units; Y, annual sales
   in millions of dollars)
3. a) $\hat{Y}_A = \$72.2181$ million   b) $\hat{Y}_B = \$53.7454$ million
5. a) $\hat{Y}_A = 45.8 + 2.64X$ b) $\hat{Y}_B = 47.4 + 0.52X$
7. b)

Quarter	I	II	III	IV
Seasonal index	86.96	76.25	101.0	135.79

9. a) $\hat{Y} = 102 + 8.8X$ (Origin, July 1, 1966; X, 1 year units; Y, annual sales in
   millions of dollars)
   b) $\hat{Y} = 24.88 + 0.55X$
11. b) $\hat{Y} = 756.365 + 52.1697X$ d) $\hat{Y} = \$1225.8921$ million
13. a) $\hat{Y} = 502.7636 + 9.8969X$ b) $\hat{Y} = 41.5189 + 0.0687X$
15.

	Quarter	I	II	III	IV
b)	Seasonal index	80.5194	96.1038	118.4414	104.935
c)	Seasonal index	87.1096	97.3271	116.1653	99.3951
e)	Seasonal index	86.0467	99.8662	116.2530	97.8169

d) $\hat{Y} = 89.4 + 32.3X$

f)

	Quarters			
Year	I	II	III	IV
1968	17.4324	22.0295	23.2252	25.5580
1969	26.7297	31.0415	33.5475	33.7365
1970	34.8648	35.0469	37.8485	40.8927
1971	45.3242	45.0603	45.5902	49.0713
1972	55.7837	52.0697	55.9125	57.2498

CHAPTER 14

7.

Year	Index number
1963	100.00
1964	110.31
1965	113.02
1966	108.50
1967	113.38
1968	126.58
1969	132.55
1970	135.62
1971	138.38
1972	139.60

9. $I_{65} = 100$, $I_{70} = 121.8$
11. $I_{65} = 100$, $I_{70} = 130.83$
13. $I_{65} = 100$, $I_{70} = 159.25$
15. $19,678.50

CHAPTER 15

1. $\chi^2 = 456.04$; the five stores do not have equal sales potential.
3. $\chi^2 = 32$; the five cities do not have same sales potential.
5. $\chi^2 = 35$; salesmen's abilities do differ significantly.
7. $\chi^2 = 0.1857$; the normal fit is good.
9. $\chi^2 = 64.11964$; the normal fit is inadequate.
11. $\chi^2 = 102.9973$; normal fit is inadequate.

13. $\chi^2 = 3.122$; Poisson fit is good.
15. $\chi^2 = 1.43776$; Poisson fit is good.
17. $\chi^2 = 8.0$; mint flavor has an effect.
19. $\chi^2 = 8.16$; type does affect ability.
21. $\chi^2 = 0.88$; the null hypothesis cannot be rejected.

CHAPTER 16

1.

Source of variation	SS	d.f.	MS	F
Between cities	37.8	3	12.6	1.52
Within cities	132.4	16	8.27	
Total	170.2	19		

No significant difference between the cities.

3.

Source of variation	SS	d.f.	MS	F
Between universities	1191.2	4	297.8	6.98
Within universities	639.75	15	42.65	
Total	1830.95	19		

Significant difference in the abilities of students from the five universities.

5.

Source of variation	SS	d.f.	MS	F
Between manufacturers	0.47	2	0.235	7.3
Within manufacturers	0.39	12	0.032	
Total	0.86	14		

Significant difference in the diameters of the parts supplied by three manufacturers.

7.

Source of variation	SS	d.f.	MS	F
Between firms	36.4	2	18.2	3.31
Within firms	66.0	12	5.5	
Total	102.4	14		

No significant difference between firms.

9.

Source of variation	SS	d.f.	MS	F
Between employees	312.50	3	88.66	8.27
Between machines	266.00	3	104.16	9.72
Error	96.50	9	10.72	
Total	675.00	15		

a) Machines affect production.
b) Employees produce at different rates.

CHAPTER 17

1. 98.76%
3. 97.74%
7. 0.0428
9. 0.2051
11. $n = 24$, $c = 2$
13. $n = 110$, $c = 6$
15. Process under control.

CHAPTER 18

1. $r = 5$; fluctuation from one bolt to another is random.
3. $r = 6$; defectives are occurring at random.
5. $T(+) = 44$; percentage has not dropped.
7. $T(1) = 134$; no difference between cities.
9. $T(1) = 67$; no difference between machines.
11. $T(1) = 48$; both drugs are equally effective.

13. $H = 1.08$; there is no difference in the earnings of the workers of the three industries.
15. $H = 4.535$; there is no difference between the mean breaking strength of cables manufactured by the three firms.
17. $S = 3.5$; all candidates appear equally good.
19. $S = 9.3$; there is a difference between employees.
21. $r_s = 0.942$ 23. $r_s = 0.58$

CHAPTER 19

1.

	Supply					
Demand	0	1	2	3	4	5
0	0	-4	-8	-12	-16	-20
1	0	3	-1	-5	-9	-13
2	0	3	6	2	-2	-6
3	0	3	6	9	5	1
4	0	3	6	9	12	8
5	0	3	6	9	12	15

3.

	Supply					
Demand	0	1	2	3	4	5
0	0	4	8	12	16	20
1	3	0	4	8	12	16
2	6	3	0	4	8	12
3	9	6	3	0	4	8
4	12	9	6	3	0	4
5	15	12	9	6	3	0

5.

Demand	Probability
0	0.056
1	0.056
2	0.166
3	0.278
4	0.388
5	0.056

7.

Actual supply	Expected loss
0	9.162
1	6.554
2	4.338
3	3.284
4	4.176
5	7.784

11. Supply three units.
13. a) $S=3$ b) $S=3$ c) $S=2$
15.

Demand	Supply			
	a) 15	b) 16	c) 17	d) 18
0	-45	-48	-51	-54
1	-38	-41	-44	-47
2	-31	-34	-37	-40
3	-24	-27	-30	-33
4	-17	-20	-23	-26
5	-10	-13	-16	-19
6	-3	-6	-9	-12
7	4	1	-2	-5
8	11	8	5	2
9	18	15	12	9
10	25	22	19	16
11	32	29	26	23
12	39	36	33	30
13	46	43	40	37
14	53	50	47	44
15	60	57	54	51
16	60	64	61	58
17	60	64	68	65
18	60	64	68	72

19. $\Pr(\text{Excellent}|D_1)=0.948$; $\Pr(\text{Good}|D_1)=0.029$; $\Pr(\text{Fair}|D_1)=0.023$

INDEX

INDEX

SUMMARIES

Distribution for Means

- use Normal Dist. when σ is given &
- t test when σ has to be estimated from sample data

	Mean	S.D
Sample	$\bar{X}$	S
Pop.	μ	σ

One Sample

(a) σ known

$$Z = \dfrac{\bar{X} - \mu}{\dfrac{\sigma}{\sqrt{n}}}$$

10-4 $\xleftarrow{\text{11-2}}_{\text{11-3}}$

(b) σ unknown

$$Z = \dfrac{\bar{X} - \mu}{\dfrac{S}{\sqrt{n}}}$$

assume normal Dist of Students

$\Leftrightarrow$ d.f. $= n - 1$ $\mu = \bar{X} \pm \dfrac{t \cdot s}{\sqrt{n}}$

Two Independent Samples [11-4]

(a) $\sigma_1 \& \sigma_2$ known

$$Z = \dfrac{(\bar{X}_1 - \bar{X}_2) - (\mu_1 - \mu_2)}{\sqrt{\dfrac{\sigma_1^2}{n_1} + \dfrac{\sigma_2^2}{n_2}}}$$

Pooled - to determine

$$S = \sqrt{\dfrac{(n_1 - 1)s_1^2 + (n_2 - 1)s_2^2}{n_1 + n_2 - 2}}$$

(b) $\sigma_1 \& \sigma_2$ unknown

assume $X_1 \& X_2$ - NORMAL

pooled
$$t = \dfrac{(\bar{X}_1 - \bar{X}_2) - (\mu_1 - \mu_2)}{S \sqrt{\dfrac{1}{n_1} + \dfrac{1}{n_2}}} \qquad \nu = n_1 + n_2 - 2$$

not pooled
$$t = \dfrac{(\bar{X}_1 - \bar{X}_2) - (\mu_1 - \mu_2)}{\sqrt{\dfrac{s_1^2}{n_1} + \dfrac{s_2^2}{n_2}}} \qquad \nu = n_1 + n_2 - 2$$

Two Paired Samples

$$t = \dfrac{\bar{d} - \mu d}{\dfrac{S_d}{\sqrt{n}}}$$

$$S_d = \sqrt{\dfrac{1}{n-1}\left(\sum d^2 - \dfrac{(\sum d)^2}{n}\right)}$$

d is the difference

$|A|B|d AB| d^2$

ANOVA

X_{ij}
$X_{\cdot j}^2$

$T_j \quad \sum X_{ij}^2$

$\bar{X}$

Totals

$H_0 \quad \mu_1 = \mu_2 = \mu_3 \cdots$
H_A - 1 is different

ASSUME: measurements are interval type
2. Normal Dist
3. pooled data

$SST \sum X_{ij}^2 - n\bar{Y}^2 = (+)$

$SSB \quad \dfrac{T_1^2}{n_1} + \dfrac{T_2^2}{n_2} \cdots$

$SSW = SST - SSB$

support H_A $\to$ reject H_0

F_{crit}

Source	d.f	S.S	M.S.	Fobs	Fcrit
B	$n-1$	SSB	$SSB/k-1$	MSB/MSW	$(k-1, n-k)$
W	$n-k$	SSW	$SSW/n-k$		

BIVARIATE DATA - LINEAR REGRESSION & CORRELATION

Predict Y given X

$$\hat{y} = a + bx$$

$$\begin{array}{|c|c|c|c|c|} x,y & x^2 & y^2 & y \cdot x \\ \hline \end{array}$$

eq. for regression line $b = \dfrac{\Sigma xy - \frac{1}{n}\Sigma x \Sigma y}{\Sigma x^2 - \frac{1}{n}(\Sigma x)^2}$

$SS_x = \Sigma x^2 - \frac{1}{n}(\Sigma x)^2$

$SS_y = \Sigma y^2 - \frac{1}{n}(\Sigma y)^2$

$SP_{xy} = \Sigma x \cdot y - \frac{1}{n}\Sigma x \Sigma y$

$$= \boxed{b = \dfrac{SP_{xy}}{SS_x}} \leftarrow \text{formula for regression line}$$

$$a = \bar{y} - b\bar{x}$$

$* = $? what is the y intercept when x=0

$(\bar{x}, \bar{y})$ + a → gives you ability to plot. R.L.

(i) Standard Error

$$S_{yx} = \left[\frac{1}{n-2}\left(SS_y - b\,SP_{xy}\right)\right]^{\frac{1}{2}}$$

$= b^2 SS_x$

(A) Significant change in slope

$H_0 \quad \beta = 0$

$H_A \quad \beta \neq 0 \qquad \alpha = 0.05 \quad$ -2 tail test

$b = $ sample slope
$\beta = $ population "

$$t = \dfrac{b - \beta}{\dfrac{S_{yx}}{SS_x^{\frac{1}{2}}}}$$

$t = \dfrac{b - \beta}{}$

$\nu = n-2$

$t_{\frac{\alpha}{2}} \quad d.f = $?

$t_{obs} < t_{crit} \therefore$ do not reject H_0

(B) Confidence Intervals

(i) 95% confidence interval for expected value of productivity - score of 16

$\hat{y} = 13.44 + b(x)$

$* \quad$ y intercept when x=0 $\qquad 16 \qquad \therefore$ expected value is equal to $\hat{y}$

$$\hat{y} \pm t_{n-2}\left(\frac{\alpha}{2}\right) S_{y \cdot x}\left[\frac{1}{n} + \frac{(x_p - \bar{x})^2}{SS_x}\right]^{\frac{1}{2}}$$

from table 6

(ii) for an individual

$$\hat{y} \pm t_{n-2}\left(\frac{\alpha}{2}\right) S_{y \cdot x}\left[1 + \frac{1}{n} + \frac{(x_p - \bar{Y})^2}{SS_x}\right]^{\frac{1}{2}}$$

(C) Coefficient "r"

$$r = \dfrac{SP_{xy}}{\sqrt{SS_x \cdot SS_y}}$$

(D) Fisher z

$z_r \pm z\left(\frac{1}{\sqrt{n-3}}\right) \qquad z = \frac{1}{2}\ln\left(\frac{1+r}{1-r}\right) - 1$

$z_r = \frac{1}{2}\ln\left(\frac{1+r}{1-r}\right) \quad z_r = ?$

$= z_r \pm z\left(\frac{1}{\sqrt{n-3}}\right)$

then convert z's using Fisher z tables